# Global Warming and Population Responses

## among Great Plains Birds

Paul A. Johnsgard

Golden Eagle

# Global Warming and Population Responses among Great Plains Birds

Paul A. Johnsgard

Foundation Professor Emeritus of Biological Sciences
School of Biological Sciences
University of Nebraska–Lincoln

Zea Books
Lincoln, Nebraska
2015

ISBN 978-1-60962-064-6  paperback
ISBN 978-1-60962-065-3  e-book

Set in Calisto and Consolas types.
Design and composition by Paul Royster.

Zea E-Books are published by the
University of Nebraska–Lincoln Libraries.

Electronic (pdf) edition available online at http://digitalcommons.unl.edu/zeabook/
Print edition can be ordered from http://www.lulu.com/spotlight/unllib

Nebraska
UNIVERSITY OF
Lincoln

# Abstract

Based on an analysis of 47 years (1967–2014) of Audubon Christmas Bird Counts (CBC), evidence for population changes and shifts in early winter (late December) ranges of nearly 150 species of birds in the Great Plains states is summarized, a region defined as including the Dakotas, Nebraska, Kansas, Oklahoma, and the Texas panhandle. The rationale for this study had its origins in Terry Root's 1988 *Atlas of North American Wintering Birds*. Root's landmark study provided a baseline for evaluating the nationwide winter distributions of 253 North American birds in the mid-20th century, using data from the National Audubon Society's annual CBC surveys from 1962–63 through 1971–72. A later summary (P. A. Johnsgard and Tom Shane, *Four Decades of Christmas Bird Counts in the Great Plains: Ornithological Evidence of a Changing Climate,* 2009) provided range maps and quantitative population data (expressed as the average number of birds tallied per party-hour) for all 210 species reported from the 1967–8 to the 2006–7 CBC, on a decade-to-decade basis. The present analysis includes all of the 40 annual CBC surveys from the 1967–8 to the 2006–7 counts, plus the results of the most recent 2013–14 CBC. The present summary quantitatively describes the early winter abundance for 147 of the most commonly encountered regional species, illustrating their temporal changes in geographic distributions and relative abundance between 1967 and 2014.

Over this 47-year period there has been a progressive winter warming trend regionally, and associated ecological changes, influencing the early winter regional abundance and geographic distributions of many birds. The great majority these changes have involved northward shifts in early winter distributions. Over this approximate half-century interval at least six species (Canada goose, mallard, black-capped chickadee, American goldfinch, and house finch) have shifted their areas of greatest early winter abundance two states northward, and the centers of maximum abundance of at least ten other species have shifted northward by at least one state. Milder and less stressful early winter temperatures, with associated extended periods of ice-free water and greater access to snow free foraging sites, are believed to be responsible. These recent population shifts have been most evident in the northern half of the region, where increases in mean January temperatures have been greatest. Nearly all of these population and distributional changes can be attributed to recent climate changes in the Great Plains. Approximately 500 literature citations are included.

Barred Owl

# Contents

8

# Foreword and Acknowledgments

## Foreword

The rationale for this book has its origins in Terry Root's 1988 *Atlas of North American Wintering Birds*, which provided a baseline landmark for evaluating the nationwide winter distributions of North American birds, using data from the National Audubon Society's annual Christmas Bird Counts (CBC) from 1962-63 through 1971-72. Tom Shane and I speculated that an updated analysis might shed light on the possible effects of more recent climatic warming trends on bird migration and wintering patterns in the Great Plains, a region known for its severe winters and also one of our continent's important migratory pathways and wintering regions. As life-long residents of the Great Plains, we have both lived long enough to have witnessed some of these changes in avian migrations and wintering patterns personally. Johnsgard tested these speculations by doing some sample species analyses during the spring of 2008, after which it appeared that a complete survey of Great Plains winter birds would be worthwhile, based on CBC data.

Following Root's precedent, it is appropriate to dedicate this study to the countless thousands of people who have participated in CBC surveys over the years. This is especially true in the northern Great Plains, where the weather in late December can be unbearable. The first Christmas Count I participated in was centered at Fargo, North Dakota, in late December of 1952. The temperature was a scant seven degrees above zero, with a mercifully unknown but memorable wind-chill, and with only 17 species to be tallied after an entire day of effort. At the time, it seemed a good reason to begin applying to graduate schools having warmer climates.

Since then, winters in North Dakota have thankfully become gradually warmer, and the number of people participating in CBC surveys has increased greatly. Considering only the 41 winters chosen for the present analysis, from 1967-68 to 2006-07, plus 2013-14, and the multi-state region extending from North Dakota through Oklahoma plus the Texas panhandle, there have been about 3,000 CBCs and collectively about 40,000 individual participations (most of which represent repeat involvements over multiple years). In a real sense, all of these people have directly contributed to this book.

The idea of doing a book on winter birds of the Great Plains was Tom Shane's, a long-time Kansas friend who introduced me to southwestern Kansas and lesser prairie-chickens. I agreed to participate and suggested patterning the book after my earlier (1979) summary on the breeding birds of the Great Plains states. I also agreed to analyze several decades of Christmas

Bird Count data, to document regional winter bird populations and to estimate possible population trends over the past half-century. My unlucky ornithology students at the University of Nebraska's Biological Field Station during 2008 were assigned the onerous job of extracting data on more than 200
species for this phase of the resulting report, and Tom Shane produced winter range maps for nearly 200 species that were adequately well-documented
by CBC data. Dr. Paul Royster very kindly agreed to deposit our report in the
University of Nebraska–Lincoln Library's Digital Commons Papers in Ornithology (P. A. Johnsgard and T. Shane, 2009, http://digitalcommons.unl.
edu/biosciornithology/46/).

To keep the present report at a reasonable length, all the range and other
maps have been eliminated, but they are archived online in the just mentioned
source. Similar maps based on CBC data through 2002 are available from the
Audubon Society's CBC website (http://audubon2.org/cbchist/). More recent Audubon studies (Weidensaul, 2007; National Audubon Society, 2014a;
Nijhuis, 2014) have predicted the effects of climate change on 403 species of
North American birds, using data from three decades (1980–2009) of CBC
surveys and Breeding Bird Surveys (http://climate.audubon.org/article/
audubon-report-glance),

It should be pointed out that the CBC surveys, which are always performed
within two weeks before or after Christmas, do not truly inventory the birds
of the coldest part of a Great Plains winter, which typically occurs around
the middle of January. They do, however, provide by far the largest and most
long-running database available for trying to judge winter bird populations
and associated population trends in North America.

Attempting to summarize older published migration data has its problems,
for most such arrival and departure information does not reflect recent climatic trends. For example, the migration data given for Nebraska are primarily based on an analysis of published arrival and departure dates published between 1933 and the late 1970s (Johnsgard, 1980). Although these dates more
closely reflect climatic conditions of the mid-1900s than those of the present,
such data provide a comparative basis for evaluating future migration information, and are the only extensive series of dates available for the Great Plains
region. Records from recent years suggest that average spring arrivals now average about two weeks earlier in Nebraska, and average autumn departures
are now several weeks later. These deviations from earlier averages appear to
be more extreme to the north of Nebraska, and less divergent southwardly, in
parallel with more rapid winter amelioration in higher latitudes during recent
decades. This climatic trend has resulted in notably longer frost-free periods in
the Great Plains, especially at higher latitudes (Kunkel *et al.*, 2013).

## Acknowledgments

For helping in the tedious and herculean job of extracting and summarizing 1967–2008 CBC data, I am deeply indebted all of the 17 unlucky students who were enrolled in my Ornithology class at the University of Nebraska's Cedar Point Biological Station during the spring of 2008. They include Laura Achterberg, Eva Anguilano, Matthew Colling, Andrew Furman, Lisa Levander, Michael McGuire, Sean O'Day, Elizabeth Olson, Gwendolyn Powell, Keenan Price, Nikaela Schernikau, Tyler Seiboldt, John Sens, Justin Tilberg, Mohammed Vhora, Charles Wessel and Kristin Williams. This report is gratefully dedicated to them, and to Linda Brown, for her heroic summarizing of the 2013–14 CBC data. Tom Shane's efforts in preparing the range maps for our 2009 report is likewise gratefully appreciated. The unlimited use of University of Nebraska computers, library, and other University facilities at Cedar Point, as well as at Lincoln, is also acknowledged. Prof. Paul Royster kindly agreed to adapt the manuscript for inclusion in the University's Digital Commons' library collection.

Wild Turkey

# Part 1: Introduction

## Methods of Study

In an exhaustive and landmark study, Terry Root (1988) analyzed the distributions of 508 species or subspecies of birds that winter throughout North America, based on her analysis of data from Audubon Christmas Bird Counts conducted over the decade of 1962-63 to 1971–72, involving a total of 1,282 sites. More recently, the annual Christmas Bird Counts have attracted far more participants and achieved surveys that are more comprehensive. For example, in 2007 nearly 58,000 Christmas Count participants at 2,005 U.S. and Canadian sites tallied 643 species. The counts also have gradually become better standardized (Arbib, 1982, 1983), making them a more reliable method for judging population changes over time (Sauer and Link, 2002). Since the 1970s there have been marked changes in avian habitats and regional winter climate patterns, and substantial changes in at least some Great Plains winter bird distributions and populations. In light of these changes it was decided that a survey of Great Plains winter bird populations over the past four decades might prove interesting, using Christmas Bird Count data mostly accrued since the time of Root's landmark study, but using her results as a basis for temporal comparisons.

The region selected for coverage by this book represents the same region as that used earlier for analyzing the breeding birds of the Great Plains region (Johnsgard, 1979). It encloses most of the Great Plains south of Canada, and includes all of five states from North Dakota to Oklahoma (hereafter the "Plains states"), plus the Texas panhandle south to 34° N. latitude. The region encompasses a maximum north-south distance of slightly more than 1,000 miles, a maximum east-west distance of about 460 miles, and includes about 406,000 square miles.

Recent efforts by the National Audubon Society to provide a convenient on-line program (http://audubon2.org/cbchist/) for extracting CBC data by species, count sites and states or regions has made analyses far more practical than had been the case previously (National Audubon Society, 2008). Using this on-line program, statewide CBC data were extracted and summarized from the count-years 1967–1968 to 2006–2007 for the five Great Plains states from North Dakota through Oklahoma, plus the Texas panhandle. This 40-year period overlaps chronologically with Root's analysis by five years, and its mid-1960s beginning corresponds to a generally accepted onset of accelerated worldwide climate change. To bring the coverage more nearly up to date, abundance data text for the most recently available (2013–2014) summary has been added in the narrative text.

In the case of the Texas panhandle, five of the six commonly active CBC sites were selected to sample that little-populated region of northwestern Texas: Amarillo, Buffalo Lake National Wildlife Refuge, Lake Meredith (west), Muleshoe National Wildlife Refuge and Quitaque. It should be noted that two of these five locations are major wildlife refuges that often support massive numbers of wetland birds, and theses sites tend to exaggerate the relative influence of wetlands in the Texas panhandle. In recent years one or both of these refuges have not been, counted, reducing the consistency of the averages of wetland bird number from this region. Otherwise, these five sites are well distributed across the panhandle, and should provide a fair sampling of most of the region's winter avifauna.

For each of the five states and the Texas panhandle the summed and averaged data (mean birds/party-hour) for all regularly observed species were grouped into four ten-year intervals and are presented in tabular form. These tables of relative abundance by decades and states, include only those 144 species having an abundance of at least 0.1 bird per party hour. When species occurred at abundance levels averaging less than 0.01 individual per party-hour of observation on these summaries, the entry was simply reported as "present" (p) in the tables. Although an earlier version of this report included tabular data for all the species tallied during the period of study, these occupied over 100 pages, and many of the species were reported too rarely to provide any useful information on their distribution and relative abundance. However, our earlier report (Johnsgard and Shane, 2009) included all the 200-plus species that were reported during the study.

National population estimates and trends are reported for as many species as possible, usually on the basis of national Breeding Bird Survey (BBS) trend data from 1966 through 2012 (Sauer *et al.*, 2014) and Partners in Flight (PIF) population estimates for the 1990s (Rich *et al.*, 2004). For those species considered by PIF to be of continental conservation significance, PIF estimates of the percentage of these species' national population wintering within the Prairie Avifaunal Biome are also presented. This grassland-dominated region consists of Bird Conservation Regions 11, 17–19, and 21–23 (North American Bird Conservation Initiative. 2000). It includes those parts of interior North America that historically supported shortgrass, mixed-grass and tallgrass prairies, as well as the prairie-hardwood transition zone, and thus occupies roughly twice the area of landmass than that represented by the region analyzed here.

## Topography, Landforms and Climate of the Great Plains

The historic prairies and plains of central North America represented on of the largest and most uniform of the continent's major ecosystems, and the "grassland biome" evokes an image of vast herds of bison amid a sea of grass that once extended across the heartland of North America. Most of these grasslands have now been converted to grain fields or else have been subjected to such grazing pressure as to degrade them almost beyond recognition. Yet major remnants remain in national parks, national grasslands, and wildlife refuges, and tiny fragments are still to be found in rural cemeteries, railroad rights-of-way, and small nature reserves. In such places essentially all the original bird life of the plains can still be found. This book documents the present wintering distributions of the birds of this region, both as a biological analysis of this major component of the North American biota, and as supporting evidence of changing winter distributions associated with global warming effects.

Although there are minor variations, the overall topography of this region is an inclined plain, which slopes downward from the west to the east at an average gradient of about ten feet per mile. The lowest point in the region is in southeastern Oklahoma, at 323 feet above sea level. The Black Hills of western South Dakota provide an isolated and distinctive montane influence, with a maximum elevation of 7,242 feet at Harney Peak. Along the region's eastern limits the only highlands of significance are Oklahoma's Ouachita Mountains, which attain a maximum height of more than 2,000 feet. Over nearly the entire region, drainage is to the southeast into the Missouri and Mississippi systems; but in North Dakota the Souris and Red rivers are part of the Hudson Bay arctic drainage system.

The pattern of rainfall throughout this region is relatively simple. In general, precipitation increases from northwest to southeast, at the approximate rate of about one inch per 40 miles at the northern edge of the region, to one inch per ten miles at the southern edge. About three-fourths of the rainfall in the Great Plains occurs during the growing season, which ranges from about 100 days in northern North Dakota to 240 days in extreme southeastern Oklahoma. The wettest locality in the region is the Ouachita Mountain range of southeastern Oklahoma, with more than 50 inches of precipitation annually, while there are fewer than 14 inches of precipitation each year in various parts of the region's western boundary.

Evaporation rates increase correspondingly as one proceeds south. The highest rates of annual evaporation occur in the Staked Plain of western Texas, characterized by evaporation rates more than four times greater than precipitation rates. Much lower evaporation rates are typical of the cooler, more northerly states, and parts of eastern North Dakota can thus support a lush tall-grass prairie vegetation with less than 20 inches of precipitation a year. In contrast, the same amount of precipitation on the Staked Plain allows only for the barest stands of buffalo grass and other xerophytic plants.

Wind has a strong accelerating influence on evaporation rates during summer, and produces devastating effects on protoplasm in conjunction with sub-freezing temperatures. The Great Plains includes five of the country's six windiest states, with North Dakota ranking first, followed in sequence by Texas, Kansas and South Dakota. Nebraska is sixth. Southerly winds are usual during late spring and summer, when warm and moist Gulf Coast air drifts north and meets cooler air masses, generating heavy rains, thunderstorms, and occasional tornadoes. In contrast, northerly wind flows are typical during winter, as cold fronts sweep southward out of Canada, often transforming autumn into winter overnight.

Winters in the Great Plains are famous for their severity, bitter wind-chills and long periods of standing snow. They are also notable for the sometimes-sudden blizzards that may materialize with little warning, and may produce periods of little or no visibility for 24 hour or more, together with bone-chilling winds. Freezing rain can be as dangerous to birds as blizzards. During the 1990s one late-winter storm in Nebraska marked by high winds and freezing rain killed hundreds if not thousands of migrant sandhill cranes. Apparently panicked by high winds, freezing rain and a partly frozen Platte River, they left their roosts and, with ice blinding their eyes, flew headlong into trees and buildings. Crippled and dead cranes littered the landscape for weeks, providing a sudden food source for predators and scavengers. Some cripples were captured and moved to the Safari Park division of Omaha's Henry Doorly Zoo, where they formed the nucleus of a new crane exhibit. A similar event was described from Minnesota (Roberts, 1932), when vast flocks of Lapland longspurs were caught in a mid-March snowstorm, and flew fatally at night into buildings, light poles and the ground. Two small lakes, with areas totaling about two square miles, had at least 750,000 dead birds scattered on their surfaces. The total area affected by this storm was about 1,500 square miles, so many millions of birds were probably killed.

Annual snowfall in the higher elevations of the Black Hills is typically well over 100 inches, while on the arid Dakota plains to the east the average is closer to 20–30 inches. A probable maximum snowfall record for the entire Plains States was 191.5 inches, which occurred during 1919 at a high-elevation Lawrence County site in South Dakota's Black Hills. North of the Black Hills standing snow is typically present for more than three months annually, but this average declines to fewer than ten days annually in regions south of a line drawn approximately from Plainview, Texas to Tulsa, Oklahoma (Wishart, 2004). Standing snow presents a special survival problem for ground-foraging birds. However, it also provides an opportunity for them to avoid extreme cold temperatures by roosting below the snow, where wind-chill effects are absent and temperatures are often only a few degrees below freezing.

## Natural Vegetation and Avifauna of the Great Plains

Although enormous changes have occurred in the vegetation of the region, numerous historical records and sufficient relict communities still exist to provide a reasonable basis for understanding the pre-settlement distribution of vegetation types through the region. Largely on the basis of the vegetation map assembled by Küchler (1964), it is possible to estimate the relative abundance of major plant communities that once covered the land surface of the region. On that basis, it seems likely that the mapped region was once 81 percent grasslands, 13 percent hardwood deciduous forest or forest-grassland mosaic, three percent sage or arid scrub grasslands, and two percent coniferous forest or arid coniferous woodland. The remaining one percent is now covered by surface water, predominantly the result of recent river impoundments.

The grassland-dominated communities in the region consist of several associations, ranging from tall-grass prairies to short-grass plains or steppe vegetation. Of these, the tallgrass prairies at the eastern edge of the region are the most fragmented and rarest of American prairies, as well as being the most species-rich. Dominated by big bluestem (*Andropogon gerardi*), these prairies historically covered the eastern Dakotas, western Minnesota and Iowa, plus portions of eastern Nebraska and Kansas, terminating in northern Oklahoma. Stewart (1975) listed four primary species of breeding birds (upland sandpiper, bobolink, western meadowlark, and Savannah sparrow) abundantly associated with tallgrass prairies in North Dakota, as well as thirteen less common or secondary species. Characteristic winter birds include the Harris's sparrow and American tree sparrow, both of which are attracted to the brushy edges where tallgrass prairie meets deciduous woods, as well as to shelterbelt plantings and woodlots. Dark-eyed juncos, white-crowned and white-throated sparrows, spotted towhees, and other seed-eaters often join them.

To the west of the now-rare bluestem prairies in the Dakotas lies the eastern mixed-grass prairie (Stewart 1975) or the wheatgrass-bluestem-needle-grass prairie (Küchler, 1964). The dominant plants are shorter than those of tallgrass bluestem prairie, but a large number of flowering forbs are also characteristic. Stewart listed eleven primary and twelve secondary species of breeding birds associated with this vegetation type in North Dakota, including Baird's sparrow, chestnut-collared longspur, Sprague's pipit, and lark bunting. All of these species except the Baird's sparrow winter to varying degrees in grasslands of the southern Great Plains. Baumgarten and Baumgarten (1992) listed 14 species of terrestrial birds as typical winter visitors of Oklahoma's variously grazed tallgrass prairies. Besides those just mentioned for North Dakota, resident or wintering species of Oklahoma include the greater prairie-chicken, rough-legged hawk, short-eared owl, American tree sparrow, Le Conte's sparrow and lark sparrow, plus Lapland and Smith's longspurs. Snowy owls, northern shrikes and snow buntings are likely to occur on snow-covered northern grassland landscapes, but rarely reach as far south as Oklahoma or Texas. Many of the tall-grass and mixed-grass species also often occupy shortgrass and grazed or cultivated grasslands during winter. The ecological differences that tend to separate grassland birds during the breeding season tend to weaken or completely disappear during winter, when snow cover and food availability conditions may be more important than other environmental variables.

The wheat-grass-needlegrass, or western mixed-grass, prairies historically occupied nearly all of North Dakota from the Missouri Valley westward and extended over more than half of South Dakota. The native vegetation is predominantly short-grass species and scattered mid-grasses, plus a moderate number of broad-leaved forbs. Stewart (1975) listed six primary species of breeding birds and nine secondary species for the western mixed-grass, prairies, all of which he indicated as also characteristic of eastern mixed-grass prairies. Primary mixed-grass species include the sharp-tailed grouse, horned lark, western meadowlark, brown-headed cowbird, grasshopper sparrow and chestnut-collared longspur. All of these species commonly winter in the Great Plains, and most rely on both cultivated grains and native plant seeds to provide winter foods.

In several areas, extensive regions of sandy soil or sand dunes have greatly affected the distributions and types of native vegetation. The largest of these is the Nebraska Sandhills, where the vegetation consists mostly of widely spaced bunchgrasses, with the intervening areas very sparsely vegetated. The most common breeding birds of the upland Sandhills grasses are the horned lark, western meadowlark and mourning dove (Johnsgard, 1995), all of which winter locally or elsewhere on the Great Plains. The Sandhills are also an important region during winter for many other grassland birds such as sharp-tailed grouse, black-billed magpie, and several sparrows, as well as for raptors such as the prairie falcon, peregrine, short-eared owl and northern shrike. Arctic-breeding migrants such as the rough-legged hawk, snowy owl and gyrfalcon also winter in the Sandhills, as well as in other areas of extensive grasslands.

The short-grass prairie, or grama-buffalo grass association, occurs on localized slopes and dry exposures in the western Dakotas and over extensive portions of the region from western Nebraska southward to the arid Staked Plains of northwestern Texas. This "high plains" biota is adapted to withstand considerable aridity, and its array of both plants and animals is somewhat restricted. Stewart (1975) listed only two bird species (horned lark and chestnut-collared longspur) as primary breeding species in this community in North Dakota. Both of these species winter in similar Great Plains grasslands varying distances to the south. Secondary shortgrass species that winter on the western Great Plains include killdeer, mourning dove, western meadowlark, brown-headed cowbird and McCown's longspur. In the southern shortgrass plains the resident lesser prairie-chicken, scaled quail, greater roadrunner and Chihuahuan raven are also often present (Baumgartner and Baumgartner (1992). Smaller numbers of variably migratory raptors such as the golden eagle, ferruginous hawk, prairie falcon, northern harrier, burrowing owl and northern shrike may also occur.

A sage-dominated community grassland occurs on clay soils in the western Dakotas, where big sagebrush (*Artemisia tridentata* and *A. cana*) grow in conjunction with shortgrass vegetation and prickly-pear cactus (*Opuntia* spp.). Stewart (1975) listed only three bird species (greater sage-grouse, lark bunting, and Brewer's sparrow) as primary breeding species for this community type. Of these, the greater sage-grouse is a distinctive but increasingly rare resident, while the lark bunting and Brewer's sparrow winter commonly in the arid grasslands of Texas and the Southwest.

In southwestern Kansas and western Oklahoma the large areas of sandy soil associated with the Cimarron and other river systems support a similar sand-adapted vegetation composed of short grasses and sand sagebrush (*Artemisia filifolia*). Rising (1974) listed the scaled quail, lesser prairie-chicken and Chihuahuan raven as typical of this habitat in Kansas. All of these are resident or near-resident species, and during winter horned larks, longspurs and other migratory sparrows supplement the resident birds. During 19631995 Christmas Counts in the Cimarron National Grassland of Kansas, horned larks were consistently the most abundant species, but there were also very large numbers of American tree sparrows, Lapland longspurs and western meadowlarks (Cable, Seltman and Cook, 1996). Baumgartner and Baumgartner (1992) listed the black-crested titmouse, verdin and curve-billed thrasher as unique residents in these grasslands of western Oklahoma.

In Oklahoma the shortgrass community likewise includes such year-around residents as the lesser prairie-chicken, scaled quail, greater roadrunner, Chihuahuan raven, and lark bunting. In the Texas panhandle and southwestern Oklahoma an arid and thorny woodland variant of this vegetation type occurs, with the inclusion of mesquite (*Prosopis juliflora*) and acacias (*Acacia* spp.). The black-crested titmouse and the ladder-backed and golden-fronted woodpeckers are resident here, in addition to those just mentioned (Sutton, 1967; Baumgartner and Baumgartner, 1992; Seyffert, 2000).

Communities dominated by deciduous or hardwood tree species are diverse and are particularly abundant in the eastern and southeastern parts of the region. The northern deciduous forest communities of eastern North Dakota are mapped as a composite of types recognized separately by Küchler (1964) as "oak savanna" and "maple-basswood forest." Additionally, a substantial area of aspen grovelands, such as those in the Turtle Mountains of north-central North Dakota, is present. The bird species associated with all these forest types are essentially those typical of the eastern deciduous forest. Green and Janssen (1975) listed seventeen breeding birds as typical of the southeastern Minnesota hardwood forests. Those that regularly winter in the Great Plains states include the red-shouldered hawk, red-bellied woodpecker, Carolina and Bewick's wrens, blue-gray gnatcatcher, tufted titmouse and northern cardinal.

Along the river systems of the Dakotas, Nebraska, and Kansas, a distinctive "gallery" or riparian forest, called by Küchler the northern floodplain forest, provides an extremely important forest corridor linking eastern and western biotas. The evolutionary significance of these river systems as gene-flow corridors has been established by a variety of field studies on hybridization between species and subspecies of eastern and western birds (*e.g.* Rising, 1974). Stewart (1975) described the associated birds of several of these riparian forests in North Dakota, and Johnsgard (2001, 2007b, 2008) described them for the Platte and Niobrara river valleys of Nebraska. Northern deciduous forest and forest-edge non-passerines that regularly winter in the Great Plains states include the red-tailed, sharp-shinned and Cooper's hawks, mourning dove, great horned owl, downy, hairy and red-bellied woodpeckers, northern flicker and yellow-bellied sapsucker. Passerines include the black-billed magpie, American crow, black-capped chickadee, white-breasted nuthatch, American robin, spotted and eastern towhees, chipping, field and song sparrows, brown-headed cowbird and American goldfinch.

From the Missouri River valley of the Nebraska-Iowa border southward, a forest community dominated by oaks and hickories (*Carya* spp.) tends to replace the northern floodplain forests along major river systems, and extends to the uplands in wetter sites. Over much of eastern Kansas the oak-hickory forest occurs as a mosaic community with bluestem prairies. In eastern Oklahoma this mosaic pattern is replaced by the "cross timbers" community of large oaks (*Quercus marilandica* and *Q. stellata*), either growing in extensive groves or scattered and interspersed with tall grasses such as bluestems and other perennial grasses. Baumgartner and Baumgartner (1992) identified 55 species as nesting birds of the post oak–blackjack oak forest communities in Oklahoma. Most of the wintering species of this habitat are those mentioned earlier as typical of more northern latitude woodlands, but additional ones include the black vulture and Carolina wren, both of which are distinctly southern in their breeding distributions.

In the wetter portions of southeastern Oklahoma the forest becomes denser, and the oaks are supplemented with hickories *(Carya* spp.) and pines *(Pinus echinata* and *P. taeda),* resulting in an extensive oak-hickory-pine community more typical of the southeastern United States and the Atlantic piedmont. Baumgartner and Baumgartner (1992) identified 24 nesting species of Oklahoma's oak-hickory and oak-pine forests, of which the brown-headed nuthatch, pine warbler, and red-cockaded and pileated woodpeckers are distinctive year-around residents.

Along the floodplains of the lower Arkansas and Red rivers a distinctive local southern floodplain forest also occurs, with oaks, tupelo *(Nyssa aquatica),* and bald cypress *(Taxodium distichum)* sharing dominance. Of Oklahoma's several breeding warblers, only the pine warbler, which shifts from insects to seed-eating during cold weather, typically winters there.

Last, along the drainage of the Canadian River of Oklahoma and across the panhandle of Texas, an arid-adapted community dominated by scrubby, sometimes shin-high (shinnery) oaks *(Quercus mohriana* and *Q. havardi)* and little bluestem *(Schizachyrum scoparius)* occurs, together with various deciduous shrubs and occasional low evergreens such as junipers. The resident western scrub-jay and plain titmouse also reach their easternmost breeding limits in this habitat type, the jay often surviving on stored acorns and pinyon pine seeds. Baumgartner and Baumgartner (1992) listed 15 species as breeders in the Oklahoma shinnery oak community. Permanent residents of this habitat include lesser prairie-chicken, scaled quail, greater roadrunner, golden-fronted and ladder-backed woodpeckers, and the rock and canyon wrens.

The coniferous-dominated communities of the Great Plains region are relatively few and distinctive. The Black Hills coniferous forest, together with the other forests of ponderosa pine *(Pinus ponderosa)* of southwestern North Dakota and western Nebraska, provides the most typically Rocky Mountain biota to be found in the entire region. Black Hills resident species geographically associated with the Rocky Mountain forests include the Lewis's, black-backed, and three-toed woodpeckers, as well as the endemic white-winged race of the dark-eyed junco. The more broadly distributed brown creeper and ruby-crowned kinglet are also typical of the region, as is the stream-dependent American dipper. These species all winter locally or at relatively short distances away from the Black Hills.

Although the "northern coniferous forest" of America's northern latitudes does not quite reach into the multi-state region covered in this book, it exerts an important ecological influence. Regional coniferous vegetation types recognized by Küchler (1964) include coniferous bogs, the Great Lakes spruce-fir forest, and the Great Lakes pine forest. With few exceptions, the birds of all three community types are the same, and are essentially those associated with Canada's transcontinental boreal forest.

Green and Janssen (1975) identified a boreal avian component among Minnesota's breeding birds. Species that they associated with Minnesota's coniferous forests and that regularly winter in the Great Plains states include the hermit thrush, golden and ruby-crowned kinglets, yellow-rumped warbler, dark-eyed junco and Lincoln's sparrow. Additionally, invasions of red and white-winged crossbills, purple finches, pine grosbeaks, evening grosbeaks, common and hoary redpolls and pine siskins sometimes appear unexpectedly from boreal or western montane coniferous forests, temporarily decorating Great Plains plantings of pines, spruces and firs. Coniferous plantings also often serve as winter roosts for northern saw-whet and long-eared owls, and if junipers are present, their berries provide prime foods for Townsend's solitaires, Bohemian and cedar waxwings, as well as hardy thrushes.

In the Black Mesa country of northwestern Oklahoma an upland woodland community type dominated by low junipers (*Juniperus* spp.) and arid-adapted pines (*Pinus edulis* and *P. monophylla*) occurs locally on mesa slopes and along dry river channels. Here the juniper titmouse, bushtit, and pinyon jay occur, and are year-around residents.

### Climate Change and its General Effects on the Great Plains

For much of the latter 20th century, and especially during the past few decades, the Great Plains have experienced a warming trend that is part of a global phenomenon. The year 2014 had the highest annual average global temperature (58.24° F. ) of any year since such records began over 130 years ago. Previous heat records were broken in 2010 and 2005 (National Oceanic and Atmospheric Administration data). The last time the Earth's record for annual cold temperatures occurred more than a century ago, in 1911.

In the Great Plains this warming trend has been most apparent in recent milder winter temperatures, which are more noticeable in the northern states such as the Dakotas than in more southern latitudes. Thus the century-long average rate of winter warming, as reflected in statewide January temperatures between 1895 and 2008, averaged 0.44° F. degree of increase per decade in North Dakota but only 0.04° F. per decade in Oklahoma, a ten-fold difference in long-term mid-winter warming rates.

In the collective five Great Plains states considered here, the rate of mean January temperature increase has averaged 1.75° F. per decade during the four-decade period 1969–2008, suggesting a substantial increase in the rate of temperature change throughout the entire region over the past four decades (Table 1). Monthly mean January temperature for the three-decade period 1971–2000 were 7.98° F. in North Dakota, 16.11° F. in South Dakota, 22.73° F. in Nebraska, 28.77° F. in Kansas, and 36.11° F. in Oklahoma (U.S. Weather Bureau data). Over a broader time-frame, the average annual 1895–1994 temperature rose between 0.5 and 1.0° F. in Oklahoma and the Texas panhandle, about 1.0° F. in Kansas and Nebraska, and from 1.5–3.0° F. per decade in the Dakotas (Cunningham and Kroeger, 1996). Heating costs during winter in the northern plains are declining as a result but air conditioning costs will continue to increase throughout the entire region.

Calculations by the National Oceanic and Atmospheric Administration (NOAA) indicate that, under a scenario of higher global emissions of greenhouse gases, future Great Plains average temperature anywhere from a few to nearly ten degrees higher by the end of the 21st century. When compared with historic regional precipitation and evaporation patterns (Fig 1), regional spring precipitation totals will be diversely affected. Spring precipitation levels are likely to decrease substantially at the southwestern edge of the region (Kunkel, 2013). Reduced precipitation in the Great Plains has resulted in increased agricultural dependence on the region's vast High Plains aquifer: from the 1950s to 2012 its average total water-level declined 41.2 feet in Texas, 12.3 feet in Oklahoma, and 25.5 feet in Kansas. The Ogallala aquifer in Nebraska has by far the greatest underground water reserves anywhere in the Great Plains; its decline there over this period averaged only 0.3 feet. However, some heavily irrigated areas in the southern plains have had aquifer declines of as much as 100–250 feet.

Predicted reduced precipitation in the Rocky Mountains will result in smaller snow packs and reduced spring runoffs, a vital source of water for the entire region. Smaller snow packs will result in reduced spring flow of

**Table 1**
Mean January Temperatures (Fahrenheit), 1969–2008*

|  | 1969–1978 | 1979–1988 | 1989–1998 | 1999–2008 | 1895–2008* | |
|---|---|---|---|---|---|---|
|  |  |  |  |  | Avg. | Change |
| N. Dakota | 2.9° | 4.25° | 8.8° | 12.3° | 6.46° | +0.44°/decade |
| S. Dakota | 11.8° | 16.9° | 17.4° | 20.7° | 15.56° | +0.19°/decade |
| Nebraska | 19.5° | 22.3° | 24.55° | 25.7° | 22.53° | +0.11°/decade |
| Kansas | 26.4° | 26.3° | 30.9° | 32.0° | 28.92° | +0.10°/decade |
| Oklahoma | 34.4° | 34.9° | 38.2° | 39.2° | 36.60° | +0.04°/decade |
| 5-state average* | 19.0° | 20.9° | 24.0° | 26.0° | 22.01° | +0.18°/decade |

* U.S. Weather Bureau data

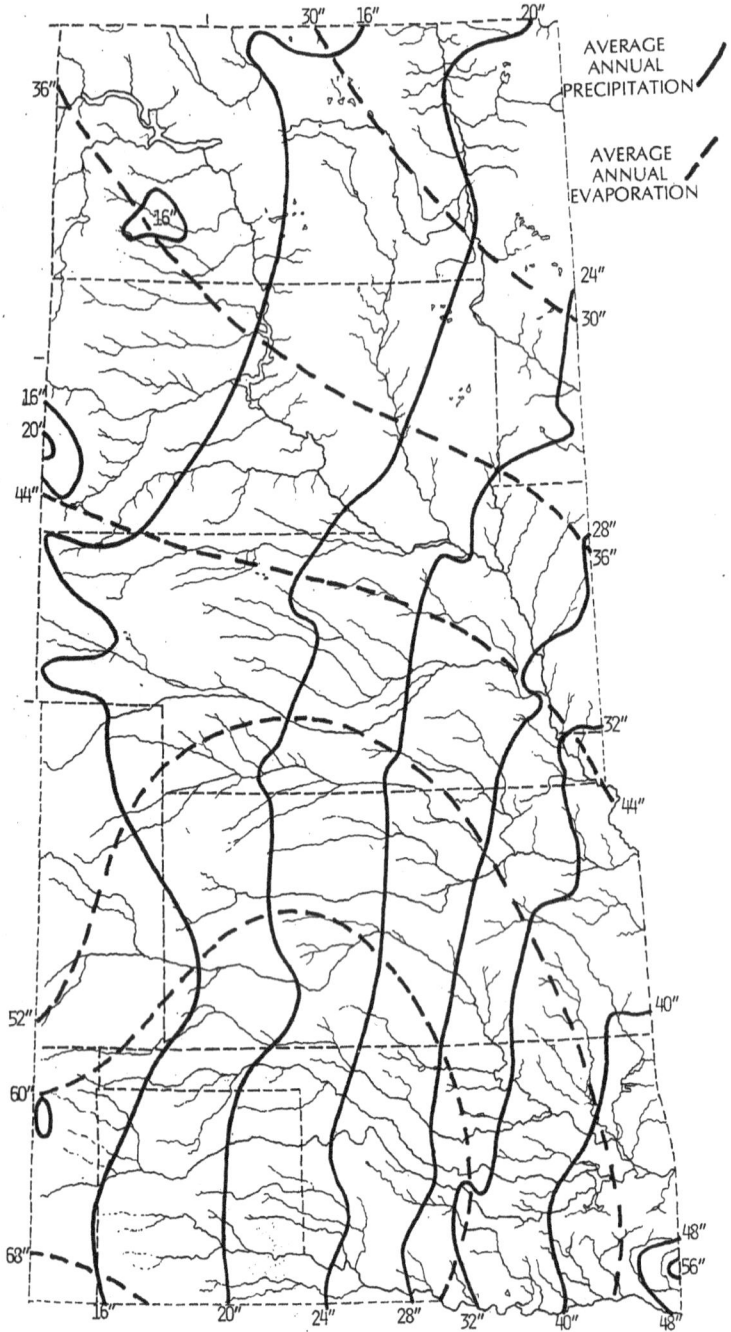

Fig. 1. Topography of the Great Plains states, showing historic mean annual precipitation and lake surface evaporation totals (inches). From Johnsgard (1979).

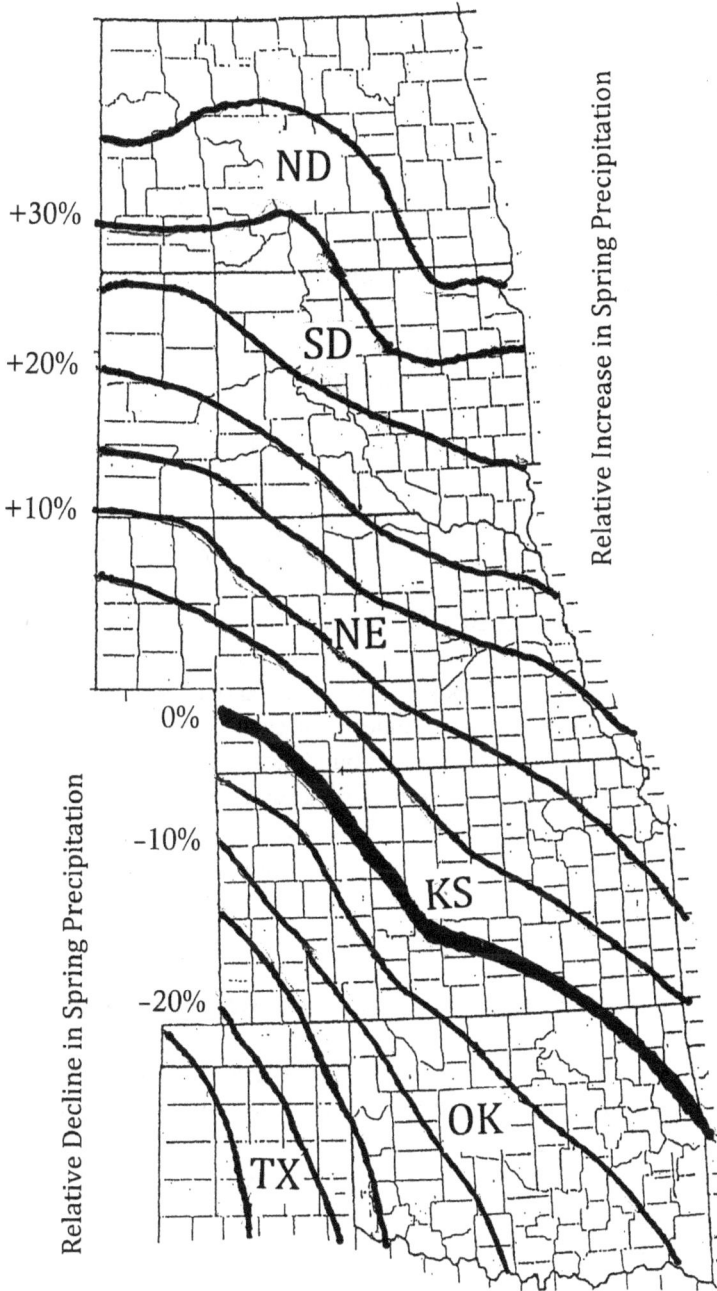

Fig. 2. Map of Great Plains states, showing projected percentage spring precipitation declines and increases by the late 2000s, under a scenario of high greenhouse gasses emissions (adapted from Karl et al., 2009).

important Plains rivers such as the Platte, affecting regional water supplies for irrigation, industry and municipal needs, and habitats for wetland-dependent species. Irrigation now accounts for nearly 95 percent of all Nebraska's water consumption; during the 2010–2012 drought, Lake McConaughy was reduced in volume to as little as about 25 percent of its holding capacity as a result of record irrigation demands.

With the expected progressively increasing levels of greenhouse gasses, the southwestern parts of the Great Plains states will potentially have reduction of 30–40 per cent in spring precipitation totals, and both increased and more severe droughts by the end of the century (Fig. 2).

In contrast, the northernmost parts of the region might have an increase of up to about 40 percent spring precipitation (Karl *et al.,* 2009). In 2011 Montana and Wyoming experienced their all-time wettest springs, and the Souris River near Minot, ND, crested at a new record of four feet above its previous high, causing two billion dollars in flood losses (U.S. Global Change Research Program, 2014). In recent decades the upper Missouri River has produced several billion-dollar floods in spite of an expensive array of dams and levees. Much of this precipitation increase in the north is likely to fall during extreme weather events, such as cloudbursts that cause severe flooding. With warmer spring and summer temperatures, the timing, duration and intensity of such catastrophic droughts and floods will be increased (Kunkel *et al.,* 2014).

There will be many associated undesirable side-effects, such as an increase in drought-related wildfires, increased abundance of drought-tolerant and rapidly-growing weeds, improved overwintering survival of insect pests, and phenological changes in seasonal phenomena, such as lengths of frost-free periods, timing of flowering and breeding, and the timing and amount of seasonal movements in migratory species. (Hogg, 2013; Rathke *et al.,* 2014).

### Climate Change and its Effects on Great Plains Birds

With regard to birds, these side-effects might also include changes in their breeding phenology or fecundity, in the composition and structure of their breeding and wintering habitats of bird, or their migration timing, routes and staging areas. Accelerating global climatic changes have already had many evident effects on birds. These include a poleward shift in avian wintering ranges (La Sorte and Thompson, 2007), northward movements in the breeding ranges of some North American birds (Hitch and Leeberg. 2007). Various other biological influences on birds and other wildlife have been reported (Peters and Lovejoy, 1992; Burton, 1995; Stavy, Dybala and Snyder, 2008; Wormworth and Mullen, undated). Less obvious indirect effects of climate change on a species might result from climate-based influences on regional parasites, diseases, competitors and predators. Other more dramatic effects of global warming, such as changes in the frequencies of severe rainstorms, hurricanes and other climatic disasters, may have massive short-term consequences on local or regional populations. During 2010 over three million acres of land were affected by wildfires in the U.S., and in Nebraska some 270,00 acres of forest lands were lost to fires during the extreme drought of 2012.

In North America the average spring arrival times of many short-distance migrants breeding in the Northeast occurred an average of nearly two weeks earlier during the second half of the 20th century than the first half (Butler, 2004). Over a 63-year period of the 20th century (1939–2001), at least 27 migratory species exhibited altered spring arrival dates at Delta Marsh, Manitoba, with 15 of the species arriving significantly earlier as the century progressed (Murphy-Klassen *et al.*, 2005). Mills (2005) found that both spring and fall migration patterns of passerines at Long Point, Ontario, were affected by global warming during the period 1975–2000. Fall migration was especially affected, with 13 of 14 species studied exhibiting delayed fall migrations.

The National Audubon Society has undertaken a major study the effects of climate change on 588 species of North American birds (http://climate. audubon.org/article/audubon-report-glance), using data from three decades of CBC surveys and Breeding Bird Surveys. More than half of the species studied will be adversely affected ("climate-threatened") by climate change, with 188 species judged likely to lose more than 50 percent of their current climatic range by 2080. The other 126 "climate-endangered" species are likely to lose more than half of their current range by 2050.

As some examples of poleward shifts in breeding ranges in the Great Plains, northern cardinals have increasingly moved north to become permanent residents of eastern South Dakota and southeastern North Dakota during the past century, and have nested west to Bismarck. Although still quite rare in winter, the red-bellied woodpecker also is moving slowly north into both eastern South Dakota and southern North Dakota along riparian woodlands. Northern mockingbirds now breed regularly in southern Nebraska, and have occasionally nested in South Dakota. Scissor-tailed flycatchers have begun nesting in southern Nebraska with some regularity. The great-tailed grackle has rapidly expanded its distribution substantially northward in the Great Plains from southern Texas (in the 1950s) through the past half-century, both in its summer and winter ranges (Dinsmore and Dinsmore, 1993). It began nesting in Oklahoma by 1959, Kansas by 1969, Nebraska by 1977 and South Dakota by the late 1990s. It has also been seen in North Dakota but, as of 2008, nesting in the state had not yet been reported. Likewise, the white-winged dove had been reported at least nine times in North Dakota by 2008

(Ron Martin, pers. comm.), and the tropically oriented Inca dove is now regular in Oklahoma, and has wandered north to North Dakota and Montana

Many migratory bird species have moved their wintering regions farther north in the Great Plains during the past 40 years, as lengths of the frost-free season have increased noticeably. The white-winged dove, Savannah sparrow, lark bunting and fish crow have all begun wintering as far north as Kansas. Inca doves have wintered in Nebraska, and it is likely that the tufted titmouse will move north into South Dakota as winters continue to ameliorate. Root reported that the American goldfinch had a winter population peak along the Texas–Oklahoma border, but by 2008 its peak mean regional abundance was in Nebraska. Snow geese and Ross's geese have shifted from wintering on the Gulf Coast of Texas to refuges as far north as northern Missouri and Kansas, while Canada geese are also commonly wintering in northern Kansas and Nebraska, and locally to the Dakotas. Double-crested cormorants, which at the time of Root's 19631972 analysis barely appeared on Oklahoma Christmas Counts, have increased a thousand-fold in average numbers seen there.

One result of such changes is that published breeding and wintering ranges of many bird species that are more than a decade or two old are no longer accurate. Likewise, average spring and fall migration arrival and departure dates are often based on data several decades old. Those reported for Nebraska by Johnsgard (1980) need to be adjusted from 2–4 weeks in both spring and fall to conform with current migration phenology patterns.

In Nebraska and Kansas, many water-dependent species are now wintering commonly on ice-free rivers and impoundments. The largest recent Christmas Count changes around Lincoln have occurred among Canada geese (from less than 1,000 in 1991 to about 12,000 birds in 2013), mallards and ring-billed gulls. All of these are water-related species that probably have benefited from increasingly later fall freeze-up and earlier spring thawing. Since 1998 three species of ducks and three sandpipers, as well as the western grebe and marsh wren, have all appeared on Lincoln Christmas Counts for the first time (Johnsgard, 2006). Similar trends have occurred in western Nebraska. Species that markedly increased in Scottsbluff's Christmas Counts during the second half of the 20th century include the Canada goose, American wigeon, blue jay and American crow, all of which have probably responded to the region's long-term warming trend (Johnsgard, 1998). Christmas Counts at Lake Mc-Conaughy continue to add new species nearly every year; the average number of species reported from the early 2000s was abut 95, but in more recent years the count has often exceeded 100, perhaps the result of better coverage.

Among terrestrial species, very large increases have occurred in Lincoln's counts of the American robin, dark-eyed junco and red-winged blackbird.

Other species that have increased to a lesser degree on Lincoln's Christmas Counts and that perhaps now winter in Nebraska in increasing numbers are the eastern bluebird, golden-crowned kinglet and yellow-rumped warbler. Some boreal species that have declined significantly on recent Lincoln counts and perhaps now regularly winter still farther north are the Bohemian waxwing, common redpoll, evening grosbeak and red crossbill (Johnsgard, 2006).

As an indication of the major shifts in species composition and relative abundance that have occurred in Christmas Counts over the past four decades, Table 2 lists the five most commonly occurring species seen in each of the Great Plains region during the period 1968 to 2007. There is a notable shift of relative abundance in the northern Plains states (especially North Dakota) from predominantly small cold-adapted passerines such as snow bunting, Harris's sparrow, Bohemian waxwing and horned lark to increasingly water-dependent birds such as mallard and Canada goose. The snow goose and greater white-fronted goose have also increased relatively, especially in the central Plains states of Kansas and Oklahoma. Although not entirely water-dependent, the red-winged blackbird is certainly closely associated with wetlands, and its numbers have greatly increased over the past four decades as the ice-free season has increased, a trend that is especially evident in the Texas panhandle.

More significant to humans than changes in avian migration patterns and breeding times is the role that global warming is having on both human and avian-borne diseases. The 1999 invasion of West Nile virus in the U.S. is an example of a previously tropical disease that in recent years has moved into temperate-climate regions and has caused great mortality both to birds and humans in the U.S. and Europe. Another potential example is highly pathogenic avian influenza (HPAI, strain H5N1), a mosquito-borne disease that originated in the Old World subtropics. By 2008 the virus had been found in some North American wild birds, but so far is not known to have measurably affected our native bird populations. If mutations occur that make this lethal virus easily transmittable to humans it could produce far more devastating consequences than has West Nile virus, both for birds and humans.

In spite of the popular positions against global warming espoused by some persons for political or religious reasons, it is now far too late to argue about whether global warming is occurring and if humans have caused it; that horse left the barn long ago. The winter conditions typical of the early to mid-1900s in the northern plains are history. The only question now is what can be done to limit the release of greenhouse gasses into the atmosphere before there is a veritable worldwide stampede of undesirable and eventually cataclysmic ecological effects.

**Table 2**

Most Abundant (in descending order) Christmas Bird Count Species, by Decades

| 1968–1977 | 1968-2007 | 1998-2007 |
|---|---|---|
| **North Dakota** | | |
| House Sparrow | House Sparrow | Canada Goose |
| Snow Bunting | Canada Goose | House Sparrow |
| Harris's Sparrow | Snow Bunting | Snow Bunting |
| Bohemian Waxwing | Harris's Sparrow | Mallard |
| Horned Lark | Mallard | Lapland Longspur |
| **South Dakota** | | |
| Mallard | Mallard | Mallard |
| House Sparrow | Canada Goose | Canada Goose |
| European Starling | House Sparrow | American Robin |
| American Robin | European Starling | European Starling |
| Canada Goose | American Robin | House Sparrow |
| **Nebraska** | | |
| Mallard | Mallard | Mallard |
| European Starling | Red-winged Blackbird | Canada Goose |
| House Sparrow | European Starling | European Starling |
| Horned Lark | Canada Goose | American Robin |
| Canada Goose | House Sparrow | American Crow |
| **Kansas** | | |
| Red-winged Blackbird | Red-winged Blackbird | Red-winged Blackbird |
| European Starling | European Starling | European Starling |
| Mallard | Mallard | Mallard |
| Canada Goose | Canada Goose | Gr. White-fronted Goose |
| Lapland Longspur | Snow Goose | Lapland Longspur |
| **Oklahoma** | | |
| Red-winged Blackbird | Red-winged Blackbird | Red-winged Blackbird |
| European Starling | Mallard | European Starling |
| Canada Goose | European Starling | Mallard |
| American Wigeon | Canada Goose | American Robin |
| Common Merganser | Common Grackle | Snow Goose |
| **Five-state average, North Dakota south through Oklahoma** | | |
| Red-winged Blackbird | Red-winged Blackbird | European Starling |
| European Starling | European Starling | Mallard |
| Mallard | Mallard | Snow Goose |
| Canada Goose | Canada Goose | Gr. White-fronted Goose |
| House Sparrow | Common Grackle | American Robin |
| **Northwest Texas (5 sites)** | | |
| Sandhill Crane | Sandhill Crane | Red-winged Blackbird |
| Northern Pintail | Red-winged Blackbird | Sandhill Crane |
| Canada Goose | Northern Pintail | Canada Goose |
| Mallard | Canada Goose | Northern Pintail |
| Red-winged Blackbird | Mallard | Mallard |

# Part 2: The Winter Birds of the Great Plains States

## FAMILY ANATIDAE: SWANS, GEESE AND DUCKS

### Greater White-fronted Goose
*Anser albifrons*

The white-fronted goose is one of the "gray geese" of the northern hemisphere; it is a close relative of the European gray-lag goose (*Anser anser*), which is the ancestor of most types of domestic geese. White-fronted geese are high-arctic breeders, nesting in lowland tundras of North America and Eurasia. The sounds of their high-pitched, laughing calls emanating from migrating birds flying unseen overhead provide one of the recurrent seasonal thrills of a naturalist's year. They are slightly later spring and earlier autumn migrants than are snow geese, and in contrast to that species tend to migrate in small flocks, occasionally in the company of mid-sized Canada geese.

*Winter Distribution.* The mid-continental population of this species largely winters along the Texas coast (Root, 1988). Its spring and autumn migration routes are centered on and largely confined to the Great Plains, with an estimated 80 percent of the population passing through the central Platte River valley each spring en route to arctic breeding grounds. Birds banded or recovered in South Dakota have been documented south to northern Mexico and north to the arctic coasts of Alaska and Canada (Tallman, Swanson and Palmer, 2002).

*Seasonality and Migrations.* The greater white-fronted goose is a widespread and common seasonal migrant, wintering southwardly. This species was reported four years during Christmas Bird Counts from 1968 to 2007 in North Dakota, 17 years in South Dakota, 26 years in Nebraska, 39 years in Kansas, 37 years in Oklahoma and one year in the Texas panhandle. Wintering in South Dakota is rare, and mostly consists of single birds (Tallman, Swanson and Palmer, 2002). Fifteen final autumn Nebraska sightings are from October 12 to December 29, with a median of November 6 (Johnsgard, 1980). It is local in Kansas during winter (Thompson *et al.*, 2011), with extreme fall-to-spring dates from September 19 to May 22 (Thompson and Ely, 1989). It is considered a winter resident in Oklahoma, with continuous monthly records extending from September to April (Woods and Schnell, 1984). There are a few December and January sightings for the

Texas panhandle (Seyffert, 2001). The average number of birds observed per party-hour for the 2013–2014 Christmas Bird Count were: N. Dakota 0, S. Dakota 0, Nebraska 2.80, Kansas 5.35, Oklahoma 5.35, northwest Texas 0. For complete species abundance data, see Appendix 1:1.

*Habitats.* Migrants are associated with large marshes, shallow lakes, wide rivers with bars and islands, and adjacent agricultural grain fields.

*National Population.* Breeding Bird Survey trend data are not available for this arctic-nesting goose. Its 2014 Pacific Coast population was estimated as 637,000 birds, while the mid-continent population was about 750,000 birds (U. S. Fish and Wildlife Service, 2014). There is also a Eurasian population that numbers in the hundreds of thousands (Kear, 2005).

*Further Reading.* Palmer, 1973; Owen, 1980; Godfrey, 1986: Johnsgard, 1975a; Ell and Dzubin, 1994 (*The Birds of North America:* No. 131); Kear, 2005; Baldassare, 2014.

## Snow Goose
### *Chen caerulescens*

The sight of immaculate white-bodied geese, projected against a cerulean sky, offers one of the best reasons for choosing to live in the Great Plains. Even in a city the size of Lincoln, where in early March countless skeins of snow geese can be seen streaming above the city for hours on end, the wonders of the natural world can be brought close to home. Their barking calls remind me of a pack of dogs in full cry, racing toward the arctic as fast as their wings will carry them. They are among the earliest of spring waterfowl migrants, following ice break-up on rivers and lakes, and likewise are late autumn migrants, increasingly wintering in the central Plains States, especially in the vicinity of wildlife refuges.

*Winter Distribution.* Root (1988) determined that the Missouri River valley, where Nebraska, Kansas, Iowa and Missouri meet (the vicinity of Squaw Creek National Wildlife Refuge) was the second-most important concentration area for snow geese (both white and blue morphs) in late December. Since then, refuge management changes, altered agricultural practices and warmer winters have had major effects on snow goose migration patterns as to both timing and major wintering sites. Additionally, increased continental snow goose populations during the past few decades have resulted in greatly relaxed hunting regulations and seasons, blurring any possible climate-induced effects. The large numbers reported on recent Kansas counts are probably a reflection of late autumn migrations rather than of wintering birds. However, certainly far more geese (especially snows and Canadas) now winter along the Missouri Valley of Kansas and Missouri than was the case during the 1960s. Over the four-decade period 1967–2006, blue-morph birds made up 27.5 percent of all the snow geese counted in the five-state Plains region. This proportion declines rapidly one moves westward across the Great Plains. Over time, the proportion of blue-morph ("blue geese") birds in the central and western states has slowly increased, for reasons that are still uncertain, but it has been speculated that there are possible differential morph responses to gradually warming arctic climates. For example, during the 1950s the proportion of blue-morph birds then nesting along the arctic coastline of the Northwest Territories was nearly zero, but the blue form was increasing at a rate of about one or two percent annually (Cooch, 1960). A few blue-morph birds have recently been seen as far west as the Pacific coast; in Oregon their frequency was estimated at about 1 in 10,000 during the 1990s (Gilligan *et al.,* 1994).

*Seasonality and Migrations.* The snow goose is a widespread and abundant seasonal migrant, wintering southwardly. This species was reported four years during Christmas Bird Counts from 1968 to 2007 in North Dakota, nine years in South Dakota, 12 years in Nebraska, 20 years in Kansas and Oklahoma and 11 years in the Texas panhandle. In South Dakota, most early winter records are autumn stragglers (Tallman, Swanson and Palmer, 2002). Thirty-eight final autumn Nebraska sightings are from October 26 to December 31, with a median of December 2 (Johnsgard, 1980). The snow goose is locally abundant in Kansas during winter (Thompson *et al.,* 2011), with extreme dates from August 14 to May 26 (Thompson and Ely, 1989). It is a winter resident in Oklahoma, with continuous monthly records extending from September to May (Woods and Schnell, 1984). Wintering is also regular in the Texas panhandle (Seyffert, 2001). The average number of birds observed per party-hour for the 2013–2014 Christmas Bird Count were: N. Dakota 0, S. Dakota 1.43, Nebraska 5.0, Kansas 545.76, Oklahoma 184.5, northwest Texas 0. The very high Kansas and Oklahoma numbers reflect the importance of a few major wildlife refuges. For complete species abundance data, see Appendix 1:2.

*Habitats.* Marshes, sloughs, river bottom meadows and croplands such as cornfields are used on migration. Lakes or reservoirs near croplands are also utilized. Snow geese in the Great Plains have increased tremendously in the past few decades; current populations of about five million birds are more than their tundra breeding grounds can support. The birds have also shifted their migration route from their traditional route along the Missouri River more than 100 miles west into Nebraska's Platte Valley during spring. This shift brings them into competitive contact with millions of migrating cackling geese, Canada geese, greater white-fronted geese and sandhill cranes.

*National Population.* Breeding Bird Survey trend data are not available for this arctic-breeding species. was estimated as about 250,000 birds, while the Western Arctic/Wrangel Island population was judged at about 1,351,000 The 2014 mid-continent population of snow and Ross's geese was about 3,800,000 birds (U. S. Fish and Wildlife Service, 2014); Several million snow/Ross's geese have been estimated to be present in Nebraska's Platte Valley and nearby Rainwater Basin during some recent springs (Johnsgard, 2010). Snow geese also breed in Greenland (the greater race) and off the coast of northeastern Siberia (Wrangel Island).

*Further Reading.* Palmer, 1973; Johnsgard, 1975a, 2010; Owen, 1980; Godfrey, 1986: Mowbray, Cooke and Ganter, 2000 (*The Birds of North America:* No. 514); Kear, 2005; Baldassare, 2014.

## Ross's Goose
### Chen rossii

The Ross's goose and cackling goose are the miniature counterparts of the snow and Canada geese. Both are high-arctic breeders whose small size allows their young to be raised to fledging in the shortest possible time (42-48 days), and thereby being able to escape the breeding grounds before the first storms of winter. Ross's geese have relatively longer wings and tend to migrate farther south than their larger relatives. They also breed in denser concentrations than snow geese, but their tiny size makes them more vulnerable to arctic foxes, and their goslings often fall prey to gulls and jaegers.

*Winter Distribution.* Like the snow goose, the Ross's goose has had an impressive population explosion since the 1960s, with a substantial expansion eastward into the Great Plains states. This increase is reflected in the increasing mean numbers reported for the five-state Plains region, from 0.0002 birds per party-hour during the 1968–1977 decade to 1.33 per party-hour during 1998–2007. Ross's geese migrate in the Great Plains with snow geese, and winter in the same regions. As considerably smaller birds, they might seem less able to winter as far north in the central Missouri Valley as snow geese have increasingly done in recent years.

*Seasonality and Migrations.* The Ross's goose is a widespread and common seasonal migrant, wintering southwardly. This species was not reported on any Christmas Bird Counts from 1968 to 2007 in North Dakota, but it appeared nine years in both South Dakota and Nebraska, 19 years in Kansas, 29 years in Oklahoma and five years in the Texas panhandle. Over the entire 40-year period of our Christmas Counts study, Ross's geese made up 2.89 percent of the combined snow/Ross's goose population in the five-state Plains region. For the recent decade 1997–2006 the percentage of Ross's goose was 3.55 percent, suggesting that the Ross's goose is still increasing relatively faster than snow geese in the Great Plains. Ross's geese first appeared on South Dakota's Christmas Count in 1981, on Nebraska's in 1995, on Kansas' in 1973, and on Oklahoma's in 1977. In Nebraska, the species has otherwise been reported regularly since about 1970, and Ross' geese made up two percent of a flock of 1,200 mixed snow and Ross's geese killed during a 1990 York County tornado. Five late autumn Nebraska records are from November 10 to December 22, with a mean of November 26 (Johnsgard, 2007a). The Ross's goose is an uncommon fall migrant and rare winter visitant in Kansas (Thompson *et al.*, 2011), having been reported from October 18 to April 22. (Thompson and Ely, 1989). It has been considered a rare winter visitor in Oklahoma, with continuous monthly records extending from November to April (Woods and Schnell, 1984). Wintering is now regular and increasingly common in the Texas panhandle (Seyffert, 2001). An extremely small proportion (under one percent) of the birds are blue-morph plumage types, of still uncertain genetic relationship to blue-morph snow geese (Thompson *et al.*, 2011; Johnsgard, 2014). The average number of birds observed per party-hour for the 2013–2014 Christmas Bird Count were: N. Dakota 0, S. Dakota 0, Nebraska 0.15, Kansas 3.73, Oklahoma 3.97, northwest Texas 0. For complete species abundance data, see Appendix 1:3.

*Habitats.* These geese occupy the same habitats as snow geese, and they are often found together.

*National Population.* Breeding Bird Survey trends and good population data are not available for this arctic-breeding goose, but the Ross's goose population seemingly had been increasing during the past decade or more at a comparable rate as the snow goose, if not more rapidly. The Western Central Flyway population of the Ross's goose was estimated in 2013 at about 75,000 birds U. S. Fish and Wildlife Service, 2013). The two largest known nesting colonies (Karrak Lake, along Queen Maud Gulf on Canada's central arctic coast and the McConnell River on Hudson's Bay's western coast) had a combined population of 767,000 birds in 2007 (U. S. Fish and Wildlife Service, 2007). Between 2005–2014 the mid-continent snow/Ross's geese populations increased about 7% annually. By 2014, the Karrak Lake colony alone held some 700,000 Ross's geese (U. S. Fish and Wildlife Service, 2014). Although these birds typically winter in California, probably all of those from the Hudson Bay region now migrate through the Great Plains.

*Further Reading.* Palmer, 1973; Johnsgard, 1975a, 2014; Owen, 1980; Godfrey, 1986: Ryder and Alisauskas, 1995 *(The Birds of North America:* No. 162); Kear, 2005; Baldassare, 2014.

## Cackling Goose
*Branta hutchinsii*

Some birders may have been initially pleased when ornithologists decided to taxonomically "split" the cackling goose from the Canada goose as a new species, as it provided one more species for their life lists. However, it has proven to be a headache in the Great Plains, where the smallest regional race of the Canada goose (*B. c. parvipes*) is nearly identical in size and appearance to the cackling goose, making the Christmas Counts for both species suspect. Because of this difficulty of visually separating cackling geese from small Canada geese, reliable winter data for the cackling goose are still lacking.

*Winter Distribution.* Because of the very recent recognition of the cackling goose as a distinct species, only three years of Christmas Count data were available, making it impossible to trace long-term wintering patterns. In the Great Plains these small geese winter from the Staked Plain of Texas and eastern New Mexico south into northern Mexico, and also along the Gulf coast of Texas and northeastern Mexico. The Alaska breeding population (*B. h. minima*) typically winters in California.

*Seasonality and Migrations.* The cackling goose is a widespread and uncommon seasonal migrant, wintering southwardly. Too few records of this recently recognized species are available to judge its migration pattern clearly. Like the Canada goose, the cackling goose is present in Nebraska from late February or early March to mid-April, and from early October to November or December; a few birds may winter locally with Canada geese in locations having open water as far north as Kansas or Nebraska. The average number of birds observed per party-hour for the 2013–2014 Christmas Bird Count were: N. Dakota 0.03, S. Dakota 0.06, Nebraska 8.14, Kansas 20.73, Oklahoma 4.08, northwest Texas 0. For complete species abundance data, see Appendix 1:4.

*Habitats.* These geese mix with and occur in the same habitats as Canada geese.

*National Population.* Breeding Bird Survey trend data are not available for this arctic-breeding goose. Cackling geese breed widely across the central and eastern Canadian arctic, and, in a separate population (the typical dark-breasted cackling goose), breeds in western Alaska and winters from Washington state south to California. The light-breasted (so-called "Hutchin's) population of this species that migrates through eastern parts of the Great Plains states is called the Tallgrass Prairie flock. It includes those *B. h. hutchinsii* breeding in the eastern Canadian arctic along the Hudson's Bay west coast, as well as others breeding on Baffin and Southampton islands. Its winter populations have increased progressively since 1970, and in 2007 were estimated as 680,300 (U. S. Fish and Wildlife Service, 2007), a record high. The 2014 cackling goose estimate was 281,000, these two widely differing estimates perhaps reflecting both population growth and the difficulties of field separation from small Canada geese.

The other major group of small white-cheeked geese, which passes through western parts of the Great Plains, is the Shortgrass Prairie flock. These birds, estimated at 190,500 in 2007 (U. S. Fish and Wildlife Service, 2007), include that part of *B. h. hutchinsii* that breeds on Canada's central arctic coasts of Northwest Territories and on Victoria and Banks Islands, and the slightly larger Canada goose race *B. c. parvipes,* which breeds throughout the subarctic interior of Canada's Northwest Territories south to northern Alberta. These two populations both migrate south through the western Great Plains, and are especially prevalent in western Nebraska. The latter population winters from southeastern Colorado and adjacent western Nebraska south to the playa lakes region of northwestern Texas and northeastern New Mexico.

The smaller *B. h. hutchinsii* winters somewhat farther south than *parvipes*, with a westerly population component being centered in western Texas, and a more easterly one mostly concentrated along coastal Texas and adjacent northeastern Mexico (Owen, 1980). Together there may be about 900,000 of these small white-cheeked geese migrating through the Great Plains. Cackling geese barely registered in some state-wide Christmas Count totals when *hutchinsii* was first recognized as a separate species in 2005, but high-count species summaries for 2004-2005 and 2007-2008 showed respective totals of 74,800 and 37,000 cackling geese at Salt Plains National Wildlife Refuge, Oklahoma.

*Further Reading.* Palmer, 1973; Owen, 1980; Godfrey, 1986; Dickson, 2000; Mowbray *et al.*, 2003 (*The Birds of North America:* No. 682); Kear, 2005; Baldassare, 2014.

## Canada Goose
*Branta canadensis*

Few changes in the bird life of the Great Plains have been more evident during the past half-century than the status of the Canada goose. It has transformed itself from a wild, rarely seen bird of distant marshes to a city-dwelling loafer, content to spend its days in parks and golf courses, barely bothering to get out of the way of moving traffic. Yet, the sight and sounds of a flight of a flock of Canada geese overhead is still magical, even if their voices can hardly be heard above the bustle of city noises. Part of the population explosion has resulted from the releases of very large (*B. c. maxima*) hand-reared Canada geese at locations across the Great Plains since the 1960s, producing many local residential or semi-residential goose populations that are very tolerant of living near humans. In Lincoln, several hundred of such "nuisance" geese were once captured at one of our city parks, and were deported by truck several hundred miles to suitable habitat in the western part of the state. Most were back at the park within a few days after their release, apparently none the worse for wear and tear. The resident 2013 population around Lincoln was about 10,000–12,000 birds.

*Winter Distribution.* Root (1988) mentioned Kansas, Missouri, western Texas, and the border region of Oklahoma and Texas as important Great Plains wintering areas for Canada geese. This description may still largely apply, but these adaptable geese are largely concentrated around wildlife refuges, large ice-free rivers, lakes and impoundments, and even city parks. Populations in the five-state Plains region have increased from about 50 individuals per party-hour, during the middle 1960s and 1970s, to nearly 150 per-party hour only four decades later. Counts in the Plains states from North Dakota to Kansas have increased, especially those in Kansas, but Oklahoma and Texas panhandle counts have declined, probably reflecting a movement to more northerly wintering areas.

*Seasonality and Migrations.* The Canada goose is an abundant Great Plains seasonal migrant and widespread breeder, wintering southwardly. This species was reported 30 years during Christmas Bird Counts from 1968 to 2007 in North Dakota, 37 years in South Dakota and Nebraska, and 38 years in Kansas, Oklahoma and the Texas panhandle. In South Dakota, several thousand have wintered locally on reservoirs and near power plants in recent years (Tallman, Swanson and Palmer, 2002). Fifty-four final autumn Nebraska sightings are from October 18 to December 31, with a median of December 10 (Johnsgard, 1980), but by 2008 well over 100,000

Canada geese were regularly wintering in Nebraska, especially in the Platte, North Platte and Republican valleys. This goose is locally common in Kansas during winter, especially on large reservoirs (Thompson *et al.*, 2011), and is a permanent resident in Oklahoma (Woods and Schnell, 1984). It is considered a common to abundant winter visitor in the Texas panhandle (Seyffert, 2001). The average number of birds observed per party-hour for the 2013–2014 Christmas Bird Count were: N. Dakota 34.68, S. Dakota 30.67, Nebraska 166.09, Kansas 91.02, Oklahoma 22.23, northwest Texas 18.92. For complete species abundance data, see Appendix 1:5.

*Habitats.* Migrant and wintering birds are found on large marshes, lakes or reservoirs, and nearby grain fields.

*National Population.* Breeding Bird Surveys between 1966 and 2012 indicate that this species exhibited a survey-wide population increase (10.0% annually) during that period. The total U.S. and Canadian 2007 population of this species was estimated as about 5.5 million birds (U. S. Fish and Wildlife Service, 2007). The populations that pass through Great Plains states include three components. The Great Plains population, consisting mostly of the high plains subspecies *B. c. moffitti,* increased greatly between 1970 and 1999. The Western Prairie population, a mixture of subspecies (from largest to smallest, *B. c. maxima, B. c. moffitti,* and *B. c. interior),* numbered about 100,000 birds by 1999 (Dickson, 2000), and by 2014 the combined Western Prairie and Great Plains populations of Canada geese were judged at 568,000 birds. The population size of the even smaller subarctic-breeding race *B. c. parvipes* is uncertain, as it is difficult if not impossible to separate it in the field from *B. hutchinsii* (see previous account). The Short Grass Prairie and Tallgrass Prairie components of these small geese were judged to number about 567,000 birds in 2014. All told, the resident and migratory Great Plains populations of all races of Canada geese averaged almost two million birds in 2009–2011 (Johnsgard, 2012a), and the entire North American Canada goose population had probably exceeded six million by 2014. There are other introduced populations elsewhere in the world, including Europe, New Zealand and Australia.

*Further Reading.* Palmer, 1973; Johnsgard, 1975a, 2012a; Owen, 1980; Godfrey, 1986; Mowbray *et al.*, 2003 (*The Birds of North America:* No. 682); Kear, 2005; Baldassare, 2014

## Tundra Swan
### *Cygnus columbianus*

My most unforgettable sight of tundra swans (then called whistling swans) occurred along the arctic coast of western Alaska. There I once watched a pair of swans flying north in front of smoky clouds along the edge of the Bering Sea, illuminated by the pale yellowish light of a near-midnight sun. It was early June, but the tundra was still winter-brown, with ice flows grudgingly making their way out to sea in a nearby river, where long-tailed ducks and spectacled eiders were engaging in last-minute courting. I later decided that tundra swan, rather than its traditional name "whistling swan," was perfect for this magnificent bird. Birds from the eastern Canadian arctic pass through North Dakota in large numbers while on autumn migration, but none remain to winter, as most continue on to winter in the Chesapeake Bay region.

*Winter Distribution.* Root (1988) reported no winter concentrations of tundra swans in the Great Plains, and that situation still applies. Those reported from the Great Plains during late December are all likely to be vagrants (most wintering occurs along the Atlantic and Pacific coasts), or some may possibly even be misidentified trumpeter swans. The vast majority of tundra swans migrating through eastern North Dakota head southeast through southern Minnesota and winter in the Chesapeake Bay region, so probably most of those seen south of North Dakota are vagrants that somehow strayed off their usual southeasterly route.

*Seasonality and Migrations.* The tundra swan is an uncommon seasonal migrant in the northern Great Plains, mainly wintering coastally. This species was reported eight years during Christmas Bird Counts from 1968 to 2007 in North Dakota, four years in South Dakota, one year in Nebraska, 17 years in Kansas, 15 years in Oklahoma and one year in the Texas panhandle. Few records exist after December in South Dakota (Tallman, Swanson and Palmer, 2002). Eleven autumn Nebraska sightings are from October 21 to December 14, with a median of November 22 (Johnsgard, 1980). It is very rare in Kansas during winter (Thompson *et al.*, 2011), with extreme dates between November 21 and April 26 (Thompson and Ely, 1989). It is considered a rare winter visitor in Oklahoma, with records for all months from October to April (Woods and Schnell, 1984). There are a several autumn and winter sightings for the Texas panhandle, mostly from November to February (Seyffert, 2001). The average number of birds observed per party-hour for the 2013–2014 Christmas Bird Count were: N. Dakota 0, S. Dakota 0.52, Nebraska 0, Kansas 0.14, Oklahoma 0, northwest Texas 0. For complete species abundance data, see Appendix 1:6.

*Habitats.* Migrants use shallow lakes, marshes and adjacent flooded fields.

*National Population.* Breeding Bird Survey trend data are not available for this arctic-nesting swan. The 2014 eastern and western populations were estimated as 105,000 and 68,200 respectively (U. S. Fish and Wildlife Service, 2014). There is also a population (previously the "Bewick's swan") breeding in arctic Eurasia that numbers in at least the tens of thousands (Kear, 2005); the two populations are now regarded as subspecies, rather than their historic recognition as separate species.

*Further Reading.* Palmer, 1973; Johnsgard, 1975a; Godfrey, 1986; Limpert and Earnst, 1994 *(The Birds of North America:* No. 89); Kear, 2005; Baldassare, 2014.


## Trumpeter Swan
*Cygnus buccinator*


Another victory in America's conservation war has been the restoration of the trumpeter swan. As a child, I thought I would never live to see one in my lifetime, as they were then thought to be almost entirely limited to a single wildlife refuge (Red Rock Lakes) west of Yellowstone National Park. The Canadian population had not been fully inventoried at that time, and the Alaskan population was still entirely unknown. Now the species is well established in the northern Great Plains as a result of re-introduction efforts, and at least in Nebraska and South Dakota it is not unusual to see nesting pairs on some of these states' larger marshes.

*Winter Distribution.* The development of a breeding stock of trumpeter swans at Lacreek National Wildlife Refuge in South Dakota during the early 1960s has produced a semi-residential Great Plains population at least 1,000 birds. Many of these birds winter south in the Nebraska Sandhills, and additionally some trumpeter swans banded in Saskatchewan have been recovered in South Dakota (Tallman, Swanson and Palmer, 2002). Birds resulting from a more recently established Minnesota population (at least 5,000 birds by 2010) often winter in Missouri, but have been seen south to the Texas panhandle (Seyffert, 2001).

*Seasonality and Migrations.* The trumpeter swan is a resident or limited seasonal migrant in the northern Great Plains, with recently increasing introduced populations. This species was reported on only one Christmas Bird Count from 1968 to 2007 in North Dakota, but was seen nine years in South Dakota, 11 years in Nebraska, six years in Kansas, five years in Oklahoma and one year in the Texas panhandle. In South Dakota, hundreds may occur during winter in and near Lacreek National Wildlife Refuge (Tallman, Swanson and Palmer, 2002). Some swans regularly winter in the Nebraska Sandhills on unfrozen rivers and creeks, such as Blue Creek in Garden County. They have been reported rarely from Kansas between November 24 and February 22 (Thompson and Ely, 1989). There are a few winter sightings for the Texas panhandle (Seyffert, 2001). The average number of birds observed per party-hour for the 2013–2014 Christmas Bird Count were: N. Dakota 0, S. Dakota 0, Nebraska 0, Kansas 0.65, Oklahoma 0, northwest Texas 0. For complete species abundance data, see Appendix 1:7.

*Habitats.* Migrants and wintering are found on rivers, lakes, large marshes, and impoundments. Breeding in the Great Plains occurs on large shallow marshes or lakes having abundant submerged vegetation, emergent plants, and stable water levels, and these habitats are also favored outside the breeding season.

*National Population.* Breeding Bird Survey trend data and continent-wide population estimates are not available, but U.S. and Canadian populations have increased enough in recent years to allow the species' removal from its earlier nationally threatened status. By the early 2000s at least 20,000 trumpeter swans existed in the wild. These included a thriving introduced restoration flock of perhaps 10,000 birds in the Midwest (mostly South Dakota, Minnesota and southern Ontario.

*Further Reading.* Palmer, 1973; Johnsgard, 1975a; Mitchell, 1994 *(The Birds of North America:* No. 105); Kear, 2005; Baldassare, 2014.

## Wood Duck

*Aix sponsa*

I have long believed that male wood ducks must be aware of their stunning beauty, as they prefer to hide in the shade of heavy vegetation during the brightest part of the day, and emerge to display their iridescent plumage toward evening. Their large eyes also make it possible for them to forage and court later in the evening and earlier in the morning than is true of most ducks. The males also tend to court females very personally, on a one-to-one basis rather than simply to join a large group of courting birds. This high degree of male attentiveness is one of the species' most endearing traits. Because of their somewhat elusive, almost crepuscular, behavior and early autumn migration, few if any are seen during Christmas Counts in the northern Plains states.

*Winter Distribution.* The wood duck, sometimes vernacularly known as the "summer duck" because of its early autumn migration, is not common anywhere during late December in the Great Plains states. Root (1988) mentioned Kirwin National Wildlife Refuge in Kansas as a local winter population, and mapped southeastern Oklahoma as another wintering area. Our more recent data suggest that Oklahoma is the region's most important wintering state, and that the state's population was increasing through the 1990s. A more recent apparent downturn may be an artifact, as national wood duck populations have shown a long-term increase.

*Seasonality and Migrations.* The wood duck is a common Great Plains seasonal migrant and widespread wetland breeder, wintering southwardly. This species was reported nine years during Christmas Bird Counts from 1968 to 2007 in North Dakota, 33 years in South Dakota, 21 years in Nebraska, 38 years in both Kansas and Oklahoma and nine years in the Texas panhandle. In South Dakota a few birds may attempt to winter in the state (Tallman, Swanson and Palmer, 2002). Thirty-five final autumn Nebraska sightings are from September 10 to December 31, with a median of October 21 (Johnsgard, 1980). It is very rare in Kansas during winter (Thompson *et al.*, 2011), with reports extending from August 22 to May 12 (Thompson and Ely, 1989). It is considered a permanent resident in Oklahoma (Woods and Schnell, 1984), and a rare winter visitor in the Texas panhandle (Seyffert, 2001). The average number of birds observed per party-hour for the 2013–2014 Christmas Bird Count were: N. Dakota 0.015, S. Dakota 0.35, Nebraska 0, Kansas 0.24, Oklahoma 0.12, northwest Texas 0.05. For complete species abundance data, see Appendix 1:8.

*Habitats.* Throughout the year this species is associated with tree-lined rivers, creeks, oxbows and lakes, and usually breeds near slow-moving rivers, sloughs or ponds where large trees are found.

*National Population.* Breeding Bird Surveys between 1966 and 2012 indicate that this species exhibited a survey-wide population increase (2.0% annually) during that period. Counts in the 1960s and 1970s suggested a continental population then of 1.6–1.7 million birds (Johnsgard, 1978). A more recent (2002) estimate for all of North America was about 3.53 million birds (Rose and Scott, 1997; Kear, 2005), reflecting this increase.

*Further Reading.* Palmer, 1973; Johnsgard, 1975a, 1978; Hepp and Bellrose, 1995 *(The Birds of North America:* No. 169); Kear, 2005; Baldassare, 2014.

## Gadwall

*Anas strepera*

The name "gadwall" is of obscure origin and meaning, and its alternative colloquial name "gray duck" describing the male's mostly gray and black plumage isn't much of an improvement. But, as Whistler proved in the painting of his mother, "An Arrangement in Gray and Black," muted colors can indeed be beautiful. In any flock of wild ducks, the inconspicuous gadwalls are often one of the last to be noticed, unless they happen to take flight and reveal their contrasting white wing speculum pattern. In spite of any color limitations, gadwalls are an important part of the Great Plains wetland fauna, and they are well adapted to the shallow and often alkaline marshes that are so common here.

*Winter Distribution.* Root (1988) mapped a broad wintering distribution of the gadwall across the southern Great Plains, with a peak in the Texas panhandle (Buffalo Lake and Muleshoe national wildlife refuges). Oklahoma now appears to be the most important state in the five-state Plains region for gadwalls, while northwestern Texas may have declined in significance. This change may reflect a northerly shift in wintering birds in recent decades. Furthermore, gadwalls are usually associated with shallow, sometimes alkaline, wetlands, and periodic drought cycles could have major effects on the distribution of favorable wintering areas.

*Seasonality and Migrations.* The gadwall is a common Great Plains seasonal migrant and widespread wetland breeder, wintering southwardly. This species was reported 17 years during Christmas Bird Counts from 1968 to 2007 in North Dakota, 40 years in South Dakota, 28 years in Nebraska, and 40 years in Kansas, Oklahoma and the Texas panhandle. In South Dakota flocks up to 150 having been seen during winter (Tallman, Swanson and Palmer, 2002). Fifty final autumn Nebraska sightings range from October 4 to December 31, with a median of November 21 (Johnsgard, 1980). It is uncommon to rare in Kansas during winter (Thompson *et al.*, 2011), with reports extending from August 3 to June 4. (Thompson and Ely, 1989). It is considered a winter resident in Oklahoma, but there are continuous monthly records throughout the year (Woods and Schnell, 1984), and is a fairly common winter visitor in the Texas panhandle (Seyffert, 2001). Average birds observed per party-hour for the 2013–2014 Christmas Bird Count were: N. Dakota 0.03, S. Dakota 0.025, Nebraska 1.51, Kansas 1.62, Oklahoma 2.24, northwest Texas 14.36. For complete species abundance data, see Appendix 1.9.

*Habitats.* Migrants are normally found in shallow marshes and sloughs, and sometimes on deeper waters such as lakes and reservoirs.

*National Population.* Breeding Bird Surveys between 1966 and 2012 indicate that this species exhibited a survey-wide population increase (2.7% annually) during that period. The long-term average (1955-2006) population estimate for the gadwall in the traditional national survey area was 1,714,000 birds; the combined 2007 U.S. and Canadian population of this species was estimated as 3,355,000 birds (U. S. Fish and Wildlife Service, 2007). The combined 2014 U.S. and Canadian population of the gadwall was estimated as 3,811,000 birds (Zimpher *et al.,* 2014). There is also a Eurasian population of nearly a million birds (Rose and Scott, 1997; Kear, 2005).

*Further Reading.* Palmer, 1973; Johnsgard, 1975a; LeSchack, McKnight and Hepp, 1997 (*The Birds of North America:* No. 283); Kear, 2005; Baldassare, 2014.

## American Wigeon
### *Anas americana*

Although gadwalls and American wigeons are often present on the same wetlands, the wigeons are far more conspicuous, both for their brighter colors and the loud whistling notes of the males, uttered both on water and during aerial courtship. During such flights their white upper wing-coverts provide an added visual treat. Wigeons are among the several duck species that hunters often call whistlers, but the others (goldeneyes and scoters) produce their whistling noises by wing feather vibrations during flight, rather than vocally. Wigeons are neither the earliest nor latest of spring and autumn migrants, but instead comprise part of the middle wave of waterfowl migration in the Plains. More than most dabbling ducks, wigeons are strongly vegetarians, and their short, tapered bills are well adapted for clipping shoreline plants. As a result, the birds are not likely to remain in an area long after local vegetation has frozen.

*Winter Distribution.* Root (188) mapped the western Great Plains as a significant wintering area for American wigeons, noting that Muleshoe (Texas) and Bitter Lake (New Mexico) national wildlife refuges represent important winter concentrations. The very high early numbers seen in northwestern Texas result from a few years of very high refuge counts in the late 1960s and early 1970s. These were sometimes followed by very low counts the following year, rather than showing any consistent pattern. In more recent years, South Dakota and Nebraska have shown small but gradually increasing late December numbers. During the same period Kansas, Oklahoma and northwest Texas have shown steep to gradual declines. Evidently this species is very responsive to water or other environmental variations in its winter distribution.

*Seasonality and Migrations.* The American wigeon is a common seasonal migrant and wetland breeder in the northern plains, wintering southwardly. This species was reported 13 years during Christmas Bird Counts from 1968 to 2007 in North Dakota. 40 years in South Dakota, 36 years in Nebraska, and 40 years in Kansas, Oklahoma and the Texas panhandle. Late autumn dates in South Dakota may represent wintering birds (Tallman, Swanson and Palmer, 2002). Fifty final autumn Nebraska sightings are from October 9 to December 31, with a median of November 18 (Johnsgard, 1980). It is uncommon to locally common in Kansas during winter (Thompson *et al.*, 2011), with reports extending from September 4 to June 6 (Thompson and Ely, 1989). It is considered a winter resident in Oklahoma, with records for all months from August to May (Woods and Schnell, 1984). It was reported all 40 years on five Texas panhandle counts, and is a common to abundant winter visitor (Seyffert, 2001). The average number of birds observed per party-hour for the 2013–2014 Christmas Bird Count were: N. Dakota 0, S. Dakota 0.42, Nebraska 0.38, Kansas 0.77, Oklahoma 1.97, northwest Texas 9.17. For complete species abundance data, see Appendix 1:10.

*Habitats.* During migration and winter, these birds are sometimes found on large lakes or reservoirs, but forage where submerged plants can easily be reached from the surface or around the shoreline in grassy meadows.

*National Population.* Breeding Bird Surveys between 1966 and 2012 indicate that this species exhibited a survey-wide population decrease (0.7% annually) during that period. The long-term average (1955–2006) population estimate for the American wigeon in the traditional national survey area was 2,608,000 birds. The 2014 U.S. and Canadian population was estimated as 3,100,000 birds (Zimpher *et al..*, 2014)

*Further Reading.* Palmer, 1973; Johnsgard, 1975a; Mowbray, 1999 (*The Birds of North America:* No. 491); Kear, 2005; Baldassare, 2014.

## American Black Duck
*Anas rubripes*

Hunters in the eastern U.S. once tended to regard black ducks as the premier and most elusive target species of all waterfowl. However, the black duck of the present day is quite different of the black duck of historic times, owing to the extensive infusion of mallard genes, and a consequent change of the species' ecology. The few that turn up on the Great Plains are likely to show the genetic influence of mallards in their plumage traits.

*Winter Distribution.* The Great Plains are well outside any significant wintering areas for this eastern species. Birds seen in this region are generally strays that are associated with mallards. Root (1988) mapped a small area of winter abundance in northwestern Missouri (Squaw Creek National Wildlife Refuge), but indicated no other wintering areas within the Great Plains.

*Seasonality and Migrations.* The American black duck is a rare to very rare seasonal migrant, mainly in the east. This species has appeared 37 out of 40 years of the five Great Plains states' Christmas Counts between 1968 and 2007. In South Dakota, some records exist for January and February (Tallman, Swanson and Palmer, 2002). There are autumn Nebraska records extending from August to December 22 (Johnsgard, 1980). It is rare during winter in Kansas (Thompson *et al.*, 2011), mainly occurring between late October and late March (Thompson and Ely, 1992), and is a rare winter visitor in Oklahoma, with monthly records extending from September to April (Woods and Schnell, 1984). It has been seen from October to February in the Texas panhandle (Seyffert, 2001).

*Habitats.* Usually found among flocks of mallards in Nebraska, and using the same habitats during migration.

*National Population.* Breeding Bird Surveys between 1966 and 2012 indicate that this species exhibited a survey-wide population decrease (2.8% annually) during that period. The 2014 eastern North American population of this species was estimated at about 600,000 birds (Zimpher *et. al..*, 2014). The population has been in almost continuous decline for the past half-century (U. S. Fish and Wildlife Service, 2007), in part because of habitat changes and frequent hybridization with the mallard (Johnsgard and DiSilvestro, 1974).

*Further Reading.* Palmer, 1973; Johnsgard, 1975a; Longcore *et al.,* 2000 (*The Birds of North America:* No. 481); Kear, 2005; Baldassare, 2014.

## Mallard
### *Anas platyrhynchos*

Like the Canada goose, the mallard has become a quasi-urbanite in recent decades, and many of the "wild" mallards that one now sees have the pot-bellied appearance of a well-fed executive who would never consider leaving the city. But mallards can be both beautiful and fascinating when seen under natural conditions, and they offer the average bird-watcher opportunities for close-up observations of behavior that are impossible for most other species of waterfowl.

*Winter Distribution.* Root (1988) illustrated important areas for wintering mallards as extending broadly from South Dakota to Oklahoma, with local peaks in southeastern South Dakota and north-central Kansas that probably reflect the distribution of major wildlife refuges. Forty-year party-hour averages for the Great Plains states show a progressive increase in count numbers from North Dakota to Kansas, and a decline from Oklahoma to northwestern Texas. There is also an apparent gradual region-wide decline over the 40-year period that contrasts with federal surveys showing a stable national population, and Breeding Bird Surveys indicating a significant national population increase. Mallards, like Canada geese, are highly adaptable and likely to respond quickly to regional environmental conditions, so the Great Plains' numbers may not be a reliable guide to national trends. The average number of birds observed per party-hour for the 2013–2014 Christmas Bird Count were: N. Dakota 14.5, S. Dakota 51.77, Nebraska 195.35, Kansas 105.5, Oklahoma 9.56, northwest Texas 7.47. For complete species abundance data, see Appendix 1:11.

*Seasonality and Migrations.* The mallard is an abundant Great Plains seasonal migrant and widespread wetland breeder, wintering southwardly. This species was reported every year during Christmas Bird Counts from 1968 to 2007 in all states from North Dakota to Oklahoma and the Texas panhandle. In South Dakota flocks of more than 100,000 birds have been reported during winter (Tallman, Swanson and Palmer, 2002). Sixty-four final autumn Nebraska sightings are from August 25 to December 31, with a median of November 27 (Johnsgard, 1980). It is locally common in Kansas during winter (Thompson *et al.*, 2011). It is considered a permanent resident in Oklahoma (Woods and Schnell, 1984), and an abundant winter visitor in the Texas panhandle (Seyffert, 2001).

*Habitats.* Breeding birds favor fairly shallow waters, either still or slowly flowing, and surrounding dry areas of non-forested vegetation. Migrants are often found on large marshes, lakes or reservoirs, especially where nearby grain fields provide food.

*National Population.* Breeding Bird Surveys between 1966 and 2012 indicate that this species exhibited a survey-wide population stability (0.0% change). Breeding Bird Surveys between 1966 and 2012 indicate that this species exhibited a survey-wide population increase (2.7% annually) during that period. The long-term average (1955–2013) population estimate for mallards in all survey areas was about 10.9 million birds (Zimpfer *et al.*, 2014). Typical mallards breed south to northernmost Mexico; a small resident central

Mexican population ("Mexican duck") is now considered to be only sub-specifically distinct (*A. p. diazi*). There is also a resident Greenland population, a multi-million Eurasian population and several introduced populations elsewhere in the world (Kear, 2005).

*Further Reading.* Palmer, 1973; Johnsgard, 1975a; Drilling, Titman and McKinney, 2002 (*The Birds of North America:* No. 658); Kear, 2005; Baldassare, 2014.


### Northern Shoveler
*Anas clypeata*


Shovelers suffer from the problem of having too large a beak to satisfy the esthetics of most people. Yet, in spring, the males are dazzling in their contrasting breeding plumages, Furthermore, few ducks are more interesting to watch as they feed is a beak-to-tail manner, sometimes forming long lines as they strain food from the water surface. Like blue-winged and cinnamon teal, they are part of a world-wide assemblage of "blue-winged" ducks having pale blue upper wing-coverts that are usually visible only in flight. All of the blue-winged ducks seem to be sensitive to cold, and the North American species are early autumn and late spring migrants.

*Winter Distribution.* Like blue-winged teal, shovelers are early autumn migrants and do not usually winter very far north of Texas. Average Great Plains counts increase progressively from North Dakota to northwestern Texas, and show a slight trend toward increasing average counts for the five Plains states over the four-decade period. These results seem to follow national Breeding Bird Survey trends.

*Seasonality and Migrations.* The northern shoveler is a common Great Plains seasonal migrant and widespread wetland breeder, wintering southwardly. This species was reported one year during Christmas Bird Counts from 1968 to 2007 in North Dakota, nine years in South Dakota, 21 years in Nebraska, 37 years in Kansas, and 40 years in Oklahoma and the Texas panhandle. Sixty-two final autumn Nebraska sightings range from September 5 to December 31, with a median of November 4 (Johnsgard, 1980). Only a few birds winter in Kansas (Thompson *et al.*, 2011). It is considered a winter resident in Oklahoma, with monthly records existing throughout the year (Woods and Schnell, 1984), and is a fairly common winter visitor in the Texas panhandle (Seyffert, 2001). The average number of birds observed per party-hour for the 2013–2014 Christmas Bird Count were: N. Dakota 0, S. Dakota 0.01, Nebraska 0.22, Kansas 01.17, Oklahoma 1.16, northwest Texas 6.99. For complete species abundance data, see Appendix 1:12.

*Habitats.* Shallow marshes with floating plant foods such as duckweed available at the water surface are preferred by this species.

*National Population.* Breeding Bird Surveys between 1966 and 2012 indicate that this species exhibited a survey-wide population increase (2.2% annually) during that period. The long-term average (1955-2006) U.S. and Canadian population estimate for the shoveler in all survey areas was about 2,206,000 birds (U. S. Fish and Wildlife Service, 2007).). The combined 2014 U.S. and Canadian population of this species was estimated as 5,279,000 birds (Zimpher *et al.*, 2014). There is also a Eurasian population numbering in the millions (Kear, 2005).

*Further Reading.* Palmer, 1973; Johnsgard, 1975a; Dubowy, 1996 (*The Birds of North America:* No. 217); Kear, 2005; Baldassare, 2014.

## Northern Pintail
### *Anas acuta*

Northern pintails are the essence of grace in flight, their long, pointed wings and extended tails would make one believe they were designed specifically for racing. They, like mallards, are among the earliest spring and latest autumn migrants, and are likely to appear in spring only a few days after marshes and lakes begin to thaw. Some of these early migrants may be headed for the high arctic tundras. They are among the most widely distributed and northerly nesting of the dabbling ducks, and are also at equally home on tundra and the shortgrass prairies of the Great Plains.

*Winter Distribution.* Root (1988) mapped Great Plains wintering areas in central Oklahoma and northwestern Texas. the latter centered on Muleshoe National Wildlife Refuge. Texas refuges account for the large numbers summarized here, especially among Christmas Counts of the 1960s. Otherwise, there is no clear population trend evident in the five-state Plains region, although an overall downward trend is suggestive.

*Seasonality and Migrations.* The northern pintail is an abundant Great Plains seasonal migrant and widespread wetland breeder, wintering southwardly. This species was reported 18 years during Christmas Bird Counts from 1968 to 2007 in North Dakota, 38 years in South Dakota, 36 years in Nebraska, 39 years in Kansas, and 40 years in Oklahoma and the Texas panhandle. In South Dakota some wintering may occur in southern locations (Tallman, Swanson and Palmer, 2002). Fifty-seven final autumn Nebraska sightings range from September 16 to December 31, with a median of November 19 (Johnsgard, 1980). It is local and uncommon in Kansas during winter (Thompson *et al.*, 2011), with extreme fall-to-spring records from September 5 to May 10 (Thompson and Ely, 1989). It is considered a winter resident in Oklahoma, with records extending throughout the year (Woods and Schnell, 1984), and an abundant winter visitor in the Texas panhandle (Seyffert, 2001). The average number of birds observed per party-hour for the 2013–2014 Christmas Bird Count were: N. Dakota 0.03, S. Dakota 0.04, Nebraska 0.95, Kansas 0.78, Oklahoma 1.06, northwest Texas 0.4. For complete species abundance data, see Appendix 1:13.

*Habitats.* While on migration nearly all wetland habitats are used, ranging from flooded fields to large lakes and reservoirs. Like mallards, feeding in grain fields is common among wintering birds.

*National Population.* Breeding Bird Surveys between 1966 and 2012 indicate that this species exhibited a survey-wide population decrease (2.7% annually) during that period The long-term average (1955–2006) population estimate for the pintail in the government's traditional national survey area was 4,098,000 birds. The 2007 U.S. and Canadian population was estimated as 3,335,000 birds (U. S. Fish and Wildlife Service, 2007). There is also a Eurasian population numbering in the millions (Kear, 2005). The combined 2014 U.S. and Canadian population of this species was estimated as 3,220,000 birds (Zimpher *et al..*, 2014).

*Further Reading.* Palmer, 1973; Johnsgard, 1975a; Austin and Miller, 1995 (*The Birds of North America:* No. 163); Kear, 2005; Baldassare, 2014.

## Green-winged Teal
*Anas crecca*

Green-winged teals are the smallest of North America's dabbling ducks, and therefore should be early autumn and late spring migrants with low tolerance for cold weather. Such is not the case, and instead the birds typically arrive in the central and northern Plains shortly after the mallards and pintails begin to appear on partly thawed wetlands in early spring. Their autumn departure is correspondingly late, especially by comparison with the other American teals.

*Winter Distribution.* According to Root (1988), the winter distribution of this species is centered in wildlife refuges, including Muleshoe National Wildlife Refuge. Central Kansas, in the vicinity of Quivera National Wildlife Refuge and Cheyenne Bottoms Wildlife Management Area, is mapped as another population locus. The influence of two Texas panhandle refuges

(Buffalo Lake and Muleshoe) is very apparent during the first half of the four-decade period analyzed, but recent late December green-winged teal populations in the southern Great Plains are relatively low, in spite of an apparently increasing national population trend. Perhaps changing water conditions in Texas and New Mexico have altered wintering patterns there.

*Seasonality and Migrations.* The green-winged teal is a common seasonal migrant and wetland breeder in the northern plains, wintering southwardly. This species was reported 11 years during Christmas Bird Counts from 1968 to 2007 in North Dakota, 40 years in South Dakota, 29 years in Nebraska, and 40 years in Kansas, Oklahoma and the Texas panhandle. In South Dakota wintering in the state is rare (Tallman, Swanson and Palmer, 2002). Forty-nine final autumn Nebraska sightings are from September 20 to December 31, with a median of November 2 (Johnsgard, 1980). It is local in Kansas during winter (Thompson *et al.*, 2011), with reports extending from August to May (Thompson and Ely, 1989). It is considered a winter resident in Oklahoma, with continuous monthly records extending from August to June (Woods and Schnell, 1984), and an abundant winter visitor in the Texas panhandle (Seyffert, 2001). The average number of birds observed per party-hour for the 2013–2014 Christmas Bird Count were: N. Dakota 0.08, S. Dakota 0.08, Nebraska 3.08, Kansas 0.39, Oklahoma 0.38, northwest Texas 0. For complete species abundance data, see Appendix 1:14.

*Habitats.* Migrants and wintering birds are associated with standing or slowly-flowing wetlands, which are often quite shallow and sometimes small.

*National Population.* Breeding Bird Surveys between 1966 and 2012 indicate that this species exhibited a survey-wide population decrease (0.2% annually) during that period. The long-term average (1955–2006) population estimate for the green-winged teal in the traditional national survey area was 1,881,000 birds; the 2007 U.S. and Canadian population was estimated as 2,890,000 birds (U. S. Fish and Wildlife Service, 2007). The combined 2014 U.S. and Canadian population of this species was estimated as 3,440,000 birds (Zimpher *et al.,* 2014). There is also a large Eurasian ("Eurasian teal") population that is often considered a separate species (Kear, 2005), with "American green-winged teal" designating the North American form. At times both names have been used simultaneously in Christmas Count summaries, confounding historic count totals.

*Further Reading.* Palmer, 1973; Johnsgard, 1975a; Johnson, 1995 (*The Birds of North America:* No. 193); Kear, 2005; Baldassare, 2014.

## Canvasback
*Aythya valisineria*

Among the largest of North American diving ducks, canvasbacks breed from the Nebraska north into the Prairie Provinces of Canada, and are one of our most unforgettable waterfowl species, either when seen on the water or in flight. They fly with great speed and power, in nearly unwavering flight lines, and traditionally wintered in estuaries along the Gulf Coast, from Louisiana to Mexico. That wintering pattern is still generally true, but the birds are now more scattered, with some wintering occurring well away from the coast.

*Winter Distribution.* Root (1988) found that the highest populations of canvasbacks occurred around Bitter Lake National Wildlife Refuge in northeastern New Mexico, with another secondary locus in southeastern Oklahoma. Long-term average Christmas Counts from Oklahoma and northwestern Texas are only slightly higher than those from farther north, and follow a progressive increase in average Christmas Count numbers from North Dakota south to Texas. The long-term counts show no upward or downward trend for the region as a whole, which fits with both Breeding Bird Surveys and federal waterfowl surveys in suggesting a probably stable national population.

*Seasonality and Migrations.* The canvasback is an uncommon seasonal migrant and local wetland breeder in the northern plains, wintering southwardly. This species was reported six years during Christmas Bird Counts from 1968 to 2007 in North Dakota, 27 years in South Dakota, 23 years in Nebraska, 39 years in Kansas, 40 years in Oklahoma and 32 years in the Texas panhandle. In South Dakota, wintering is rare (Tallman, Swanson and Palmer, 2002). Thirty-nine final autumn Nebraska sightings are from October 12 to December 31, with a median of November 14(Johnsgard, 1980). It is very rare in Kansas during winter (Thompson *et al.*, 2011). It is considered a winter resident in Oklahoma, with continuous monthly records extending from September to May (Woods and Schnell, 1984), and is a fairly common winter visitor in the Texas panhandle (Seyffert, 2001). The average number of birds observed per party-hour for the 2013–2014 Christmas Bird Count were: N. Dakota 0, S. Dakota 0.04, Nebraska 0.34, Kansas 0.57, Oklahoma 1.03, northwest Texas 0.25. For complete species abundance data, see Appendix 1:15.

*Habitats.* On migration, this species uses marshes, rivers and shallow lakes rich in submerged pondweeds and similar vegetation. Prairie marshes with abundant emergent vegetation and some areas of open water are favored for nesting, while migrants and wintering birds are more likely to use both larger and deeper bodies of water.

*National Population.* Breeding Bird Surveys between 1966 and 2012 indicate that this species exhibited a survey-wide population increase (0.5% annually) during that period. The long-term average (1955–2006) population estimate for the canvasback in the traditional federal survey area was 565,000 birds. The combined 2014 U.S. and Canadian population was estimated as 685,000 birds (Zimpher *et al.,* 2104).

*Further Reading.* Palmer, 1973; Johnsgard, 1975a; Mowbray, 2002 (*The Birds of North America:* No. 659); Kear, 2005; Baldassare, 2014.

## Redhead
### *Aythya americana*

Redheads are highly social during migration and wintering, often forming large "rafts" that include canvasbacks and other diving ducks. Redheads and canvasbacks are much dependent on vegetative foods than are, for example, scaups, goldeneyes or mergansers, and concentrate in wetlands where plant life is abundant.

*Winter Distribution.* Root (1988) reported that the vicinity of Bitter Lake National Wildlife Refuge in New Mexico, and adjacent portions of the Texas panhandle are important wintering areas of redheads in the Great Plains. Like the canvasback, there is a progressive increase in redhead Christmas Count numbers from North Dakota south to Texas. There is no clear long-term trend evident in counts from Oklahoma or Texas. However, progressively increasing numbers were seen over the four decades in Kansas and Nebraska, perhaps suggestive of later autumn migrations in recent decades. Christmas Counts from the entire Great Plains region also support a possibly increasing population trend over the past three decades, as also suggested by Breeding Bird Survey and federal waterfowl survey data.

*Seasonality and Migrations.* The redhead is an uncommon seasonal migrant and local wetland breeder in the northern plains, wintering southwardly. This species was reported 13 years during Christmas Bird Counts from 1968 to 2007 in North Dakota, 34 years in South Dakota, 21 years in Nebraska, 40 years in Kansas and Oklahoma, and 32 years in the Texas panhandle. In South Dakota wintering on the Missouri River or elsewhere is rare (Tallman, Swanson and Palmer, 2002). Fifty-six final autumn Nebraska sightings are from October 9 to December 1, with a median of November 9 (Johnsgard, 1980). It is uncommon in Kansas during winter (Thompson *et al.,* 2011) , with extreme dates from September 26 to May 20 (Thompson and Ely, 1989). It is considered a winter resident in Oklahoma (Woods and

Schnell, 1984), and a fairly common winter visitor in the Texas panhandle (Seyffert, 2001). The average number of birds observed per party-hour for the 2013–2014 Christmas Bird Count were: N. Dakota 0, S. Dakota 0.17, Nebraska 0.51, Kansas 0.36, Oklahoma 0.36, northwest Texas 0.34. For complete species abundance data, see Appendix 1:16.

*Habitats.* Migrants and wintering birds are found on large prairie marshes, lakes and reservoirs, especially where submerged vegetation is abundant.

*National Population.* Breeding Bird Surveys between 1966 and 2012 indicate that this species exhibited a survey-wide population increase (0.8% annually) during that period. The long-term average (1955–2006) population estimate for the redhead in the traditional federal survey area was 630,000 birds; the 2007 U.S. and Canadian population was estimated as 1,109,000 birds (U. S. Fish and Wildlife Service, 2007). The combined 2014 U.S. and Canadian population of this species was estimated as 1,279,000 birds (Zimpher *et al.,* 2014). There was also an historic resident Mexican population of unknown current status.

*Further Reading.* Palmer, 1973; Johnsgard, 1975a; Woodin and Michot, 2003 (*The Birds of North America:* No. 695); Kear, 2005; Baldassare, 2014.

### Ring-necked Duck
*Aythya collaris*

Although ring-necked ducks are sometimes confused with scaups, they are much more closely related to redheads and canvasbacks. The relationship is revealed by the pale lemon-yellow pattern of their downy young, and by the similarities of the adult female plumages. Ring-necked ducks are prone to nest in coniferous bogs and muskegs, rather than on the prairie marshes favored by redheads and canvasbacks. During migration ring-necked ducks are likely to be seen on the same larger and deeper wetlands that are used by these other species of diving ducks, and often mix with them to form multi-species rafts.

*Winter Distribution.* Root (1988) reported a concentration of ring-necked ducks occurring at Bitter Lake National Wildlife Refuge in northeastern New Mexico, and also along the eastern edge of the Texas panhandle. Like the redhead and canvasback, this species exhibited gradually increasing Christmas Count numbers from North Dakota south to Texas. It exhibited irregular population trends in Oklahoma and Texas over the four-decade period, but showed generally increasing long-term averages in counts in the more northern states from Kansas to at least South Dakota. An overall increasing population trend, as suggested by Breeding Bird Surveys, is thus supported.

*Seasonality and Migrations.* The ring-necked duck is an uncommon seasonal migrant and local wetland breeder in the northern plains, wintering southwardly. This species was reported five years during Christmas Bird Counts from 1968 to 2007 in North Dakota, 30 years in South Dakota, 20 years in Nebraska, 39 years in Kansas, 35 years in Oklahoma and 36 years in the Texas panhandle. In South Dakota, wintering is rare (Tallman, Swanson and Palmer, 2002). Twenty-three final autumn Nebraska sightings are from October 27 to December 31, with a median of November 17 (Johnsgard, 1980). It is very rare in Kansas during winter (Thompson *et al.*, 2011), with reports extending from October 3 to May 31 (Thompson and Ely, 1989). It is considered a winter resident in Oklahoma, with monthly records extending from October to June (Woods and Schnell, 1984), and an uncommon to fairly common winter visitor in the Texas panhandle (Seyffert, 2001). The average number of birds observed per party-hour for the 2013–2014 Christmas Bird Count were: N. Dakota 0.03, S. Dakota 0.19, Nebraska 2.83, Kansas 0.69, Oklahoma 1.26, northwest Texas 0. 46. For complete species abundance data, see Appendix 1:17.

*Habitats.* Migrants are found on large prairie marshes, lakes and reservoirs. Rather acidic swamps and bogs, surrounded by shrubby cover, are the primary breeding habitat.

*National Population.* Breeding Bird Surveys between 1966 and 2012 indicate that this species exhibited a survey-wide population increase (2.3% annually) during that period. The combined 2014 U.S. and Canadian population of this species was estimated as 600,000 birds (Zimpher *et al.*, 2014).

*Further Reading.* Palmer, 1973; Johnsgard, 1975a; Hohman and Eberhardt, 1998 (*The Birds of North America:* No. 329); Kear, 2005; Baldassare, 2014.

## Greater Scaup
### *Aythya marila*

The scaups are diving ducks that more often feed on mussels and other invertebrates than do redheads, ring-necked ducks or canvasbacks; the word "scaup" is a variant of "scalp," which in earlier English referred to mussels. The greater scaup is prone to use larger and deeper waters than lesser scaup, which breed on shallow wetlands of the boreal forests and often use relatively shallow wetlands during migration. Because of their deep-water preferences, greater scaups in the Great Plains tend to occur on larger reservoirs. They often remain fairly far from shore, where they may be hard to distinguish from lesser scaups.

*Winter Distribution.* The Great Plains states are an insignificant wintering area for greater scaups, which are primarily coastal birds in winter. In the last fifteen years the birds have appeared regularly in Nebraska, Kansas and Oklahoma, but generally in numbers too small to be meaningfully measured. They seem to be increasing very slowly in the region, especially in Oklahoma.

*Seasonality and Migrations.* The greater scaup is an occasional to rare seasonal migrant, wintering southwardly and coastally. This species was reported three years during Christmas Bird Counts from 1968 to 2007 in North Dakota, 13 years in South Dakota, 16 years in Nebraska, 13 years in Kansas, 21 years in Oklahoma and two years in the Texas panhandle. In South Dakota wintering is extremely rare (Tallman, Swanson and Palmer, 2002). There are autumn Nebraska records extending from October 27 to December 30 (Johnsgard, 2007a), suggesting only rare wintering. It is rare during winter in Kansas (Thompson *et al.*, 2011), with extreme fall-to-spring records from October 27 to April 16 (Thompson and Ely, 1992), and a rare winter visitor in Oklahoma, with continuous monthly records extending from October to May (Woods and Schnell, 1984). There are several winter records for the Texas panhandle, extending from October to March (Seyffert, 2001). The average number of birds observed per party-hour for the 2013–2014 Christmas Bird Count were: N. Dakota 0, S. Dakota 0.26, Nebraska 0.19, Kansas 0.17, Oklahoma 0.13, northwest Texas 0. For complete species abundance data, see Appendix 1:18.

*Habitats.* Migrants and wintering birds utilize lakes and reservoirs in the interior, but most birds winter coastally.

*National Population.* See the lesser scaup account for a combined-species population estimate. Breeding Bird Survey trend data are not available for this boreal species. Summer surveys of "tundra scaup" (nearly all greaters) in Alaska from 1961–2001 ranged from 340,000–642,000, but have been in decline. Winter U.S. estimates of greater scaups from 1961–2000 have ranged from 140,000–699,000, and also show a consistent long-term declining trend (Kessel, Rocque and Barclay, 2002). The 2013 U.S. and Canadian population of both scaup species was estimated as 4,611,000 birds (Zimpfer *et al.,* 2014), with the great majority being the lesser species. There are also Icelandic and Eurasian populations totaling about 500,000 birds (Kear, 2005).

*Further Reading.* Palmer, 1973; Johnsgard, 1975a; Godfrey, 1986; Kessel, Rocque and Barclay, 2002 (*The Birds of North America:* No. 650); Kear, 2005; Baldassare, 2014.

## Lesser Scaup
*Aythya affinis*

Lesser scaups are sometimes called "little bluebills," for the bills of both sexes tend to be fairly uniformly bluish in hue, lacking the pale ring near the tip that is present in ring-necked ducks. Lesser scaups often nest in prairie marshes of the northern Great Plains, but are more abundant in the coniferous wetlands of boreal Canada. Because of their invertebrate-oriented diet, they are not highly rated by hunters as table fare, and their other values are thus also under-estimated. They are fairly late autumn migrants, the last flocks sometimes barely escaping the first winter blizzards on the northern plains.

*Winter Distribution.* During the winter this species is likely to be found in fairly deep freshwater wetlands across the southern Great Plains. Root (1988) indicated centers of abundance along the Missouri Valley of South Dakota and northeastern Nebraska, in south-central Kansas, and in much of Oklahoma. Our data show a progressive increase in numbers from North Dakota to the Texas panhandle, and also a progressive long-term increase in numbers over the five-state Plains region from the 1960s onward. This runs counter to the possible downward population trend suggested by Breeding Bird Surveys, and so our apparent upward trend may not be typical of the national population.

*Seasonality and Migrations.* The lesser scaup is a common seasonal migrant and local wetland breeder in the northern plains, wintering southwardly. This species was reported 16 years during Christmas Bird Counts from 1968 to 2007 in North Dakota, 38 years in South Dakota, 29 years in Nebraska, 40 years in Kansas and Oklahoma, and 37 years in the Texas panhandle. In South Dakota wintering is rare (Tallman, Swanson and Palmer, 2002). Thirty-one final autumn Nebraska sightings are from November 22 to December 31, with a median of December 14(Johnsgard, 1980). It is a very rare in Kansas during winter (Thompson *et al.*, 2011) , with extreme dates of September 3 and May 29 (Thompson and Ely, 1989). It is considered a winter resident in Oklahoma, although monthly records exist throughout the year (Woods and Schnell, 1984), and is an uncommon to common winter visitor in the Texas panhandle (Seyffert, 2001). The average number of birds observed per party-hour for the 2013–2014 Christmas Bird Count were: N. Dakota 0, S. Dakota 0.13, Nebraska 0.35, Kansas 1.79, Oklahoma 0, northwest Texas 0.58. For complete species abundance data, see Appendix 1:19.

*Habitats.* Migrating and wintering birds commonly use deep marshes, reservoirs, and lakes.

*National Population.* Breeding Bird Surveys between 1966 and 2012 indicate that this species exhibited a survey-wide population decrease (1.3% annually) during that period. Estimated U.S. and Canada breeding populations of both scaups averaged about 5.5 million from 1955 to 1995 (Custer and Afton, 1998). The combined 2014 U.S. and Canadian population of this species and the greater scaup was estimated as 4,611,000 birds (Zimpher *et al.*, 2014). Lesser scaups are believed to comprise nearly 90 percent of the combined North American scaup population (Bellrose, 1980), so that species may have numbered about four million birds in 2014.

*Further Reading.* Palmer, 1973; Johnsgard, 1975a; Austin, Custer and Afton, 1998 (*The Birds of North America:* No. 338); Kear, 2005; Baldassare, 2014.


## White-winged Scoter
### *Melanitta fusca*

During the early part of the 20$^{th}$ century the white-winged scoter nested in North Dakota, but that nesting population disappeared, and now it is limited to Canada and Alaska as a continental breeder. Scoters are invertebrate-eaters, and during migration they are likely to be found on deeper lakes and wetlands, far from shore, The word "scoter" is of uncertain origin, but is a probable variant of "coot" or "scoot." Along the East Coast, scoters are at times referred to as coots by hunters, who consider them to provide fine sporting targets but virtually inedible. In contrast to their generally dull blackish plumages, all male scoters have wonderfully colorful bills while in breeding condition, and their courtship consists of spectacular aquatic chases and flights.

*Winter Distribution.* Of the three species of scoters, the white-winged is the most likely species to turn up in the Great Plains states; during the 19$^{th}$ century, it nested in northern North Dakota, and in Manitoba until the mid-20$^{th}$ century. However, like other scoters, it winters coastally, and many of the individuals that show up in the Great Plains are females or immatures.

*Seasonality and Migrations.* The white-winged scoter is a very rare seasonal migrant, mainly wintering coastally. This species was reported two years during Christmas Bird Counts from 1968 to 2007 in North Dakota, one year in South Dakota, four years in Nebraska, six years in Kansas, four years in Oklahoma, and none in the Texas panhandle. In South Dakota, the latest reported record is January 3 (Tallman, Swanson and Palmer, 2002). Twenty-one autumn Nebraska records are from October 7 to December 10, with a median of November 10 (Johnsgard, 2007a). It is a very rare during winter in Kansas (Thompson *et al.*, 2011) , with December and February

records (Thompson and Ely, 1992), and is a rare winter visitor in Oklahoma, with continuous monthly records extending from October to May
(Woods and Schnell, 1984). It is a very rare migrant during autumn and
winter in the Texas panhandle (Seyffert, 2001).

*Habitats.* Lakes, reservoirs and larger rivers are used by migrants in the Great
Plains.

*National Population.* Breeding Bird Survey trend and population data are not
available for this boreal and arctic species, but collective estimated populations for all three scoters from the 1950s to the 1970s averaged about 1–2
million birds (Brown and Fredrickson, 1997). The North American white-
winged scoter population has been estimated at a million birds (Rose and
Scott, 1997). There is also a large Eurasian ("velvet scoter") population that
has two recognized subspecies (Kear, 2005).

*Further Reading.* Palmer, 1973; Johnsgard, 1975a; Godfrey, 1986; Brown and
Fredrickson, 1997 (*The Birds of North America:* No. 274); Kear, 2005; Baldassare, 2014.

### Black Scoter

*Melanitta americana*

Rarely seen in the Great Plains, the black scoter is very thinly scattered across
Canada as a breeder, and it probably appears here only as rare vagrants. Females and immatures are nondescript and could be easily overlooked among
flocks of diving ducks, so some individuals may be overlooked here. However,
adult males are very rarely seen in the continental interior.

*Winter Distribution.* This sub-arctic-breeding species is only a very rare visitor
to the Great Plains states, and was seen only slightly more frequently than
is the surf scoter, with totals of seven and five Christmas Count records respectively during the four-decade period studied.

*Seasonality and Migrations.* The black scoter is a very rare seasonal migrant,
mainly wintering coastally. This species was not reported during Christmas
Bird Counts from 1968 to 2007 in North Dakota, but was seen one year in
South Dakota, two years in Nebraska, Kansas and Oklahoma, and none in
the Texas panhandle. Wintering in South Dakota is considered extremely
rare (Tallman, Swanson and Palmer, 2002). Seven autumn Nebraska records range from September 28 to December 10, with a mean of October
28 (Johnsgard, 2007a). It is a very rare during winter in Kansas (Thompson
*et al.*, 2011). There are apparently no winter records for Oklahoma. It is an

extremely rare autumn migrant (November and December records) in the Texas panhandle (Seyffert, 2001).

*Habitats.* Lakes, reservoirs and larger rivers are used by migrants in the Great Plains.

*National Population.* Breeding Bird Survey trend and population data are not available for this arctic-breeding species. During the 1980s there were an estimated 235,000 birds in Alaska and 315,000 elsewhere in North America (Bordage and Savard, 1995). There are no reliable estimates of the North American population, but the closely related European common scoter population may number about 1.6 million birds (Kear, 2005).

*Further Reading.* Palmer, 1973; Johnsgard, 1975a; Godfrey, 1986; Bordage and Savard, 1995 (*The Birds of North America*: No. 177); Kear, 2005; Baldassare, 2014.


## Long-tailed Duck
*Clangula hyemalis*

The name "long-tailed duck" is a far better one than its earlier North American name "oldsquaw." Anyone who has heard the wild and raucous calls of this duck echoing over the arctic tundra might wonder if the earlier name based of a racist view of Native American women. Occasionally a migrating or wintering long-tailed duck will also call, but it somehow seems out of place anywhere except in the high arctic.

*Winter Distribution.* Long-tailed ducks do not winter in any numbers in the Great Plains states, and the lack of deep, ice-free lakes in the region makes them infrequent winter visitors. None of the states had mean numbers high enough to rise above the minimal "present" category for any single ten-year period.

*Seasonality and Migrations.* The long-tailed duck is a rare seasonal migrant, mainly wintering coastally and on the Great Lakes. This species was reported on 16 Christmas Bird Counts from 1968 to 2007 in North Dakota, four years in South Dakota, seven years in Nebraska, 11 years in Kansas and Oklahoma, and one year in the Texas panhandle. In South Dakota, very rare wintering occurs on the Missouri River (Tallman, Swanson and Palmer, 2002). Ten autumn Nebraska records extend from October to December 11, with a median of November 27 (Johnsgard, 2007a). There are only scattered migration and winter records for Kansas (Thompson *et al.*, 2011)., extending from October 31 to April 18 (Thompson and Ely, 1992). The species is a rare winter visitor in Oklahoma, with continuous monthly records extending from November to April (Woods and Schnell, 1984), and is a very rare winter visitor (October to March records) in the Texas panhandle (Seyffert, 2001).

*Habitats.* Lakes, reservoirs and larger rivers are used by migrating birds.

*National Population.* Breeding Bird Survey trend and population data are not available for this abundant arctic-breeding species. The North American population may be about 2.7 million birds (Rose and Scott, 1997). There are also populations in Greenland, Iceland and Eurasia that probably total over five million birds (Kear, 2005).

*Further Reading.* Palmer, 1973; Johnsgard, 1975a; Godfrey, 1986; Robertson and Savard, 2002 (*The Birds of North America:* No. 651); Kear, 2005; Baldassare, 2014.

## Bufflehead
*Bucephala albeola*

The word "sprite" immediately comes to mind at the thought of buffleheads. Their tiny size, dainty movements, and rapid if not frenetic courtship all combine to produce a nearly perfect vision of beauty in a small package. The enormously crested head of the male, for which the original English name buffalo-head derives, can be seen for nearly as great a distances as the rest of its body. The smaller and much more inconspicuous female is often overlooked until she is seen in the presence of the male. In spite of their tiny size, buffleheads are fairly early spring migrants, often appearing in the central and northern plains only a few weeks after the notably cold-tolerant goldeneyes and common mergansers.

*Winter Distribution.* According to Root (1988), the Great Plains states support moderate numbers of buffleheads from South Dakota south to Texas, with larger numbers along the New Mexico, Texas border. Our data support that view, with long-term mean numbers increasing progressively from North Dakota to Texas. There has also been generally increasing numbers in most states from the 1960s onward. Like most waterfowl species, the high numbers that were reported in northwestern Texas during the 1968–1977 decade do not fit the long-term pattern.

*Seasonality and Migrations.* The bufflehead is a common seasonal migrant, and a very local wetland breeder in northern North Dakota, wintering southwardly. This species was reported 20 years during Christmas Bird Counts from 1968 to 2007 in North Dakota, 35 years in South Dakota, 23 years in Nebraska, 39 years in Kansas, 40 years in Oklahoma and 35 years in the Texas panhandle. In South Dakota wintering is uncommon (Tallman, Swanson and Palmer, 2002). Thirty-one final autumn Nebraska sightings are from October 29 to December 31, with a median of November 24 (Johnsgard, 1980). It is a very rare in Kansas during winter (Thompson *et al.*, 2011), with extreme dates of September 19 and May 24 (Thompson and Ely, 1989). It is considered a winter resident in Oklahoma, with continuous monthly records extending from October to May (Woods and Schnell, 1984), and is a fairly common to common winter visitor (October to March records) in the Texas panhandle (Seyffert, 2001). The average number of birds observed per party-hour for the 2013–2014 Christmas Bird Count were: N. Dakota 0.3, S. Dakota 0.61, Nebraska 1.47, Kansas 1.09, Oklahoma 0.79, northwest Texas 0.72. For complete species abundance data, see Appendix 1:20.

*Habitats.* Migrating birds use lakes, reservoirs and deeper marshes.

*National Population.* Breeding Bird Surveys between 1966 and 2012 indicate that this species exhibited a survey-wide population increase (2.7% annually) during that period Current rangewide population estimates are not available, but the continental 1992 population was estimated at 1,390,000 (Gauthier, 1993). More recent studies suggest a stable population of about one million birds (Kear, 2005).

*Further Reading.* Palmer, 1973; Johnsgard, 1975a; Gauthier, 1993 (*The Birds of North America:* No. 67); Kear, 2005; Baldassare, 2014.

## Common Goldeneye
### *Bucephala clangula*

Together with common merganser, the appearance of common goldeneyes on just-thawing lakes of the northern plaice provides one of the first signs that the grip of winter has finally been broken. The mostly white plumaged males of both species contrasts strongly with a deep blue of the lakes that they appear on, and the females are often greatly outnumbered by males among these earliest of waterfowl migrants. Goldeneyes are a bird-watcher's delight, as migrants spend long hours performing complex courtship displays that show great gymnastic-like abilities, not only on the water but also during spirited underwater and aerial chases.

*Winter Distribution.* Root (1988) indicated a rather irregular wintering pattern in the Great Plains, with no clear peaks. Our data show a geographic increase in average numbers from North Dakota to Kansas, but a decline in numbers from there to northwestern Texas. There have also been long-term numerical increases in at least Nebraska and Kansas, but declines in Oklahoma and Texas, It seems likely that a warming climate has allowed goldeneyes to remain as far north as Kansas and Nebraska for longer periods. Wintering numbers have apparently increased substantially in the five-state Plains region since the 1960s, which is consistent with significant national population increases as judged by Breeding Bird Surveys.

*Seasonality and Migrations.* The common goldeneye is a common seasonal migrant, and a very local wetland breeder in northern North Dakota, wintering southwardly. This species was reported 35 years during Christmas Bird Counts from 1968 to 2007 in North Dakota, 40 years in South Dakota, 37 years in Nebraska, and 40 years in Kansas and Oklahoma, and 27 years in the Texas panhandle. In South Dakota wintering is common in the Black Hills, Missouri River and elsewhere (Tallman, Swanson and Palmer, 2002). Thirty-one final autumn Nebraska sightings are from November 22 to December 31, with a median of December 14 (Johnsgard, 1980). It is uncommon in Kansas during winter(Thompson *et al.*, 2011), with reports ranging from November 4 to May 4 (Thompson and Ely, 1989). It is considered a winter resident in Oklahoma, with continuous monthly records extending from October to May (Woods and Schnell, 1984), and is an uncommon to common winter visitor (November to May records) in the Texas panhandle (Seyffert, 2001). Average number of birds observed per party-hour for the 2013–2014 Christmas Bird Count were: N. Dakota 1.53, S. Dakota 1.16, Nebraska 1.94, Kansas 14.48, Oklahoma 3.75, northwest Texas 0.29. For complete species abundance data, see Appendix 1:21.

*Habitats.* Deeper marshes, rivers, lakes and reservoirs are used during migration.

*National Population.* Breeding Bird Surveys between 1966 and 2012 indicate that this species exhibited a survey-wide population increase (1.0% annually) during that period. The 2014 U.S. and Canadian survey of the two goldeneye species collectively estimated 500,000 birds (Zimpfer *et al.*, 2014) but more realistic North American estimates for the common goldeneye are of 1.25–1.5 million (Bellrose, 1980; Kear, 2005). There is also a Eurasian population of about 300,000–450,000 birds (Kear, 2005).

*Further Reading.* Palmer, 1973; Johnsgard, 1975a; Eadie, Mallory and Lumsden, 1995 (*The Birds of North America:* No. 170); Kear, 2005; Baldassare, 2014.

## Barrow's Goldeneye
*Bucephala islandica*

Sometimes bird-watchers in the Great Plains are treated to the sight of a single oddly-plumaged goldeneye among the many flocks of common goldeneyes that can be seen on lakes and deeper marshes. Barrow's goldeneyes (or rarely their hybrids with common goldeneyes) are just frequent enough as to make it worthwhile checking every flock of goldeneyes to determine if it might contain a Barrow's that has strayed in from the Rocky Mountains.

*Winter Distribution.* Root (1988) indicated no wintering areas for this western species in the Great Plains. The only Great Plains location that consistently has Barrow's goldeneyes on its Christmas Counts is in southwestern South Dakota, where the birds are regularly seen at Canyon Lake, Pennington County (Tallman, Swanson and Palmer, 2002). It has also occasionally been seen in the North Platte valley of Nebraska.

*Seasonality and Migrations.* The Barrow's goldeneye is a rare seasonal migrant, mainly in the west. This species was reported one year during Christmas Bird Counts from 1968 to 2007 in North Dakota, 29 years in South Dakota, six years in Nebraska, one year in Kansas, six years in Oklahoma, and none in the Texas panhandle. In South Dakota this species is rare but regular during winter (Tallman, Swanson and Palmer, 2002). Autumn Nebraska records are from November 26 to December 21 (Johnsgard, 2007a). There are a few Kansas records between September and February (Thompson *et al.*, 2011, and Oklahoma records exist for November, January and February (Woods and Schnell, 1984). There are two December sightings for the Texas panhandle (Seyffert, 2001).

*Habitats.* While on migration and during winter this species uses the same habitats as the common goldeneye, but is more prone to winter in coastal or brackish waters.

*National Population.* Breeding Bird Surveys between 1966 and 2012 indicate that this species exhibited a survey-wide population decrease (0.9% annually) during that period Estimates of the Pacific coast population of this species have ranged from 100,000–150,000, and the eastern population has been estimated about 4,000 (Eadie, Savard and Mallory, 2000). There is also a resident Icelandic population of a few thousand birds (Kear, 2005).

*Further Reading.* Palmer, 1973; Johnsgard, 1975a; Eadie, Savard and Mallory, 2000 (*The Birds of North America:* No. 548); Kear, 2005; Baldassare, 2014.

## Hooded Merganser
*Lophodytes cucullatus*

The hooded merganser is all too rare in the Great Plains, although it does seem to be gradually moving west as our prairie rivers gradually become more heavily forested. Males in spring are fascinating to watch, as their crests can be raised to a seemingly impossible height as they assiduously court females. Oddly, the male's courtship calls are more frog-like than avian in their timbre, and his soft rattling notes are among the few vocalizations one is likely to hear from these quietly beautiful birds. Unlike common mergansers, the birds never occur in large flocks, at least within the Great Plains region.

*Winter Distribution.* Root (1988) indicated small wintering populations of this species from South Dakota southwards to Oklahoma and Texas, with a few small areas of higher density in eastern and southern Oklahoma. Our data also shows a peak density in Oklahoma, but very few in northwestern Texas. There is a gradual upward trend in numbers for the five-state Plains region since the 1960s, but the birds are still distinctly uncommon to rare over most parts of the Plains region.

*Seasonality and Migrations.* The hooded merganser is an uncommon seasonal migrant and very local wetland breeder in the northern and eastern plains, wintering southwardly. This species was reported 29 years during Christmas Bird Counts from 1968 to 2007 in North Dakota, ten years in South Dakota, 16 years in Nebraska, 39 years in Kansas, 40 years in Oklahoma and 13 years in the Texas panhandle. In South Dakota, wintering is rare (Tallman, Swanson and Palmer, 2002). Nineteen final autumn Nebraska sightings are from November 6 to December 17, with a median of November 22 (Johnsgard, 1980). It is a very rare in Kansas during winter (Thompson et al., 2011)., with reports ranging from October 10 to June 14 (Thompson and Ely, 1989). It is considered a winter resident in Oklahoma, with continuous monthly records extending throughout the year (Woods and Schnell, 1984), and is an uncommon to fairly common winter visitor in the Texas panhandle (Seyffert, 2001). Average number of birds observed per party-hour for the 2013–2014 Christmas Bird Count were: N. Dakota 0, S. Dakota 0.09, Nebraska 0.09, Kansas 0.92, Oklahoma 3.98, northwest Texas 0.56. For complete species abundance data, see Appendix 1:22.

*Habitats.* Migrants are found on clear-water rivers, lakes, reservoirs and deeper marshes. In contrast, breeding is usually done on rivers, creeks and oxbows bordered by woods and supporting good populations of fish.

*National Population.* Breeding Bird Surveys between 1966 and 2012 indicate that this species exhibited a survey-wide population increase (4.3%

annually) during that period. Range wide population estimates are not available, but hunter harvest data suggest a population during the 1990s of at least 270,000–385,000 (Dugger, Dugger and Fredrickson, 1994).

*Further Reading.* Palmer, 1973; Johnsgard, 1975a; Dugger, Dugger and Fredrickson, 1994 (*The Birds of North America:* No. 98); Kear, 2005; Baldassare, 2014.

## Red-breasted Merganser
*Mergus serrator*

This middle-sized merganser is the most rakish-looking of the American mergansers; both sexes have rather shaggy crests that fit their generally exotic appearances. They are regrettably rather rare on the Great Plains, as they are very prone to migrate directly to a coastline and winter on salt water. Only there is one likely to be able to watch their equally bizarre courtship behaviors, during which males often resemble puppets, being pulled quickly into impossible shapes by invisible strings. Sadly, fisherman and some hunters fail to appreciate the exotic beauty of mergansers, and often shoot them on sight, only to discard them.

*Winter Distribution.* Compared with the common and hooded mergansers, very few red-breasted mergansers winter in the Great Plains, but instead concentrate along coastal regions. The largest numbers are found in Oklahoma, but even there they are less than half as common as hooded mergansers. Our numbers are too low to judge any population trends with confidence.

*Seasonality and Migrations.* The red-breasted merganser is an uncommon seasonal migrant, wintering southwardly and coastally. This species was reported seven years during Christmas Bird Counts from 1968 to 2007 in North Dakota, 13 years in South Dakota, 16 years in Nebraska, 20 years in Kansas, 37 years in Oklahoma and two years in the Texas panhandle. In South Dakota wintering occurs rarely along the Missouri River (Tallman, Swanson and Palmer, 2002). Sixteen autumn Nebraska sightings are from September 21 to December 31, with a median of November 18 (Johnsgard, 1980). It is reported to winter occasionally in Kansas (Thompson *et al.*, 2011), and is considered a rare winter visitor in Oklahoma, with continuous monthly records extending from October to June (Woods and Schnell, 1984). It is a very rare winter visitor (with November to February records) in the Texas panhandle (Seyffert, 2001). Average number of birds observed per party-hour for the 2013–2014 Christmas Bird Count were: N. Dakota 0, S. Dakota 0.04, Nebraska 0.17, Kansas

0.08, Oklahoma 6.5, northwest Texas 0. For complete species abundance data, see Appendix 1:22.

*Habitats.* Lakes, reservoirs and large rivers are used by migrants; wintering mostly occurs coastally.

*National Population.* Breeding Bird Surveys between 1966 and 2012 indicate that this species exhibited a survey-wide population increase (3.1% annually) during that period Rangewide population estimates are not available, but winter and breeding estimates for Canada and Alaska have approached 250,000 birds (Titman, 1999). There is also a similarly large Eurasian population (Kear, 2005).

*Further Reading.* Palmer, 1973; Johnsgard, 1975a; Titman, 1999 (*The Birds of North America:* No. 443); Kear, 2005; Baldassare, 2014.

## Common Merganser
*Mergus merganser*

These spectacular birds are the largest of all mergansers; they are called "goosanders" in the British Isles. On the water they remind one of sleek naval warships, as they rapidly maneuver about, especially during winter or spring courtship. Then the males' contrasting plumages are especially visible, as are their orange feet as they splash water behind them in their efforts to attract the attention of females.

*Winter Distribution.* Root (1988) documented that the common merganser has some major wintering areas in the Great Plains, with concentrations in southern South Dakota, central Kansas and central Oklahoma, but with marked year-to-year variations in total numbers seen. Our data indicate that Kansas is now probably the most important Plains state for wintering, followed by Nebraska. Oklahoma numbers are apparently declining, perhaps because the birds are now wintering increasingly farther north, and relatively few are now wintering there or in northwestern Texas.

*Seasonality and Migrations.* The common merganser is an abundant Great Plains seasonal migrant and a very local breeder, wintering southwardly. This species was reported 35 years during Christmas Bird Counts from 1968 to 2007 in North Dakota, 40 years in South Dakota, 38 years in Nebraska, 40 years in Kansas and Oklahoma, and 35 years in the Texas panhandle. In South Dakota, wintering is locally common (Tallman, Swanson and Palmer, 2002). Thirty-six final autumn Nebraska sightings are from November 20 to December 31, with a median of December 17 (Johnsgard, 1980). It is common in Kansas during winter (Thompson *et al.*, 2011), with extreme dates between August 22 and June 8 (Thompson and Ely, 1989). It is considered a winter resident in Oklahoma, with continuous monthly records extending from October to June (Woods and Schnell, 1984, and a common to fairly common winter visitor in the Texas panhandle (Seyffert, 2001). Average number of birds observed per party-hour for the 2013–2014 Christmas Bird Count were: N. Dakota 1.9, S. Dakota 5.95, Nebraska 12.66, Kansas 23.63, Oklahoma 4.18, northwest Texas 2.78. For complete species abundance data, see Appendix 1:24

*Habitats.* Migrants and wintering birds are found on rivers, lakes, reservoirs, and any other large water areas supporting fish populations. Most nesting occurs on forest-lined lakes and ponds near rivers, but rarely breeding occurs in treeless areas, the nests in rock crevices or other natural cavities.

*National Population.* Breeding Bird Surveys between 1966 and 2012 indicate that this species exhibited a survey-wide population decrease (2.1% annually) during that period. Rangewide population estimates are not available, but Mallory and Metz (1999) suggested a North American population of nearly 1.1 million birds. There is also a Eurasian population (consisting of two subspecies) that numbers in the hundreds of thousands of birds (Kear, 2005).

*Further Reading.* Palmer, 1973; Johnsgard, 1975a; Mallory and Metz, 1999 (*The Birds of North America:* No. 442); Kear, 2005; Baldassare, 2014.

### Ruddy Duck
*Oxyura jamaicensis*

Students in ornithology classes who have the good fortune to witness a male ruddy duck in spring plumage and in full display often quickly decide that it is their favorite duck. It is hard to resist such a seemingly pompous performance by a rusty-brown bird having a cocked-up tail, white cheeks, a sky-blue bill, an incongruously swollen neck, and two Satan-like feather tufts projecting above its black crown. With this unlikely combination of traits, the ruddy duck is a bird apart, although the grayish females and males in winter plumage can easily be overlooked among wintering flocks of waterfowl.

*Winter Distribution.* Root (1988) indicated no areas of winter concentrations of ruddy ducks in the Great Plains, but does show a concentration in adjacent northeastern New Mexico at Bitter Lake National Wildlife Refuge. That pattern fits with our data showing relatively few birds in the Plains States, except for large numbers in northwestern Texas wildlife refuges during the late 1960s and 1970s. Our data are too few to judge any population trends.

*Seasonality and Migrations.* The ruddy duck is a common seasonal migrant and local wetland breeder in the central and northern plains, wintering south-wardly. This species was reported four years during Christmas Bird Counts from 1968 to 2007 in North Dakota, 12 years in South Dakota, 15 years in Nebraska, 34 years in Kansas, 39 years in Oklahoma and 33 years in the Texas panhandle. In South Dakota. wintering is apparently rare (Tallman, Swanson and Palmer, 2002). Fifty-nine final autumn Nebraska records are from August 30 to December 31, with a median of November 27 (Johnsgard, 1980). It is a very rare in Kansas during winter (Thompson *et al.*, 2011), with extreme dates between August 31 and June 21 (Thompson and Ely, 1989). It is considered a permanent resident in Oklahoma (Woods and Schnell, 1984), and is a rare to uncommon winter visitor in the Texas panhandle (Seyffert, 2001). Average number of birds observed per party-hour for the 2013–2014 Christmas Bird Count were: N. Dakota 0, S. Dakota 0, Nebraska 0.03, Kansas 0.21, Oklahoma 0.21, northwest Texas 0.23. For complete species abundance data, see Appendix 1:25.

*Habitats.* Migrants may be found on lakes, reservoirs, larger marshes and similar habitats offering considerable open water and mud-bottom feeding areas. Breeding occurs on prairie marshes having stable water levels and an abundance of emergent vegetation, as well as some areas of open water.

*National Population.* Breeding Bird Surveys between 1966 and 2012 indicate that this species exhibited a survey-wide population increase (3.1% annually) during that period. Current range wide population estimates are not available, but there is an average 1995–2000 spring estimate of nearly 410,000 birds for the U.S.A. and Canada (Brua, 2003). There are also resident Mexican and Caribbean populations, some Andean populations of still-disputed taxonomic status, and a small but increasing introduced European population (Johnsgard and Carbonell, 1996).

*Further Reading.* Palmer, 1973; Johnsgard, 1975a; Johnsgard and Carbonell, 1996; Brua, 2003 (*The Birds of North America:* No. 696); Kear, 2005; Baldassare, 2014.

# FAMILY PHASIANIDAE:
## PHEASANTS, GROUSE AND TURKEYS

### Gray Partridge
*Perdix perdix*

Gray partridges are still moderately common in the Dakotas, but will never again reach the abundance that they one enjoyed before the Second World War, when weedy fields and waste grain were both plentiful. Somehow these introduced European birds seem to still survive the brutal winters of the Dakotas; I remember flushing coveys from snow-blanketed North Dakota wheat fields in below-zero temperatures, where it seemed that neither grain nor any protective weed cover was present.

*Winter Distribution.* Gray partridges were long-time favorite targets for hunters in the Prairie Provinces and northern Plains states, but they have greatly declined in numbers in recent decades, and now barely enter northern Nebraska. The Christmas Count numbers are too low to judge population trends with much assurance, and in general suggest a low but stable population in the northern plains. The Nebraska population seems to hover constantly on the brink of extirpation, but Dakota populations still seem secure.

*Seasonality and Migrations.* This species is a permanent resident throughout its range. Average number of birds observed per party-hour for the 2013–2014 Christmas Bird Count were: N. Dakota 0.68, S. Dakota 0, Nebraska 0, Kansas 0, Oklahoma 0, northwest Texas 0. For complete species abundance data, see Appendix 1:26.

*Habitats.* This introduced species seems to favor highly fertile soils associated with natural grasslands in the northern Great Plains, especially those that are associated with small grain crops. Brushy cover may be used during winter (Johnsgard, 1973).

*National Population.* Gray partridges were reported all 40 years during Christmas Bird Counts from 1968 to 2007 in North Dakota, 38 years in South Dakota, and seven years in Nebraska. Breeding Bird Surveys between 1966 and 2012 indicate that this species exhibited a survey-wide population decrease (1.4% annually) during that period. There are no range-wide population estimates, but there was an annual harvest of 970,000 birds in the 1970s (Johnsgard, 1973), which would suggest a fall population then of at least 3–4 million. Gray partridges are legal game in nearly two dozen states and provinces, and the 1986–88 annual kills in three of these (North Dakota, Saskatchewan and Iowa) totaled about 329,000 birds (Carroll, 1993). There is also a large but generally declining native Eurasian population.

*Further Reading.* Johnsgard, 1973; Carroll, 1993 (*The Birds of North America:* No. 58); del Hoyo, Elliott and Sargatal, 1994.

### Ring-necked Pheasant
*Phasianus colchicus*

Perhaps the most successful of all upland game bird introductions in North America, ring-necked pheasant have to a large degree replaced our native grouse and quails as primary prey for sport hunters. They do not directly compete with these natives, and from that standpoint have been ecologically successful. They also provide prey for large raptors and mammalian predators. And, to some degree they serve as a gene repository for the original stocks, which in many parts of their Asian ranges have become very rare, if not extirpated.

*Winter Distribution.* Pheasants have long been the prime upland game bird species in the Dakotas, and the populations of the 1940s and 1950s were amazingly high. Since then there has been a sharp national decline, but our data suggest that, at least North Dakota has experienced a recent increase, perhaps reflecting positive Conservation Reserve Program effects on pheasant populations since the 1980s. The trend pattern in South Dakota is less clear, as are those of Nebraska and Kansas.

*Seasonality and Migrations.* This species is a permanent resident throughout its range. Average number of birds observed per party-hour for the 2013–2014 Christmas Bird Count were: N. Dakota 4.64, S. Dakota 3.6, Nebraska 0.34, Kansas 0.21, Oklahoma 0.24, northwest Texas 0. For complete species abundance data, see Appendix 1:27

*Habitats.* Throughout the year a combination of small grain croplands and adjacent heavier covers such as weedy ditches, sloughs, wooded areas or shelterbelts provide optimum pheasant habitats.

*National Population.* This species was reported all 40 years during Christmas Bird Counts from 1968 to 2007 from North Dakota through Oklahoma, and 31 years in the Texas panhandle. Breeding Bird Surveys between 1966 and 2012 indicate that this species exhibited a survey-wide population decrease (0.7% annually) during that period. During the 1960s pheasant harvests in the United States averaged about 12 million birds annually (Johnsgard, 1975b), suggesting a population then of more than 20 million birds. However populations have since declined greatly, so that by the mid-1980s there may have been less than half this number. No recent range-wide population estimates of the North American population are available. There are also several native Asian populations (of several sub-species) and various other introduced populations in Europe, New Zealand and elsewhere.

*Further Reading.* Hill and Robertson, 1988; del Hoyo, Elliott and Sargatal, 1994; Guidice and Ratti, 2001 (*The Birds of North America:* No. 572).

### Ruffed Grouse
*Bonasa umbellus*

This bird of the eastern deciduous forests once had a wider distribution in the Great Plains states, occurring along the Missouri valley of southeastern Nebraska and formerly northeastern Kansas. Starting in 1983 it was reintroduced into several eastern Kansas counties, but without success, and the species is now gone from Kansas.

*Winter Distribution.* Now limited to isolated populations in South Dakota's Black Hills, the Turtle Mountains the Pembina Hills (Pembina County) of northeastern North Dakota, and the Killdeer Mountains (Dunn County) in west-central North Dakota

*Seasonality and Migrations.* This species is a permanent resident throughout its range.

*Habitats.* In North Dakota, ruffed grouse are associated with northern deciduous forests having scattered openings, the common trees being quaking aspen (*Populus tremuloides*), paper birch (*Betula papyrifera*), green ash (*Fraxinus pennsylvanica*), bur oak (*Quercus macrocarpa*) and balsam popular (*Populus balsamifera)* (Stewart, 1975). In South Dakota they are associated with young to medium-age stands of aspens, other hardwoods and open pine forests. In 1983 they were re-introduced into seven eastern Kansas counties, with apparently only limited success.

*National Population.* This species was reported during 23 of 40 years of Christmas Bird Counts from 1968 to 2007 from North Dakota and two years in South Dakota. It never achieved abundance levels above the "present" category in either state, but has been reported nearly every year in North Dakota counts since 1981. Breeding Bird Surveys between 1966 and 2012 indicate that this species exhibited a survey-wide population decrease (0.4%) annually. Breeding Bird Surveys between 1966 and 2012 indicate that this species exhibited a survey-wide population increase (2.7% annually) during that period. The estimated 1990s continental population was about 8.3 million birds (Rich *et al.*, 2004).

*Further Reading.* Johnsgard, 1973, del Hoyo, Elliott and Sargatal, 1994; Rusch *et al.*, 2000 (*The Birds of North America:* No. 515).

## Greater Sage-Grouse
### *Centrocercus urophasianus*

This is the largest of North American grouse, and probably until the middle of the past century one of the more abundant species. However, it is closely associated with sagebrush (especially *Artemisia tridentata*), and the massive clearing of sagebrush for irrigation or other purposes in the past half-century has had dire consequences for the sage-grouse. Like the other grassland grouse it is slowly retreating from the all perimeters of its original range, It is again being considered a candidate for threatened or endangered status, an action that was blocked in 2005 for purely political reasons by the G. W. Bush administration.

*Winter Distribution.* Greater sage-grouse barely enter the western edge of the Great Plains states, being confined to a few counties in southwestern North Dakota and three counties in northwestern and southwestern South Dakota. They once also occurred in northwestern Nebraska, but were extirpated by about the start of the 20th century. The species is now threatened in South Dakota by sagebrush destruction (Tallman, Swanson and Palmer, 2002), and hunting is now limited to two counties. Its North Dakota range has been reduced from four to two counties since Stewart's (1975) summary of that state' breeding birds.

*Seasonality and Migrations.* This species is a permanent resident, with only limited seasonal movements.

*Habitats.* Throughout the year, sagebrush shrublands are this species' primary habitat, although in summer meadows, grasslands and aspen (*Populus*) thickets adjacent to sagebrush are frequently also used (Andrews and Righter, 1992), Paothong, 2012.

*National Population.* This grouse was reported two years during Christmas Bird Counts from 1968 to 2007 in North Dakota. Breeding Bird Surveys between 1966 and 2012 indicate that this species exhibited a survey-wide population decrease (2.4% annually) during that period. Most state surveys have indicated recent significant national population declines (Johnsgard, 2002). It has been designated a species of continental conservation importance, with the Prairie Avifaunal Biome supporting about 20 percent of the continental winter population (Rich *et al.*, 2004). The sage-grouse's estimated 1990s continental population was 150,000 birds (Rich *et al.*, 2004), but it has continued to decline and deserves federal listing as a threatened if not endangered, species. This species was yellow-listed in the National Audubon Society's 2007 WatchList of rare and declining birds (Butcher et al., 2007).

*Further Reading.* Johnsgard, 1973, 2002a; Bergerud and Gratson, 1988; del Hoyo, Elliott and Sargatal, 1994; Schroeder, Young and Braun, 1999 *(The Birds of North America:* No. 425).

### Sharp-tailed Grouse
*Pedioecetes phasianellus*

As the grassland grouse with the largest historic North American range, involving several subspecies that are adapted to quite varied habitats and climates, this grouse has the best prospects for long-term survival. If the greater prairie-chicken is the iconic species of the tallgrass prairies, this is the counterpart species of the mixed-grass and shortgrass grasslands of the Great Plains. Like the other prairie grouse, this species is so well camouflaged that it is nearby impossible to see before it flushes, so the only way to survey the birds by locating the males' spring communal display sites, or leks.

*Winter Distribution.* This species was reported all 40 years during Christmas Bird Counts from 1968 to 2007 in North Dakota and South Dakota, and 37 years in Nebraska. Root's (1988) map of sharp-tailed grouse distribution indicates a major peak in western North Dakota and northwestern South Dakota, but extending south to northwestern Kansas. The Kansas population is now extirpated, as is that of Nebraska's south of the centrally located Sandhills region. Our data suggest that North Dakota's population has been increasing in recent decades, as too has South Dakota's smaller population. In Nebraska the population may be stable.

*Seasonality and Migrations.* This species is a permanent resident throughout its range. Average number of birds observed per party-hour for the 2013–2014 Christmas Bird Count were: N. Dakota 1.54, S. Dakota 1.8, Nebraska 0.07, Kansas 0, Oklahoma 9, northwest Texas 0. For complete species abundance data, see Appendix 1:28

*Habitats.* Open grassland, where trees are absent or nearly so, is the typical Nebraska habitat for sharp-tailed grouse. Brushy cover covering from five to 30 percent of the land is used in more northerly areas, especially where winter snow accumulation is considerable.

*National Population.* Breeding Bird Surveys between 1966 and 2012 indicate that this species exhibited a survey-wide population increase (0.3%

annually) during that period. It has been designated a species of continental conservation importance, with the Prairie Avifaunal Biome supporting an estimated 86 percent of the 1990s continental population of 1,2 million birds (Rich *et al.*, 2004). However, Johnsgard (2002) reported an estimated autumn North Dakota population of 500,000–600,000 birds, South Dakota's at least 400,000, and Nebraska's of about 250,000. Thus, the continental population must be substantially higher than 1.2 million birds, considering the additional populations in other states and the geographically large but numerically unknown Canadian and Alaskan components. However, some populations, such as the Columbian race of the western states, are in severe decline and are locally endangered.

*Further Reading.* Johnsgard, 1973, 2001b, 2002a; Bergerud and Gratson, 1988; del Hoyo, Elliott and Sargatal, 1994; Connelly, Gratson and Reese, 1998 *(The Birds of North America:* No. 354) Paothong, 2012.

### Greater Prairie-Chicken
*Tympanuchus cupido*

If there is an avian spirit of the true prairie, the greater prairie-chicken probably is best qualified. Its range once included the entire tallgrass prairie, east to the coastal prairies of New England where (known as the now-extinct "heath hen") it nourished the early European settlers, and originally extended west to the eastern edges of the Great Plains states. With the advent of small grain agriculture prairie-chickens expanded their range northwest well into the Canadian prairie provinces, only to rather quickly decline and finally disappear as the native tallgrass and mixed-grass prairies also disappeared.

*Winter Distribution.* Root (1988) did not provide a range map for the greater prairie-chicken, but said that the area of apparent highest abundance was near the Kansas-Missouri border, presumably in the Flint Hills of Kansas. Since the 1960s nearly all Great Plains populations of prairie-chickens have declined or barely remained stable, although the federal Conservation Reserve Program has help preserve grouse habitat in some areas. Christmas Count data for the region are not adequate to judge population trends. Johnsgard (2002) judged that less than 1,000 birds were then present in North Dakota, plus 25,000–50,000 in South Dakota, 100,000 in Nebraska, 60,000 in Kansas and 1,000–10,000 in Oklahoma. These figures would mean that the Great Plains states might have had a total autumn population of about 190,000–225,000 birds at the turn of the 21st century, plus a few thousand in some states with relict populations. If history provides any guide, all of our prairie-chickens will all be gone before the turn of the next century.

*Seasonality and Migrations.* This species is a permanent resident throughout its range. Average number of birds observed per party-hour for the 2013–2014 Christmas Bird Count were: N. Dakota 0.13, S. Dakota 0.56, Nebraska 0.45, Kansas 1.0, Oklahoma 0.14, northwest Texas 0. For complete species abundance data, see Appendix 1:29.

*Habitats.* Greater prairie-chickens are primarily associated with native grasslands, and where native grasslands and grain croplands interdigitate. Nesting usually occurs in grassy open habitats such as ungrazed meadows or hayfields, usually in rather dry sites. Sometimes nests are placed in brushy vegetation, in open woods, or at the edge of woods.

*National Population.* This species was reported 11 years during Christmas Bird Counts from 1968 to 2007 in North Dakota, 28 years in South Dakota, 30 years in Nebraska, and 39 years in Kansas. Breeding Bird Surveys between 1966 and 2012 indicate that this species exhibited a survey-wide population increase (2% annually) during that period However, most recent state surveys have indicated alarming national population declines during the past half-century (Johnsgard, 2002a). It has been designated a species of continental conservation importance, with the Prairie Avifaunal Biome supporting 97 percent of the continental winter population (Rich *et al.,* 2004). The estimated 1990s continental population was 690,000 birds (Rich *et al.,* 2004), almost three times more than the roughly quarter-million calculated by Johnsgard (2002a) to be alive at the turn of the 21$^{st}$ century. This species was red-listed in the National Audubon Society's 2007 WatchList of rare and declining birds (Butcher *et* al., 2007).

*Further Reading.* Johnsgard, 1973, 2001b, 2002a; Bergerud and Gratson, 1988; Schroeder and Robb, 1993 (*The Birds of North America:* No. 36); del Hoyo, Elliott and Sargatal, 1994 Paothong, 2012.

## Lesser Prairie-Chicken
*Tympanuchus pallidicinctus*

It seems a shame to give any bird as wonderful as this prairie-chicken a name that belittles it as a "lesser" species. It is indeed smaller than the greater species, but what it lacks in size it makes up for in toughness and character. The males display each spring with an enthusiasm unmatched by any of the larger grouse, and have somehow managed to survive over much of their original sandsage (*Artemisia filifolia*) and shinnery oak (*Quercus havardi*) range in spite of a harsh, arid climate and loss of much of their prime habitat.

*Winter Distribution.* No map was provided for this species by Root (1988).

Johnsgard's (2002) map shows the species' range to be completely limited to the Great Plains states except for small areas in southeastern Colorado and eastern New Mexico. Our data for Kansas, Oklahoma and northwestern Texas are too limited to judge possible population trends. There are no recent and reliable estimates of lesser prairie-chicken numbers for any of the states in which the species occurs, but several have reported apparent population declines. It is clear that the species should be listed as nationally threatened, but political pressures are unlikely to allow that to happen.

*Seasonality and Migrations.* This species is a permanent resident throughout its range.

*Habitats.* Sandsage and sandsage grasslands are favored habitats in northern parts of this species' range, while shinnery oak grasslands are the primary habitat farther south (Johnsgard, 1973).

*National Population.* The lesser prairie-chicken was reported 28 years during Christmas Bird Counts from 1968 to 2007 in Kansas, 23 years in Oklahoma and one year in the Texas panhandle. National Breeding Bird Survey trend data are not available, but state surveys have indicated significant national population declines (Johnsgard, 2002a), including an approximate 80 percent decline in Oklahoma in the last two decades of the 20th century (Reinking, 2004). Kansas and Texas populations have also been declining. It has been designated a species of continental conservation importance, with the Prairie Avifaunal Biome supporting 100 percent of the species' population, and an estimated 1990s continental population of 32.000 birds (Rich *et al.,* 2004). My 2000 national estimate was of 10,000-20,000 birds (Johnsgard, 2002a), clearly indicating that this species belongs in the threatened or endangered category, in spite of strongly opposing political pressures from the states and federal government. This species was red-listed in the National Audubon Society's 2007 WatchList of rare and declining birds (Butcher *et* al., 2007).

*Further Reading.* Johnsgard, 1973, 2001b, 2002a; del Hoyo, Elliott and Sargatal, 1994; Giesen, 1998 (*The Birds of North America:* No. 364) Paothong, 2012.

## Wild Turkey
### *Meleagris gallopavo*

Like the trumpeter swan and a few other species, the conservation efforts of state and federal wildlife agencies toward the survival of wild turkeys has been little short of amazing. The species now occurs in all the contiguous 48 states, including some that were outsides the species' original ranges, as well as in several Canadian provinces. Like Canada geese, wild turkeys have also adapted to a quasi-urban existence, especially around smaller towns of the Great Plains. They have become so common in such locations that it is often impossible to decide if one is seeing wild or domestic turkeys, and the two sometimes interbreed.

*Winter Distribution.* This species' distribution and overall abundance have greatly expanded since Root's (1988) analysis. She mapped a major population center in western Kansas and adjacent Oklahoma, and much smaller numbers from North Dakota south through western South Dakota to western Nebraska. In addition to a substantial long-term population increase throughout the Great Plains region, the highest recent (1998–2007) counts have been in the Dakotas, followed by Kansas, Nebraska and Oklahoma. Except in northwestern Texas, populations everywhere appear to be increasing, in accordance with national trends.

*Seasonality and Migrations.* This species is a permanent resident throughout its range. Average number of birds observed per party-hour for the 2013–2014 Christmas Bird Count were: N. Dakota 1.88, S. Dakota 0, Nebraska 0.93, Kansas 2.04, Oklahoma 2.65, northwest Texas 0.26. For complete species abundance data, see Appendix 1:30.

*Habitats.* Although various races of turkeys differ greatly in habitats utilized, the birds of the Great Plains region are mostly found in floodplain forests having a variety of hardwood trees, especially those bearing acorns or other large and edible seeds. In more western areas the birds are also associated with ponderosa pines (*Pinus ponderosa*), other conifers, running water and a fairly rugged topography.

*National Population.* This species was reported all 40 years during Christmas Bird Counts from 1968 to 2007 in North Dakota, 38 years in South Dakota, 30 years in Nebraska, 34 years in Kansas, 40 years in Oklahoma and 38 years in the Texas panhandle. Breeding Bird Surveys between 1966 and 2012 indicate that this species exhibited a survey-wide healthy population increase (8.0% annually) during that period. The estimated 1990s continental population north of Mexico was 1,170,000 birds (Rich *et al.,* 2004). The species breeds south to central Mexico.

*Further Reading.* Kilpatrick, Husband and Pringle, 1988; Eaton, 1992 (*The Birds of North America:* No. 22); del Hoyo, Elliott and Sargatal, 1994.

# FAMILY ODONTOPHORIDAE:
## NEW WORLD QUAIL

### Scaled Quail
*Callipepla squamata*

The scaled quail is a desert-grassland species, having a tendency to run rather than fly when threatened, and able to hide under only scant cover. It survives best where there is some surface water, but is not so dependent on a reliable source of water as the northern bobwhite. Both are highly social species, with strong pair and covey bonds, large clutch-sizes, and high annual mortality rates.

*Winter Distribution.* This Great Plains and southwestern desert species extends south into Mexico, but its U.S. distribution centered in northeastern New Mexico, and east into the Texas panhandle (Root, 1988). It occurs north into southwestern Kansas, although it s relatively rare there. Our data suggest that is now common only in northwestern Texas, and has been on the decline there since the 1970s, as is apparently also the case in Oklahoma. Survey data from Texas between 1978 and 2003 also indicate a gradual statewide population decline (Brennen, 2007).

*Seasonality and Migrations.* This quail is a permanent resident throughout its range. Average number of birds observed per party-hour for the 2013–2014 Christmas Bird Count were: N. Dakota 0, S. Dakota 0, Nebraska 0, Kansas 0, Oklahoma 0, northwest Texas 1.33. For complete species abundance data, see Appendix 1:31.

*Habitats.* Winter habitats of the scaled quail are often associated with soapweed (Y*ucca*) or cacti (*Opuntia*) in arid grassland habitats, often fairly far from water. Shrubs are commonly associated with scaled quail occurrence, as are various artificial structures that are used for escape or loafing cover and shade. Mid-day shade and loafing cover is important, but the cover must be sufficiently open as to allow for running escapes (Johnsgard, 1973).

*National Population.* This species was reported 29 years during Christmas Bird Counts from 1968 to 2007 in 34 years in Kansas, 35 years in Oklahoma and 40 years in the Texas panhandle. Breeding Bird Surveys between 1966 and 2012 indicate that this species exhibited a survey-wide population decrease (3.1% annually) during that period. It has been designated a species of continental conservation importance, with the Prairie Avifaunal Biome supporting 13 percent of the continental winter population (Rich *et al.,* 2004). The estimated 1990s continental population north of Mexico was about 600,000 birds (Rich *et al.,* 2004). The species breeds south to central Mexico.

*Further Reading.* Johnsgard, 1973; Scheminitz.1994 (*The Birds of North America:* No. 106); del Hoyo, Elliott and Sargatal, 1994.

## Northern Bobwhite
### *Colinus virginianus*

The northern bobwhite is easily the most popular upland game bird in North America; in the 1970s I calculated that about 35 million were being shot each year by sport hunters, or nearly ten times the number of ruffed grouse. Both the bobwhite and ruffed grouse have the reputation of being high-class game birds probably because they are largely woodland and woodland-edge birds that tend to "hold" well to a trained dog. But bobwhites have many charms in addition to being considered prized targets for hunters; their social behavior is complex, their vocalizations are correspondingly diverse, and their adaptations to cooperative covey living are crucial to their winter survival on the Great Plains.

*Winter Distribution.* The northern bobwhite reaches the western edge of its historic range in the Great Plains, and its western boundary is rather fluid, shifting from year to year under the possible influences of changes in habitat, water availability and snow cover, or winter temperatures. In Root's analysis, the highest Great Plains populations were in the Oklahoma–Texas panhandle border region, and secondarily in Kansas. Our data suggest the populations in Nebraska, Kansas and Oklahoma are in significant long-term declines, while that in the Texas panhandle is of variable abundance but does not show a clear trend. Long term surveys in the Texas panhandle and high plains also show an irruptive population pattern without a clear trend (Brennen, 2007). Greater livestock density, increased cropland acreage and suboptimal autumn and winter precipitation have all been hypothesized as possible causes of bobwhite declines in western Texas. The average number of birds observed per party-hour for the 2013–2014 Christmas Bird Count were: N. Dakota 0, S. Dakota 0, Nebraska 0.49, Kansas 0, 39 Oklahoma 0.8 northwest Texas 2.56. For complete species abundance data, see Appendix 1:32.

*Seasonality and Migrations.* The bobwhite is a permanent resident throughout its range.

*Habitats.* Throughout the year bobwhites are normally found where there is a combination of grassy nesting cover, cultivated crops, and brushy cover or woodlands with a brushy understory. Nesting is typically done in open herbaceous cover consisting of rather short vegetation that does not obstruct easy entry and exit, but sufficient to provide concealment from above.

*National Population.* This species was reported three years during Christmas Bird Counts from 1968 to 2007 in North Dakota, 19 years in South Dakota, and all 40 years in Nebraska, Kansas, Oklahoma and the Texas panhandle. Breeding Bird Surveys between 1966 and 2012 indicate that this species exhibited a survey-wide population decrease (4.2% annually) during that period. The estimated 1990s continental population north of Mexico was about 7,550,000 birds (Rich *et al.,* 2004). The species breeds south to central Mexico, and has been introduced widely throughout the world.

*Further Reading.* Johnsgard, 1973; del Hoyo, Elliott and Sargatal, 1994; Brennan, 1999 (*The Birds of North America:* No. 397).

FAMILY GAVIIDAE: LOONS

## Pacific Loon
### *Gavia pacifica*

Pacific loons are considerably smaller than common loons, and thereby are able to take flight from wetlands having shorter available "runways." This is a useful ability on small tundra wetlands in the arctic, but during autumn and winter the birds are likely to appear only on large bodies of water, wherever fish populations are easily available.

*Winter Distribution.* Root (1988) mapped the Pacific loon as occurring only along the Pacific Coast. This pattern still applies, with vagrants only rarely occurring in the interior states.

*Seasonality and Migrations.* The Pacific loon is a rare to very rare Great Plains seasonal migrant or winter vagrant. This species was not reported during Christmas Bird Counts from 1968 to 2007 in North Dakota or South Dakota, but was seen two years in Nebraska, five years in Kansas, five years in Oklahoma, and none in the Texas panhandle. South Dakota records extend to late November. Several of the 14 Nebraska sightings obtained from the mid-1930s to the late 1970s are also for November. There are a few late autumn records (through December 15 from Kansas (Thompson *et al.*, 2011), and Oklahoma records exist for October and November (Woods and Schnell, 1984). There are a few December and January sightings for the Texas panhandle (Seyffert, 2001).

Habitats Reservoirs, lakes and large rivers are used by this rare species on migration and during winter.

*National Population.* Breeding Bird Survey trend and population data are not available for this arctic-breeding species. The Alaska population may number about 100,000–125,000 birds (Russell, 2002), but the size of the presumably much larger Canadian population is unknown.

*Further Reading.* Palmer, 1962; Godfrey, 1986; Johnsgard, 1987; Russell, 2002 (*The Birds of North America:* No. 657).

## Common Loon
### *Gavia immer*

Of all the American loons, only the common loon is likely to appear during winter in the Great Plains states. It is a large, cold-tolerant bird, and even as far north as Nebraska has been observed every month of the year. Large and deep bodies of water with good fish populations are its favored winter habitat.

*Winter Distribution.* Root (1988) mapped the winter distribution of the common loon as primarily coastal, with the only notable Great Plains occurring in central and eastern Oklahoma. Our data suggest that even in Oklahoma the species is rare, but has shown a slight long-term increase in numbers seen during Christmas Counts. This apparent increase corresponds with national Breeding Bird Survey trend estimates.

*Seasonality and Migrations.* The common loon is an uncommon seasonal migrant, rarely wintering southwardly. This species was reported one year during Christmas Bird Counts from 1968 to 2007 in North Dakota and South Dakota, five years in Nebraska, 18 years in Kansas, 40 years in Oklahoma, and seven years in the Texas panhandle. In South Dakota, wintering birds are occasionally reported (Tallman, Swanson and Palmer, 2002). Seventeen final autumn Nebraska sightings are from October 25 to December 7 with a median of November 2. In a total of 135 Nebraska sightings, reports exist for all months except February (Johnsgard, 1980). Autumn records from Kansas extend to December (Thompson and Ely, 1989, Thompson *et al.*, 2011). It is considered a rare winter visitor in Oklahoma, with continuous monthly records extending throughout the year (Woods and Schnell, 1984). It is a rare winter visitor in the Texas panhandle, but has been seen during every month (Seyffert, 2001).

*Habitats.* Larger rivers, lakes and reservoirs are used while on migration and during winter.

*National Population.* Breeding Bird Surveys between 1966 and 2012 indicate that this species exhibited a survey-wide population increase (0.7% annually) during that period. The world loon population may be about 500,000–600,000 individuals, of which the majority is in Canada (McIntyre and Barr, 1997). I (1987) estimated the population south of Canada at no more than 15,000 birds. There are also a small Icelandic population and a European population ("great northern diver") of uncertain size.

*Further Reading.* Palmer, 1962; Godfrey, 1986; Johnsgard, 1987; McIntyre and Barr, 1997 (*The Birds of North America:* No. 313).

## FAMILY PODICIPEDIDAE: GREBES

### Pied-billed Grebe
*Podilymbus podiceps*

Compared with the western grebe, for example, the pied-billed grebe is a rather dumpy bird, reminding one of a Volkswagen beetle rather than a Ferrari, and a bird that would seem hardly able to fly on its short, rounded wings. Like other grebes it needs a watery "runway" to take flight, and it sometimes becomes stranded after accidentally landing on ground rather than water. Its foods are diverse, including especially crayfish, other crustaceans, fish, and insects. Wintering habitats often include brackish waters, but true marine environments are used very little. Like many other grebes the birds migrate nocturnally, and thus may appear and disappear from wetlands quite unexpectedly.

*Winter Distribution.* Unlike most grebes, the pied-billed mostly winters on freshwater wetlands. Root (1988) indicated winter Great Plains concentrations mostly located in central Texas, but with some wintering extending north to northern Kansas. Our data suggest that populations in Oklahoma are now larger than those in northwestern Texas, although there is an apparent slight increasing long-term population trend from Kansas south to Texas. This apparent upward trend runs counter to the national downward trend indicated by national Breeding Bird Surveys. The average number of birds observed per party-hour for the 2013–2014 Christmas Bird Count were: N. Dakota 0, S. Dakota 0, Nebraska 0.2, Kansas 3.1, Oklahoma 0.86, northwest Texas 3.14. For complete species abundance data, see Appendix 1:33.

*Seasonality and Migrations.* The pied-billed grebe is a common Great Plains seasonal migrant and wetland breeder, wintering southwardly. This species was reported two years during Christmas Bird Counts from 1968 to 2007 in North Dakota, 13 years in South Dakota, eight years in Nebraska, 40 years in Kansas and Oklahoma and 39 years in the Texas panhandle. South Dakota records extend to early winter (Tallman, Swanson and Palmer, 2002). Eighty-four final autumn Nebraska sightings are from August 21 to December 6, with a median of November 6 (Johnsgard, 1980). There are a few winter records for Kansas (Thompson *et al.*, 2011), with extreme dates of August 25 and May 29 (Thompson and Ely, 1989). It is considered a permanent resident in Oklahoma (Woods and Schnell, 1984). It is a common to fairly common winter visitor in the Texas panhandle, and breeds there uncommonly (Seyffert, 2001).

*Habitats.* Ice-free ponds, rivers, lakes and reservoirs are used while on migration and during winter.

*National Population.* Breeding Bird Surveys between 1966 and 2012 indicate that this species exhibited a survey-wide population increase (0.4% annually) during that period. No national population estimates are available, as is also true of the central and South American populations. The species also breeds in the West Indies and south locally to Chile and Argentina.

*Further Reading.* Palmer, 1962; Johnsgard, 1987; Muller and Storer, 1999 (*The Birds of North America:* No. 410).

## Horned Grebe
*Podiceps auritus*

Horned grebes are primarily fish-eaters, but can shift to other foods such as insects and other small aquatic invertebrates as needed. They tend to forage in fairly shallow water, less than about five feet deep, so they are typically seen in shallower wetlands, at least during he breeding season. During winter they are likely to be found on estuaries and even at sea, sometimes fairly far from shore.

*Winter Distribution.* Root's (1988) map of the horned grebe's winter distribution indicates moderate numbers of birds wintering in eastern Oklahoma and adjacent southeastern Kansas. However, she pointed out that these are perhaps mostly ephemeral clusters of migrants concentrating in ice-free rivers and lakes, and most birds winter coastally. Our data also show small numbers of birds in Oklahoma, following a similar distribution pattern to that of the pied-billed grebe.

*Seasonality and Migrations.* The horned grebe is an uncommon seasonal migrant, and a local wetland breeder in North Dakota, wintering southwardly. This species was reported two years during Christmas Bird Counts from 1968 to 2007 in North Dakota and South Dakota, nine years in Nebraska, 18 years in Kansas, 40 years in Oklahoma and three years in the Texas panhandle. In South Dakota wintering is considered to be extremely rare (Tallman, Swanson and Palmer, 2002). Seventeen final autumn Nebraska sightings are from October 9 to November 27, with a median of November 11 (Johnsgard, 1980). It is rare in Kansas during winter (Thompson *et al.*, 2011), with extreme dates of October 3 and May 19 (Thompson and Ely, 1989). It is considered a winter resident in Oklahoma, with continuous monthly records extending from September to June (Woods and Schnell,

1984). It is a rare winter visitor in the Texas panhandle, with records extending from September to May (Seyffert, 2001). The average number of birds observed per party-hour for the 2013–2014 Christmas Bird Count were: N. Dakota 0, S. Dakota 0, Nebraska 0, Kansas 0, Oklahoma 0.04, northwest Texas 0.06. For complete species abundance data, see Appendix 1:34.

*Habitats.* Rivers, lakes and reservoirs are used while on migration and during winter.

*National Population.* Breeding Bird Surveys between 1966 and 2012 indicate that this species exhibited a survey-wide population decrease (1.8% annually) during that period. A continental estimate of 100,000 pairs is the only one available (Stedman, 2000). There is also a Eurasian population (the "Slavonian grebe") of unknown size.

*Further Reading.* Palmer, 1962; Johnsgard, 1987; Stedman, 2000 (*The Birds of North America:* No. 505).

## Eared Grebe
### *Podiceps nigricollis*

The nonbreeding habitats of eared grebes are diverse, and include freshwater and saline lakes, as well as coastal waters such as estuaries, bays and channels. Unlike the horned grebe, which consumes a preponderance of fish, the eared grebe is largely insectivorous. However, in marine environments where insects are lacking crustaceans may provide the species' major food source.

*Winter Distribution.* The eared grebe winters mostly on the Pacific Coast and the Salton Sea, so counts in the Great Plains are not informative of overall winter distributions (Root, 1988). Our data indicate that southern Oklahoma does attract some eared grebes in late December, but the numbers barely exceed the "present" threshold level, and provide no clues as to this species' populations or trends.

*Seasonality and Migrations.* This grebe is an uncommon seasonal migrant and local wetland breeder in the northern plains, wintering southwardly. It was reported one year during Christmas Bird Counts from 1968 to 2007 in Nebraska, 18 years in Kansas, 30 years in Oklahoma and three years in the Texas panhandle. In South Dakota, wintering is considered extremely rare (Tallman, Swanson and Palmer, 2002). Twenty-three final autumn Nebraska sightings are from August 23 to November 15, with a median of October 16 (Johnsgard, 1980). There are a few winter records for Kansas

(Thompson *et al.*, 2011), and reports extending from August through May in Oklahoma (Woods and Schnell, 1984). It is a very rare winter visitor in the Texas panhandle, with records extending from September to May (Seyffert, 2001).

*Habitats.* Rivers, lakes and reservoirs are used by eared grebes during migration and winter. Breeding occurs on ponds, marshes, and shallow impoundments that are usually rich in submerged aquatic plants.

*National Population.* Breeding Bird Surveys between 1966 and 2012 indicate that this species exhibited a survey-wide population increase (0.5% annually) during that period. There is a fall, 1997, continental estimate of 4.1 million birds (Cullen, Jehl and Nuechterlein, 1999). The species breeds south to central Mexico, and there are also Eurasian ("black-necked grebe") and African populations of unknown size.

*Further Reading.* Palmer, 1962; Johnsgard, 1987; Cullen, Jehl and Nuechterlein, 1999 (*The Birds of North America:* No. 433).

# FAMILY PHALACROCORACIDAE: CORMORANTS

## Double-crested Cormorant
*Phalacrocorax auritus*

Even more than the American white pelican, this fish-eating species has improved its population status in recent years, probably in part as a result of the proliferation of fish farms in the southern states, where cormorants are increasingly frequent if not welcome guests. Their foods are comprised almost entirely of fish, with very small quantities of crustaceans, amphibians, reptiles, mollusks, aquatic insects and other invertebrates. Compared with pelicans the birds are swift fliers, but like them they tend to nest colonially in areas with abundant fish populations. During migration the birds fly at night or during daylight, often following river systems, and avoiding mountain ranges. First-year birds tend to winter farther from their natal colony than do adults, and their spring migration occurs about a month later than that of adults.

*Winter Distribution.* Winter populations of this cormorant are largely coastal-oriented, although in recent years the birds have sometimes been attracted to commercial fish farms. Root's (1988) map shows a minor concentration in southeastern Oklahoma, and one near the Texas border in northeastern New Mexico, at Bitter Lake National Wildlife Refuge. Our data show a significant Oklahoma population that has had a phenomenal numerical change since the 1960s, namely an approximate thousand-fold increase. Our count results suggest a much more rapid rate of regional population increase than is indicated by national Breeding Bird Surveys.

*Seasonality and Migrations.* This cormorant is a common seasonal migrant and local wetland breeder in the central and northern plains, wintering southwardly. It was reported one year during Christmas Bird Counts from 1968 to 2007 in North Dakota, four years in South Dakota, ten years in Nebraska, 36 years in Kansas, 37 years in Oklahoma and ten years in the Texas panhandle. In South Dakota there are a few January and February records (Tallman, Swanson and Palmer, 2002). Thirty-one final autumn Nebraska sightings are from September 17 to December 14, with a median of October 23 (Johnsgard, 1980). It is a very rare during winter in Kansas, with migrants reported from August 8 to June 3 (Thompson and Ely, 1989). It is considered a permanent resident in Oklahoma (Woods and Schnell, 1984). It is a rare winter visitor in the Texas panhandle, with records for every month (Seyffert, 2001). The average number of birds observed per party-hour for the 2013–2014 Christmas Bird Count were: N. Dakota 0, S. Dakota 0, Nebraska 0, Kansas 1.3, Oklahoma 47.91, northwest Texas 16.58. For complete species abundance data, see Appendix 1:35.

*Habitats.* Migrating and wintering cormorants use deeper marshes, lakes, rivers and reservoirs. Fish farms and fish hatcheries are especially favored *Habitats.*

*National Population.* Breeding Bird Surveys between 1966 and 2012 indicate that this species exhibited a survey-wide population increase (3.7% annually) during that period In 1990 there were an estimated 1–2 million birds nationally, or about 350,000 breeding pairs (Hatch and Wesley, 1999). The species breeds south to Belize.

*Further Reading.* Palmer, 1962; Johnsgard, 1993; Hatch and Wesley, 1999 (*The Birds of North America:* No. 442).

## FAMILY PELECANIDAE: PELICANS

### American White Pelican
*Pelecanus erythroryhnchos*

The white pelican is one of America's largest water birds, weighing up to as much as 16–17 pounds and is entirely dependent upon fish for its diet. Although slow and seemingly ponderous in its flight, its massive wing area results in a fairly low wing-loading (the ratio of body weight to total wing area). As a result, in contrast to tundra swans of about the same weight, it is able to take advantage of thermals to gain altitude and facilitate its migrations or even local foraging flights. Additionally, the birds may undertake long round-trip foraging flights of up almost 40 miles, which would not be practical if much energy loss were required. Partly as a result of better protection on breeding grounds, the white pelican population has been gradually increasing, and it never experienced the massive pesticide-related die-offs that occurred with the coastal brown pelican (*Pelecanus occidentalis*).

*Winter Distribution.* White pelicans that breed in the northern Plains winter along the Gulf Coast, so those appearing on Christmas Counts are likely to be in-transit individuals on their way to the Texas coast. Yet, a surprising number remain in late December as far north as North Dakota, The numbers seen during the decade 1998–2007 are certainly greater than those of earlier decades, so perhaps some wintering in Oklahoma is becoming regular. The average number of birds observed per party-hour for the 2013–2014 Christmas Bird Count were: N. Dakota 0, S. Dakota 0, Nebraska 0, Kansas 0.06, Oklahoma 3.2, northwest Texas 1.38. For complete species abundance data, see Appendix 1:35.

*Seasonality and Migrations.* This pelican is a common seasonal migrant and local wetland breeder in the northern plains, wintering southwardly. It was reported two years during Christmas Bird Counts from 1968 to 2007 in North Dakota, six years in South Dakota, nine years in Nebraska, 28 years in Kansas, 37 years in Oklahoma and one year in the Texas panhandle. In South Dakota wintering birds are probably cripples (Tallman, Swanson and Palmer, 2002). Twenty-eight final autumn Nebraska sightings are from September 16 to November 10, with a median of October 16 (Johnsgard, 1980). It is a very rare during winter in Kansas (Thompson *et al.*, 2011). Oklahoma records extend throughout the year (Woods and Schnell, 1984), and it is a very rare winter visitor in the Texas panhandle, with records for every month (Seyffert, 2001).

*Habitats.* Deeper marshes, lakes and reservoirs rich in fish life are used by migrating and wintering birds.

*National Population.* Breeding Bird Surveys between 1966 and 2012 indicate that this species exhibited a survey-wide population increase (5.1% annually) during that period. During 1981–82 there were more 22,000 nests in the U.S.A., and over 53,000 nests in Canada during 1985–86 (Evans and Knopf, 1993). Given the recent rate of increase, the species' total population may now be double these early estimates (Johnsgard, 2013).

*Further Reading.* Palmer, 1962; Johnsgard, 1993, 2013; Evans and Knopf, 1993 (*The Birds of North America:* No. 57).

FAMILY ARDEIDAE: HERONS AND BITTERNS

## Great Blue Heron
*Ardea herodias*

This is the largest and most common of Great Plains herons, and a species that seems able to tolerate freezing temperatures so long as there is enough open water left to allow for foraging. With its bayonet-like beak and powerful neck-thrusting ability, a wide variety of animals may fall prey to this predator. Not only are fish, amphibians, reptiles, small mammals, birds, and crustaceans taken, but also domestic fowl, pets, and even brown pelicans are known to have paid with their lives for letting themselves get within this bird's striking range. As a result, great blue herons can occur along the shorelines of almost any body of water, preying on whatever may be available, until being forced out by freezing conditions.

*Winter Distribution.* This heron requires open water and a supply of fish for winter survival, so it concentrates in ice-free regions of the southern Great Plains. Like the double-crested cormorant and white pelican, its largest Christmas Count numbers occur in Oklahoma. Like them, this heron's count numbers have increased substantially during the past four decades. Root's (1988) map shows an increased population density in northwestern Oklahoma and adjacent Kansas. Our data show a maximum long-term density for Oklahoma, followed by northwestern Texas and Kansas. The apparent gradual population increase in the Great Plains evident in our data is supported by national Breeding Bird Survey trend results.

*Seasonality and Migrations.* The great blue heron is a common Great Plains seasonal migrant and widespread wetland breeder, wintering southwardly. It was reported seven years during Christmas Bird Counts from 1968 to 2007 in North Dakota, 26 years in South Dakota, 30 years in Nebraska, 40 years in Kansas and Oklahoma and 37 years in the Texas panhandle. This species was reported all 40 years during Christmas Bird Counts from 1968 to 2007 in North Dakota, 38 years in South Dakota, 30 years in Nebraska, 34 years in Kansas, 40 years in Oklahoma and 38 years in the Texas panhandle. In South Dakota wintering is rare (Tallman, Swanson and Palmer, 2002). Of 103 final autumn Nebraska sightings dating from the mid-1930s to the late 1970s, the range is August 8 to December 30, and the median is October 13 (Johnsgard, 1980). It is rare to locally common along some rivers during winter in Kansas (Thompson and Ely, 1989, Thompson *et al.*, 2011). It is considered a permanent resident in

Oklahoma (Woods and Schnell, 1984), and is an uncommon winter visitor in the Texas panhandle, with records for every month (Seyffert, 2001). The average number of birds observed per party-hour for the 2013–2014 Christmas Bird Count were: N. Dakota 0, S. Dakota 0.5, Nebraska 0.7, Kansas 0.36, Oklahoma 0.81, northwest Texas 0.26. For complete species abundance data, see Appendix 1:37.

*Habitats.* Migrants and wintering birds are found around all water areas supporting a fish population and having shallows for foraging.

*National Population.* Breeding Bird Surveys between 1966 and 2012 indicate that this species exhibited a survey-wide population increase (0.5% annually) during that period There are no national population estimates, although various counts from the Atlantic and Gulf coasts suggest populations may be in the tens of thousands of pairs. The species breeds south to Ecuador.

*Further Reading.* Bent, 1926; Palmer, 1962; Butler, 1992 (*The Birds of North America:* No. 25).

# FAMILY CATHARTIDAE: AMERICAN VULTURES

## Black Vulture
*Coragyps atratus*

One usually doesn't think of the American vultures as winter birds; all of our species have featherless heads and necks that seem to make the birds quite sensitive to cold, and seemingly prone to early autumn migration. The black vulture is even more of a warmth-loving bird than the turkey vulture, never straying north of its permanent ranges in the southern Great Plains. Both species are preeminent soarers. The black vulture has shorter but broader wings than the turkey vulture, and holds its wings horizontally rather than at a slight upward angle when gliding or soaring, which probably results in somewhat poorer aerial stability, but also presumably provides greater soaring efficiency.

*Winter Distribution.* Root's (1988) map shows only two minor population centers in the Great Plains states, one in southeastern Oklahoma and the other in southwestern Oklahoma. Our data show no birds from the Texas panhandle, but a gradually increasing number have been seen in Oklahoma. There was a roughly eight-fold increase in Oklahoma averages over the four-decade study period, supporting the national long-term population increase documented by Breeding Bird Surveys.

*Seasonality and Migrations.* The black vulture is largely non-migratory within its subtropical breeding range, in contrast to the more migratory turkey vulture. Average birds observed per party-hour for the 2013–2014 Christmas Bird Count were: N. Dakota 0, S. Dakota 0, Nebraska 0, Kansas 0, Oklahoma 1.79, northwest Texas 0. For complete species abundance data, see Appendix 1:38.

*Habitats.* Like the turkey vulture, this vulture is a species of open country, but it is more social than the turkey vulture, often becoming established near villages or towns.

*National Population.* This species was reported 39 of 40 years during Christmas Bird Counts from 1968 to 2007 in Oklahoma. Breeding Bird Surveys between 1966 and 2012 indicate that this species exhibited a survey-wide population increase (4.7% annually) during that period. The estimated 1990s continental population north of Mexico was about 200,000 birds (Rich *et al.,* 2004). The species breeds south to Chile and Argentina.

*Further Reading.* Bent, 1937; Palmer, 1988; del Hoyo, Elliott, and Sargatal, 1994; Buckley, 1999 (*The Birds of North America:* No. 411).

## Turkey Vulture
*Cathartes aura*

Turkey vultures have increased in numbers in the Great Plains states during recent decades. In some areas they have learned to roost in cities, where night-time temperatures are warmer than in the country, and where city-related traffic provides a good supply of road-killed animals around the town perimeters. Such is the case in Lincoln, Nebraska, where some years ago a flock of about 60 turkey vultures began to roost in an upscale part of the city having many tall, old-growth shade trees, which didn't enhance property values there. Some residents chose to cut down their trees in an attempt to rid themselves of these messy guests. These birds scavenge road-kills in the nearby countryside, and often nest in abandoned farm buildings.

*Winter Distribution.* The winter range of the turkey vulture is very similar to that of the black vulture, according to Root's (1988) map, with eastern Oklahoma probably having the highest winter abundance within the Great Plains region. Our data for the two species show strong similarities, with the numbers for the turkey vulture in Oklahoma being generally about twice as high as those for the black vulture. Our turkey vulture data suggest an approximate five-fold increase in Oklahoma over the four-decade period, which follows Breeding Bird Survey results that also suggest a long-term national population increase.

*Seasonality and Migrations.* The turkey vulture is a seasonal migrant and local breeder, wintering southwardly. This species was reported one year each during Christmas Bird Counts from 1968 to 2007 in North Dakota and South Dakota, four years in Nebraska, 14 years in Kansas, and all 40 years in Oklahoma. Thirty-five final autumn Nebraska sightings are from August 6 to December 30, with a median of September 26 (Johnsgard, 1980). This date is about three weeks prior to the first average frost in Nebraska, so the birds seem to prefer to leave the state while thermal updrafts still allow for migrational soaring. The species is occasional during winter in Kansas (Thompson et al., 2011), with scattered records between December 8 and January 22 (Thompson and Ely, 1989). It is considered a permanent resident in Oklahoma (Woods and Schnell, 1984), and is a an extremely rare winter visitor in the Texas panhandle (Seyffert, 2001). Average birds observed per party-hour for the 2013–2014 Christmas Bird Count were: N. Dakota 0, S. Dakota 0, Nebraska 0, Kansas 0, Oklahoma 1.63, northwest Texas 0. For complete species abundance data, see Appendix 1:39.

*Habitats.* Migrants or wintering birds are found widely over open plains, sand-hills or other areas offering visual foraging.

*National Population.* Breeding Bird Surveys between 1966 and 2012 indicate that this species exhibited a survey-wide population increase (2.4% annually) during that period. The estimated 1990s continental population north of Mexico was about 1,3 million birds (Rich *et al.*, 2004). The species breeds south to Chile and Argentina.

*Further Reading.* Bent, 1937; del Hoyo, Elliott, and Sargatal, 1994; Palmer, 1988; Kirk and Mossman, 1998 (*The Birds of North America:* No. 339).

# FAMILY ACCIPITRIDAE:
# KITES, HAWKS AND EAGLES

## Bald Eagle
*Haliaeetus leucocephalus*

Like the trumpeter swan, whooping crane, peregrine and osprey, the bald eagle has been a species for which conservation investments have reaped large rewards in recent years. After more than a century of no nesting in the state of Nebraska, for example, bald eagles began to nest again in the early 1970s, and as of 2007 there were more than 50 active nests in the state. Additionally, by that time close to 1,000 eagles were wintering yearly in Nebraska, primarily along major rivers and large reservoirs, especially Lake McConaughy. These large winter populations have developed while wintering populations of waterfowl have also increased, and the birds have also shifted from winter diets of fish to one comprised a mixture of fish and waterfowl that have been killed or weakened by gunshot or disease.

*Winter Distribution.* Root (1988) determined that one of the highest populations of wintering bald eagles occurred along the section of the Missouri River marking the South Dakota–Nebraska border during her period of study, and that a second concentration occurred along the northern Kansas-Missouri border. These concentration areas still exist and, since the continental population of bald eagles has greatly increased in the past four decades, wintering numbers in the Great Plains have increased correspondingly. Winter numbers in the Great Plains states have roughly doubled during this time, with Kansas and Nebraska Christmas Counts increasing most conspicuously. These increases in the central Plains region probably reflect the larger numbers of geese wintering farther north along the Missouri Valley and on major ice-free reservoirs, since sick and wounded waterfowl provide important foods for bald eagles during autumn and winter periods, supplementing their basic fish diets.

*Seasonality and Migrations.* The bald eagle is an uncommon Great Plains seasonal migrant and very local breeder, wintering southwardly. This species was reported 33 years during Christmas Bird Counts from 1968 to 2007 in North Dakota, 39 years in South Dakota and Nebraska, all 40 years in Kansas and Oklahoma, and 39 years in the Texas panhandle. It is a local breeder (since 1965) and winter visitor in North Dakota, a local breeder (since 1992) and common winter visitor in South Dakota, and a local breeder (since 1973) and common winter visitor in Nebraska. It is a local breeder (since 1989) and uncommon during winter in Kansas(Thompson *et al.*, 2011). It is a local breeder (since 1978) and permanent resident in Oklahoma (Woods and Schnell, 1984), and is an uncommon to common winter visitor in the Texas panhandle, with records for every month except July (Seyffert, 2001). Average birds observed per party-hour for the 2013–2014 Christmas Bird Count were: N. Dakota 0.28, S. Dakota 0.15, Nebraska 0.58, Kansas 1.03, Oklahoma 0.42, northwest Texas 0.23. For complete species abundance data, see Appendix 1:40.

*Habitats.* Bald eagles utilize ice-free areas of larger tree-lined rivers and reservoirs during winter periods. Perching is usually done in tall cottonwoods near water.

*National Population.* Breeding Bird Surveys between 1966 and 2012 indicate that this species exhibited a survey-wide population increase (4.3% annually) during that period The estimated 1990s continental population was about 330,000 birds (Rich *et al.*, 2004).The species breeds south to northern Mexico.

*Further Reading.* Bent, 1937; Palmer, 1988; Russock, 1979; Griffen, 1981; Johnsgard, 1990; del Hoyo, Elliott and Sargatal, 1994; Buehler, 2002 (*The Birds of North America:* No. 506).

## Osprey
### *Pandion haliaetus*

The osprey is a species that once apparently had a wider breeding distribution in the Great Plains, but was extirpated in the late 1800s from Nebraska, South Dakota and perhaps elsewhere. This disappearance probably resulted from various factors, including persecution by fisherman, mindless killing by hunters with an antipathy for raptors, and possibly even the efforts of egg-collectors, since osprey eggs have been historically prized for their beauty and rarity. Re-introduction efforts have begun in South Dakota, to try supplementing a few recent nesting efforts around the Black Hills, and in 2008 a nesting effort occurred in western Nebraska, the first definite record in the state's history.

*Winter Distribution.* The winter range of the osprey is limited northwardly by the ice-free zone, and Root (1988) indicated no significant wintering within the Great Plains states. A clustering of the few Christmas Counts records during the last 15–20 years suggests that birds are now remaining the central and southern plains progressively longer into the winter. Breeding areas are probably also going to be expanding northward, and nesting has recently (2008) begun in Nebraska.

*Seasonality and Migrations.* The osprey is an uncommon seasonal migrant and a very rare local breeder in the northern plains, wintering southwardly. There are apparently no winter records for North Dakota, but during Christmas Bird Counts from 1968 to 2007 it was seen one year in South Dakota, four years in Nebraska, 12 years in Kansas, 20 years in Oklahoma, and three years in the Texas panhandle. Most of these records have occurred since the 1990s. There are apparently no other winter records for South Dakota (Tallman, Swanson and Palmer, 2002). Seventeen final autumn Nebraska sightings are from September 17 to December 26, with a median of October 9 (Johnsgard, 1980). It is occasional during winter in Kansas (Thompson *et al.*, 2011), with fall-to-spring records between September 1 and June 6 (Thompson and Ely, 1992), and a winter resident in Oklahoma, with continuous monthly records extending from September to June (Woods and Schnell, 1984). It is an extremely rare winter visitor in the Texas panhandle (Seyffert, 2001).

*Habitats.* Reservoirs, lakes and large rivers are used by this species on migration and during winter.

*National Population.* Breeding Bird Surveys between 1966 and 2012 indicate that this species exhibited a survey-wide population increase (2.3% annually) during that period The estimated 1990s continental population north of Mexico was about 212,000 birds (Rich *et al.,* 2004). The species breeds south to Belize. Ospreys also occur in several widely separated populations elsewhere in the world (Ferguson-Lees and Christie, 2001; del Hoyo, Elliott, and Sargatal, 1994).

*Further Reading.* Bent, 1937; Palmer, 1988; Johnsgard, 1990; del Hoyo, Elliott, and Sargatal, 1994; Poole, Bierregaard and Martell, 2003 (*The Birds of North America:* No. 683).

### Northern Harrier
*Circus hudsonicus*

The northern harrier provides a perfect symbol of the Great Plains whenever one sees a lone bird, gliding low over grasslands that stretch to the horizon, patiently making its way back and forth until it suddenly drops down on an unseen rodent in the grass. The silvery-gray adult males typically persist in the northern plains longer into late autumn than do the chocolate-toned females and immatures, and are the first to appear in late winter. Harrier mostly prey on small voles (*Microtis*), although on rare occasions they have been known to attack prey as large as ring-necked pheasants or ducks, especially already wounded or disabled ones.

*Winter Distribution.* Root (1988) reported high winter concentrations of northern harriers around Muleshoe, Buffalo Lake and Bitter Lake refuges of the Texas panhandle and adjacent New Mexico, as well as around Quivira National Wildlife Refuge in central Kansas, and the vicinity of Sequoya National Wildlife Refuge in eastern Oklahoma. Like bald eagles, northern harriers often are attracted to sources of sick and wounded waterfowl. Our data show a progressive increase in count numbers for the five-state Plains region, with Nebraska having the greatest actual and relative increase, followed by Kansas. These increases run counter to national population declines indicated by Breeding Bird Survey data, and may relate to the influence of Conservation Reserve Program (CRP) lands that has been planted to grass in the Great Plains since it began in 1985.

*Seasonality and Migrations.* The northern harrier is an uncommon Great Plains seasonal migrant and local breeder, wintering southwardly. It was reported 18 years during Christmas Bird Counts from 1968 to 2007 in North Dakota, 38 years in South Dakota, all 40 years in Nebraska, Kansas, Oklahoma, and the Texas panhandle. In South Dakota, wintering is uncommon (Tallman, Swanson and Palmer, 2002). Thirty-six final autumn Nebraska records are from September 14 to December 31, with a median of December 9 (Johnsgard, 1980). It is common during winter in Kansas (Thompson *et al.*, 2011), mostly reported from early September to late April (Thompson and Ely, 1989). It is considered a permanent resident in Oklahoma (Woods and Schnell, 1984), and is a common winter visitor in the Texas panhandle, with records for every month of the year (Seyffert, 2001). Average birds observed per party-hour for the 2013–2014 Christmas Bird Count were: N. Dakota 0, S. Dakota 0.09, Nebraska 0.25, Kansas 0.59, Oklahoma 0.42, northwest Texas 0.23. For complete species abundance data, see Appendix 1:41.

*Habitats.* This species occupies open habitats such as native grasslands, prairie marshes and wet meadows.

*National Population.* Breeding Bird Surveys between 1966 and 2012 indicate that this species exhibited a survey-wide population decrease (1.0% annually) during that period. The estimated 1990s continental population was about 455,000 birds (Rich *et al.*, 2004). There is also a Eurasian population ("hen harrier") that is often considered a separate species (Ferguson-Lees and Christie, 2001; del Hoyo, Elliott and Sargatal, 1994).

*Further Reading.* Bent, 1937; Hammerstrom, 1986; Palmer, 1988; Johnsgard, 1990, 2001b; del Hoyo, Elliott and Sargatal, 1994; MacWhirter and Bildstein, 1996 (*The Birds of North America:* No. 210).

## Sharp-shinned Hawk
*Accipiter striatus*

Probably when most people set out bird feeders, the last thing they want to attract is an accipiter hawk, and even less do they anticipate stealth attacks on their songbirds from these elusive and silent predators. At least in the Great Plains the incidence of sharp-shinned and Cooper's hawks has greatly increased at urban feeders in recent decades. On any cold, overcast winter day in Lincoln, Nebraska, one can almost count on a visit to a backyard feeder by one of these species. Dunn and Tessaglia-Hymes (1999) reported that most attacks occur in early winter, and common prey associated with feeders include mourning doves, blue jays, European starlings, dark-eyed juncos, pine siskins, house finches and house sparrows.

*Winter Distribution.* Root (1988) doubted the reliability of her map of the winter distribution of this evasive species, which shows an area of concentration in east-central Oklahoma. Our data also are based on small sample sizes, with little if any indication of regional concentrations. However, all states show slowly increasing numbers, with Kansas and northwestern Texas attaining the highest average counts. The increasing popularity of bird-feeding nationally has progressively attracted sharp-shinned hawks into towns and suburbs, which may be the reason for the apparently general upward trend in sharp-shin populations, as is also suggested by national Breeding Bird Survey data.

*Seasonality and Migrations.* The northern harrier is a common Great Plains seasonal migrant and local breeder, wintering southwardly. It was reported 35 years during Christmas Bird Counts from 1968 to 2007 in North Dakota, 38 years in South Dakota, 37 years in Nebraska, all 40 years in Kansas and Oklahoma, and 37 years in the Texas panhandle. In South Dakota this species is a permanent resident in the Black Hills and a winter resident elsewhere (Tallman, Swanson and Palmer, 2002). Forty-one initial autumn Nebraska sightings dating from the mid-1930s to the late 1970s are from July 26 to December 30, with a median of September 16. A total of 142 spring records range from January 1 to June 1, with a median of March 29. Half the records fall within the two periods January 1–9 and March 17–April 27, suggesting that this species is probably mainly a winter visitor and early spring migrant in Nebraska (Johnsgard, 2007a). It is uncommon during winter in Kansas (Thompson *et al.*, 2011), with extreme dates between August 1 and June 9 (Thompson and Ely, 1989). It is considered a winter resident in Oklahoma, with continuous monthly records extending from August to May (Woods and Schnell, 1984). It is a fairly common winter visitor in the Texas panhandle, with records for every month (Seyffert, 2001).

*Habitats.* Throughout the year this species is associated with fairly dense forests, especially mixed woods with some coniferous trees. During winter it often enters wooded yards and hides near feeders to ambush prey.

*National Population.* Breeding Bird Surveys between 1966 and 2012 indicate that this species exhibited a survey-wide population increase (1.0% annually) during that period. The estimated 1990s continental population north of Mexico was about 580,000 birds (Rich *et al.,* 2004). The species breeds south to Argentina.

*Further Reading.* Bent, 1937; Palmer, 1988; Johnsgard, 1990; del Hoyo, Elliott and Sargatal, 1994; Dunn and Tessaglia-Hymes, 1999; Bildstein and Meyer, 2000 (*The Birds of North America:* No. 482).

## Cooper's Hawk
### *Accipiter cooperii*

This medium-sized accipiter hawk is now nearly as common at urban and suburban bird-feeders in the Great Plains as is the sharp-shinned hawk, and in contrast to the sharp-shin is present throughout the Plains States all year long. Like the sharp-shin, it is most likely visit to feeding stations during the coldest winter period, and it too is not very selective also what it attacks. Dunn and Tessaglia-Hymes (1999) mentioned that at least 22 species have been reported as prey, with mourning doves being especially frequent.

*Winter Distribution.* The map that Root (1988) provided for this species suggests a widely spread and lower-density distribution than she reported for the sharp-shinned hawk, with no notable peaks in the Great Plains. Our data suggest a regional abundance almost as great as that of the sharp-shinned hawk and, like it, an increasing abundance over the four-decade study period. Nebraska and Kansas seem to support the largest numbers of Cooper's hawks around Christmas. Like the sharp-shin, they have adapted to life the suburbs and even to city life. For example, nesting has occurred in some of the Lincoln, Nebraska, and city parks during recent years. The gradual population increase indicated by our data is supported by an upward population trend documented by national Breeding Bird Surveys.

*Seasonality and Migrations.* The Cooper's hawk is an uncommon Great Plains seasonal migrant and local breeder, wintering southwardly. It was reported 17 years during Christmas Bird Counts from 1968 to 2007 in North Dakota, 31 years in South Dakota, 34 years in Nebraska, all 40 years in Kansas and

Oklahoma, and 27 years in the Texas panhandle. In South Dakota this species is a local summer resident and rare winter resident. (Tallman, Swanson and Palmer, 2002). Thirty-five final autumn Nebraska sightings from the mid-1930s to the late 1970s are from September 8 to December 31, with a median of October 30 (Johnsgard, 1980), but recent evidence indicates the species is increasingly a year-around breeding resident in Nebraska. It is uncommon during winter in Kansas (Thompson *et al.*, 2011), with extreme dates for non-breeders of August 5 and May 21 (Thompson and Ely, 1989). It is considered a permanent resident in Oklahoma (Woods and Schnell, 1984), and is an uncommon to fairly common winter visitor in the Texas panhandle, with records for every month (Seyffert, 2001).

*Habitats.* Throughout the year this species is associated with mature forests, especially hardwood forests.

*National Population.* Breeding Bird Surveys between 1966 and 2012 indicate that this species exhibited a survey-wide population increase (2.3% annually) during that period. The estimated 1990s continental population north of Mexico was about 550,000 birds (Rich *et al.*, 2004). The species breeds south to northern Mexico.

*Further Reading.* Bent, 1937; Palmer, 1988; Johnsgard, 1990; Rosenfield and Bielefeldt, 1993 (*The Birds of North America:* No. 75); del Hoyo, Elliott and Sargatal, 1994; Dunn and Tessaglia-Hymes, 1999.

### Northern Goshawk
*Accipiter gentilis*

Although most birders would probably drive a long ways to see a wild goshawk, Great Plains bird populations are probably very lucky that this species is such a rare winter visitor to the region. A forest-adapted bird, the goshawk at times it will enter cities as large as Lincoln, NE, where it occasionally is killed by flying headlong into glass windows while in full chase. Hunting along woodland edges is preferred, and hunts started from tree perches are most successful than those begun while in flight. Although medium-sized birds such as grouse are favored winter prey, rabbits and squirrels are also taken, so its winter distribution is probably more dependent on general prey availability than on temperature or amount of snow cover.

*Winter Distribution.* This generally rare and elusive winter migrant was seen too infrequently during Christmas Counts to offer much information on its real abundance. Our data indicate that only in North Dakota, and secondly

South Dakota, are northern goshawks common enough to be considered regular winter visitors. In Nebraska only five goshawks were received by Raptor Recovery Nebraska in 25 years, as compared with 34 Cooper's hawks and 76 sharp-shins (Johnsgard, 2002c). Stragglers sometimes appear south to Oklahoma and even rarely to northern Texas, but the goshawk has not adapted to city and suburb foraging in the manner of sharp-shinned and Cooper's hawks. The numbers seen are too few to judge if the Great Plains population is increasing.

*Seasonality and Migrations.* The northern goshawk is an occasional to rare wintering migrant, most common northwardly. It was reported 29 years during Christmas Bird Counts from 1968 to 2007 in North Dakota, 31 years in South Dakota, 18 years in Nebraska, 24 years in Kansas, 16 years in Oklahoma, and two years in the Texas panhandle. In South Dakota this species is a rare winter visitor (Tallman, Swanson and Palmer, 2002). Twenty-two total autumn Nebraska sightings dating from the mid-1930s to the late 1970s are from September 16 to December 31, with half of the records occurring during the two periods September 21–October 17 and December 25–31. Forty-eight spring records range from January 1 to June 1, with a median of March 15, suggesting this species is both a winter visitor and late spring migrant in Nebraska (Johnsgard, 2007a). It is rare and irregular during winter in Kansas (Thompson *et al.*, 2011), with confirmed records between October 3 and February 24 (Thompson and Ely, 1989). It is considered a rare winter visitor in Oklahoma, with continuous monthly records extending from September to May (Woods and Schnell, 1984), and is an extremely rare winter visitor in the Texas panhandle, with scattered records extending from September to April (Seyffert, 2001).

*Habitats.* Throughout the year this species is rarely found far from wooded to heavily forested areas.

*National Population.* Breeding Bird Surveys between 1966 and 2012 indicate that this species exhibited a small survey-wide population increase (0.1% annually) during that period. The estimated 1990s continental population was about 240,000 birds (Rich *et al.,* 2004). The species breeds south to southern Mexico. There is also a Eurasian population, possibly numbering in the hundreds of thousands of birds (Ferguson-Lees and Christie, 2001; del Hoyo, Elliott, and Sargatal, 1994).

*Further Reading.* Bent, 1937; Godfrey, 1986; Palmer, 1988; Johnsgard, 1990; del Hoyo, Elliott and Sargatal, 1994; Squires and Reynolds, 1997 (*The Birds of North America:* No. 298).

## Red-shouldered Hawk
*Buteo lineatus*

This is a forest-adapted buteo, and one that has disappeared from the extreme western edge of its range. There is little habitat in the Great Plains states that is sufficiently wooded for this species, which prefers mature lowland hardwood forest with water and clearings nearby.

*Winter Distribution.* The red-shouldered hawks winter range barely enters the Great Plains, and Root's (1988) map shows it as largely limited to Oklahoma, eastern Kansas and the Missouri Valley of Nebraska, probably being mostly confined to riverbottom forests. The Nebraska population is now virtually extirpated, and only in Oklahoma are the numbers on Christmas Counts large enough to allow a population analysis. Although Breeding Bird Surveys indicate that the national population is increasing, this is evidently not the case in the Great Plains.

*Seasonality and Migrations.* The red-shouldered hawk is an uncommon to occasional Great Plains seasonal migrant and local breeder, wintering southwardly. It was reported one year each during Christmas Bird Counts from 1968 to 2007 in North Dakota and South Dakota, eight years in Nebraska, 33 years in Kansas, and all 40 years in Oklahoma. Forty-nine initial spring Nebraska sightings dating from the mid-1930s to the late 1970s (Johnsgard, 1980) have a nearly random temporal distribution, suggesting that this species is a resident in its limited Nebraska range. Eleven final autumn sightings also show no clear indication of migratory movements. It is rare during winter in Kansas (Thompson *et al.*, 2011), with only a few records (Thompson and Ely, 1989). It is considered a permanent resident in Oklahoma (Woods and Schnell, 1984), and is a very rare winter visitor in the Texas panhandle, with records scattered from August to May (Seyffert, 2001). However, it is numerous in eastern Texas, where migrant birds from northern areas mix with residents (Palmer, 1988).

*Habitats.* Throughout the year this species is found in relatively moist woodlands, especially floodplain forests, with adjacent open country for foraging.

*National Population.* Breeding Bird Surveys between 1966 and 2012 indicate that this species exhibited a survey-wide population increase (2.9% annually) during that period, although during most of the 20th century the species was in marked decline. The estimated 1990s continental population north of Mexico was about 830,000 birds (Rich *et al.*, 2004). The species breeds south to northern Mexico.

*Further Reading.* Bent, 1937; Palmer, 1988; Johnsgard, 1990; Crocoll, 1994 (*The Birds of North America:* No. 107); del Hoyo, Elliott and Sargatal, 1994.

## Red-tailed Hawk
*Buteo jamaicensis*

This common and widespread buteo is easily the most abundant hawk of the Great Plains states, both in summer and in winter. By late September red-tails from farther north are starting to become apparent in Nebraska, just as Swainson's hawks are flocking and starting to move south. For the next month or two it sometimes seems that a red-tailed hawk can be seen on telephone poles every mile or less, and as autumn progresses the chances of seeing dark-colored (melanistic) Harlan's hawks (*B. j. harlani)* type increase. Unusually pale (leucistic) birds, so-called Krider's hawks, are less common. During 2004–2007 autumn raptor counts at Hitchcock Nature Center, near Council Bluffs, Iowa, the Harlan's morphs totaled 0.76 percent, whereas Krider's morphs comprised only 0.16 percent of 16,718 total red-tails (Mark Orsag, pers. comm.). As Palmer (1988) concluded, leucistic red-tails cannot be assigned to a specific breeding range, whereas Harlan's-morph birds breed in interior and western Alaska. During winter they are most commonly seen in northwestern parts of the Great Plains, especially Montana and North Dakota. During Christmas Counts from 1968 to 2007 they comprised 6.2 percent of red-tails seen in North Dakota, 1.9 percent in South Dakota 1.0 percent in Nebraska, 1.2 percent in Kansas and 2.1 percent in Oklahoma. They were not reported from the Texas panhandle. Average number of red-tailed hawks observed per party-hour for the 2013–2014 Christmas Bird Count were: N. Dakota 0.94, S. Dakota 0.33. Nebraska 0.93, Kansas 1.72, Oklahoma 1.13., northwest Texas 0.33. For complete species abundance data, see Appendix 1:42.

*Winter Distribution.* The red-tailed hawk is the classic winter hawk of the Great Plains. Root's (1988) map shows concentrations in eastern Kansas, eastern Oklahoma and along the Oklahoma–Texas panhandle border region. Our data show a gradually increasing population across the entire five-state Plains region, as well as the Texas panhandle, with overall concentrations highest in Kansas. This upward trend is consistent with national Breeding Bird Survey data.

*Seasonality and Migrations.* The red-tailed hawk is a common permanent resident in the Great Plains, supplemented by winter influxes of birds from farther north. It was reported on 32 years during North Dakota Christmas Bird Counts between 1968 and 2007, and on all 40 years in South Dakota, Nebraska, Kansas and Oklahoma. Twenty-three final autumn Nebraska sightings dating from the mid-1930s to the late 1970s and 32 initial spring Nebraska sightings have a nearly random temporal distribution, suggesting that this species is essentially a permanent resident in Nebraska (Johnsgard,

1980). It is a permanent resident in Kansas (Thompson *et al.*, 2011) and Oklahoma(Woods and Schnell, 1984), with winter numbers augmented by migrants from farther west and north, and is a fairly common to common resident in the Texas panhandle (Seyffert, 2001). Average birds observed per party-hour for the 2013–2014 Christmas Bird Count were: N. Dakota 0.09, S. Dakota 0.33, Nebraska 0.93, Kansas 1.72, Oklahoma 1.13, northwest Texas 0.33.

*Habitats.* A combination of extensive open habitat for visual hunting and scattered clumps or groves of tall trees for nesting provide the year-round needs for this species.

*National Population.* Breeding Bird Surveys between 1966 and 2012 indicate that this species exhibited a survey-wide population increase (1.7% annually) during that period. The estimated 1990s continental population north of Mexico was about 1,960,000 birds (Rich *et al.*, 2004). The species breeds south to Panama.

*Further Reading.* Bent, 1937; Palmer, 1988; Johnsgard, 1990; Preston and Beane, 1993 (*The Birds of North America:* No. 52); del Hoyo, Elliott and Sargatal, 1994.

### Ferruginous Hawk
*Buteo regalis*

The classic buteo hawk of the high plains, the ferruginous hawk never fails to stir the heart of birders when it is seen. Although recent Breeding Bird Surveys indicate an increasing national population, it has almost entirely disappeared from some parts of its range, such as southern Canada. To a large degree this decline parallels the national decline of large grassland rodents, especially prairie dogs.

*Winter Distribution.* This is a shortgrass and high plains species whose distribution and population are closely tied to those of prairie dogs and other large mammals such as ground squirrels and rabbits. Root (1988) mapped the highest densities of this species in the Texas panhandle and adjacent New Mexico (especially around Amarillo and Dalhart), with a smaller concentration in western Kansas. Our data show the birds to be concentrated almost entirely in the Texas panhandle, and suggests that they are slowly increasing in that region. Given the total failure of all the Great Plains states to protect prairie dogs as a keystone species of the shortgrass ecosystem, and a major prey of ferruginous hawks, this is a surprising but welcome trend, and is supported by national Breeding Bird Survey results.

*Seasonality and Migrations.* The ferruginous hawk is a occasional to rare permanent resident over of the western Great Plains, with some southward migration in northern areas. It was reported 36 years during Christmas Bird
Counts from 1968 to 2007 in South Dakota, 26 years in Nebraska, 39 years
in Kansas, 38 years in Oklahoma, and all 40 years in the Texas panhandle.
In South Dakota this species is rare during winter (Tallman, Swanson and
Palmer, 2002). Twenty final autumn Nebraska records dating from the mid-
1930s to the late 1970s are widely spread between August 26 and December
31. Seventy initial spring Nebraska sighting range from January 1 to May
25, with a median of March 1. The wide spread of the records suggests that
this species is residential in Nebraska (Johnsgard, 1980). It is considered a
winter resident in eastern Kansas (late September to mid-April) and a permanent resident in western Kansas. Oklahoma records extend throughout
the year (Thompson *et al.*, 2011; Woods and Schnell, 1984). It is a fairly
common to common winter visitor in the Texas panhandle, and has been
seen in every month (Seyffert, 2001). Average birds observed per party-hour
for the 2013–2014 Christmas Bird Count were: N. Dakota 0, S. Dakota
0.04, Nebraska 0.09, Kansas 0.14, Oklahoma 0.03, northwest Texas 0.23.
For complete species abundance data, see Appendix 1:43.

*Habitats.* This species is normally found in grassland habitats having scattered trees, buttes or bluffs for roosting sites, and in areas having a reliable
food supply in the form of prairie dogs, ground squirrels or other abundant rodents.

*National Population.* Breeding Bird Surveys between 1966 and 2012 indicate
that this species exhibited a survey-wide population increase (1.1% annually) during that period, making it one of the few grassland endemic
birds to show probable population increases (Johnsgard, 2001b). The estimated 1990s continental population was only about 23,000 birds (Rich
*et al.,* 2004). If accurate, this would make it the rarest of North American
hawks, and one in dire need of conservation efforts, especially in the Great
Plains, the heart of its winter range.

*Further Reading.* Bent, 1937; Palmer, 1988; Johnsgard, 1990; del Hoyo, Elliott
and Sargatal, 1994; Bechard and Schmutz, 1995 (*The Birds of North America:* No. 172); Dechant *et al.,* 1999c.

## Rough-legged Hawk
*Buteo lagopus*

One of the few ornithological bright spots in a Great Plains winter is the possibility of seeing an influx of arctic-breeding birds, such as snowy owls and rough-legged hawks. Rough-legged hawks are much more common than snowy owls in the Great Plains states, and they are also more likely to reach the southern plains. Dark-morph rough-legs are common in the Great Plains, and during fall and winter they pose challenges in field separation from dark-morph red-tailed hawks, which they sometimes closely resemble. During 2004–2007 autumn hawk counts at Hitchcock Nature Center, near Council Bluffs, Iowa, dark-morph (melanistic) individuals comprised 13.7 percent of 183 field-identified rough-legs (Mark Orsag, pers. comm.).

*Winter Distribution.* Root's (1988) map shows western Kansas as the highest concentration of rough-legged hawk in the Great Plains states during the 1960s and early 1970s, with smaller numbers extending north into western Nebraska and even fewer reaching western South Dakota. Although the overall 40-year pattern in our data might support that general pattern persists, the most recent counts would suggest that Kansas now supports fewer birds, and that they are now increasingly concentrated in Nebraska and South Dakota. The five-state long-term averages suggest a stable Great Plains population

*Seasonality and Migrations.* The rough-legged hawk is an uncommon wintering migrant throughout the Great Plains. It was reported 39 years during Christmas Bird Counts from 1968 to 2007 in North Dakota, 38 years in South Dakota, all 40 years in Nebraska, Kansas and Oklahoma, and 39 years in the Texas panhandle. In South Dakota this common wintering species occurred at least once on 73 percent of Christmas Bird Count locations between 1949 and 1998 (Tallman, Swanson and Palmer, 2002). Eighty-five initial autumn Nebraska sightings dating from the mid-1930s to the late 1970s range from September 30 to December 30, with a median of November 2. A total of 73 final spring Nebraska sightings range from January 8 to May 20, with a median of March 26 (Johnsgard, 1980). It is considered a regular winter resident in Kansas (early October to late April) and Oklahoma, with Oklahoma records extending from October to May (Thompson *et al.*, 2011; Woods and Schnell, 1984). It is an uncommon to fairly common winter visitor in the Texas panhandle, and has been seen from September to April (Seyffert, 2001). Average birds observed per party-hour for the 2013–2014 Christmas Bird Count were: N. Dakota 0.13, S. Dakota 0.57, Nebraska 0.56, Kansas 0.20, Oklahoma 0.12, northwest Texas 0.15. For complete species abundance data, see Appendix 1:44.

*Habitats.* Open prairies, plains and other grassland habitats are used by this species while on migration and during winter.

*National Population.* Breeding Bird Survey trend data are not available. The estimated 1990s continental population was about 260,000 birds (Rich *et al.*, 2004). There is also a Eurasian population of unknown size (Ferguson-Lees and Christie, 2001; del Hoyo, Elliott and Sargatal, 1994).

*Further Reading.* Bent, 1937; Godfrey, 1986; Palmer, 1988; Johnsgard, 1990, 2001b; del Hoyo, Elliott and Sargatal, 1994; Bechard and Swem, 2002 (*The Birds of North America:* No. 641).

## Golden Eagle
*Aquila chrysaetos*

Golden eagles are the largest raptors of the high plains region, roughly encompassing the western halves of the Great Plains states. Although sometimes found soaring over creeks and rivers in search of sick or wounded waterfowl, they are more often seen well away from water, hunting in the open countryside for rabbits and rodents.

*Winter Distribution.* Root (1988) judged that the largest Great Plains concentration of golden eagles occurred along the Nebraska–South Dakota, in the general vicinity of Fort Niobrara and Lake Andes National Wildlife Refuges. A smaller concentration was associated with central Kansas. Our data does not have a consistent pattern of abundance, but seem to show a declining if substantial population in the Texas panhandle, and a secondary population in South Dakota that might be increasing slightly. Our data also does not indicate a clear upward trend in the Great Plains population, as is suggested by national Breeding Bird Survey data.

*Seasonality and Migrations.* The golden eagle is an uncommon to occasional and local permanent resident in the western parts of the region, and a migrant and winter visitor elsewhere. It was reported 38 years during Christmas Bird Counts from 1968 to 2007 in North Dakota, all 40 years in South Dakota, 31 years in Nebraska, 40 years in Kansas and Oklahoma, and 38 years in the Texas panhandle. It is a year-around resident in western Nebraska and a winter visitor elsewhere in the state. It is uncommon during winter in Kansas (Thompson *et al.*, 2011), with non-breeders present from early October through early April (Thompson and Ely, 1989). It is considered a permanent resident in Oklahoma (Woods and Schnell, 1984), and an uncommon resident in the Texas panhandle (Seyffert, 2001). Average birds observed per party-hour for the 2013–2014 Christmas Bird Count were: N. Dakota 0.1, S. Dakota 0.09, Nebraska 0.08, Kansas 0.07, Oklahoma 0.13, northwest Texas 0.12. For complete species abundance data, see Appendix 1:45.

*Habitats.* Throughout most of the year this species is associated with arid, open country, often areas having buttes, mountains or canyons that offer secure nesting sites and large areas of grassland vegetation for foraging.

*National Population.* Breeding Bird Surveys between 1966 and 2012 indicate that this species exhibited a survey-wide population stability (0.0% annually) during that period The estimated 1990s continental population north of Mexico was about 80,000 birds (Rich *et al.,* 2004). The species breeds south to northern Mexico. There is also a Eurasian population of unknown total size (Ferguson-Lees and Christie, 2001; del Hoyo, Elliott and Sargatal, 1994).

*Further Reading.* Bent, 1937; Palmer, 1988; Johnsgard, 1990; del Hoyo, Elliott and Sargatal, 1994; Kocher *et al.,* 2003 (*The Birds of North America:* No. 684).

# FAMILY RALLIDAE:
# RAILS, GALLINULES AND COOTS

## Virginia Rail
### *Rallus limicola*

One does not normally think of rails when contemplating winter birds, but the Virginia rail seems to be relatively cold-tolerant, and at least some individuals in the Great Plains are prone to linger late into autumn before heading south. Its foods during cold weather probably shift from the usual diet of insects and earthworms to crayfishes and the seeds of sedges and smartweeds. Like other rails, it is so elusive that little is known of its cold-weather ecology or behavior.

*Winter Distribution.* Root (1988) mapped a minor winter concentration of this rail in the Texas panhandle and adjoining Oklahoma, but identified no other Great Plains concentrations. She judged that unfrozen freshwater marshes provide winter habitat needs. Our data show a long-term use of Oklahoma sites on Christmas Counts, and a more recent apparent late autumn or early winter lingering in both Kansas and Nebraska, where unfrozen marshes and actual wintering are probably very rare. Our data supports a probable long-term regional population increase, as is also indicated by national Breeding Bird Survey data.

*Seasonality and Migrations.* The Virginia rail is an uncommon and inconspicuous Great Plains seasonal migrant and local wetland breeder, with little evidence of wintering. It was not reported during Christmas Bird Counts from 1968 to 2007 in North Dakota, but was seen three years in South Dakota, 12 years in Nebraska, 15 years in Kansas, 31 years in Oklahoma, and five years in the Texas panhandle. The species is considered extremely rare in South Dakota during winter (Tallman, Swanson and Palmer, 2002). Thirteen final autumn sightings in Nebraska are from July 21 to October 13, with a median of September 16 (Johnsgard, 1980). It is a very rare during winter in Kansas (Thompson and Ely, 1989; Thompson *et al.*, 2011), and is evidently rare during winter in Oklahoma, with but monthly records extend from autumn to February (Woods and Schnell, 1984), and is an uncommon to fairly common resident in the Texas panhandle (Seyffert, 2001). The average number of birds observed per party-hour for the 2013–2014 Christmas Bird Count were: N. Dakota 0, S. Dakota 0, Nebraska 0.03, Kansas 0.08, Oklahoma 0.14, northwest Texas 0. For complete species abundance data, see Appendix 1:46.

*Habitats.* Primary habitats are marshes with extensive stands of emergent vegetation such as taller grasses, bulrushes and sedges.

*National Population.* Breeding Bird Surveys between 1966 and 2012 indicate that this species exhibited a survey-wide population increase (1.1% annually) during that period. No national population estimates are available for this elusive species. The species breeds south to southern Mexico.

*Further Reading.* Bent, 1926; Tacha and Braun, 1994; Conway, 1995 (*The Birds of North America:* No. 173).

## American Coot
*Fulica americana*

In contrast to rails, coots are easily visible, and might be seen in almost any wetland of large to moderate size. They need to take flight from a running start, and seem to avoid very small wetlands where takeoffs might be difficult. They are rather weak fliers, and tend to delay autumn migration as long as possible. Thus, at times they are left swimming around small openings in the ice, leaving them vulnerable to predation by hawks, owls, or coyotes.

*Winter Distribution.* Root (1988) mapped the winter distribution of coots as mostly south of the Texas border, but including a minor concentration in central Kansas along the Arkansas River (Quivira National Wildlife Refuge and the Cheyenne Bottoms Wildlife Management Area. Our data suggest that Oklahoma and northwest Texas are the region's most important wintering regions, with declining numbers in northwest Texas but increasing numbers in Oklahoma and Kansas, There have also been recent small increases in Nebraska and even South Dakota, reflecting increasingly later freeze-ups in those states.

*Seasonality and Migrations.* The American coot is a common Great Plains seasonal migrant and local wetland breeder, wintering southwardly. It was reported one year during Christmas Bird Counts from 1968 to 2007 in North Dakota, 39 years in South Dakota, 20 years in Nebraska, 40 years in Kansas and Oklahoma, and 39 years in the Texas panhandle. In South Dakota, it is rare to uncommon during winter (Tallman, Swanson and Palmer, 2002). Eighty-two final autumn Nebraska records are from July 25 to December 31, with a median of November 2 (Johnsgard, 1980). It is rare during winter in Kansas (Thompson and Ely, 1989; Thompson *et al.*, 2011). It is considered a winter resident in Oklahoma, with continuous monthly records extending throughout the year (Woods and Schnell, 1984), It is a fairly common to abundant resident in the Texas panhandle (Seyffert, 2001). The average number of birds observed per party-hour for the 2013–2014 Christmas Bird Count were: N. Dakota 0, S. Dakota 0, Nebraska 6.7, Kansas 1.04, Oklahoma 4.75, northwest Texas 6.9. For complete species abundance data, see Appendix 1:47.

*Habitats.* A wide variety of wetlands, ranging from small ponds or large lakes and reservoirs are used throughout the year, but those that are fairly shallow and rich in submerged aquatic plants are favored.

*National Population.* Breeding Bird Surveys between 1966 and 2012 indicate that this species exhibited a survey-wide population increase (1.1% annually) during that period U.S. Fish and Wildlife Service surveys of the late 1990s suggest a national population of about three million birds (Brisbin, Pratt and Mobray, 2003). The species breeds south to Costa Rica.

*Further Reading.* Bent, 1926; Tacha and Braun, 1994; Brisbin, Pratt and Mobray, 2003 (*The Birds of North America:* No. 697).

# FAMILY GRUIDAE: CRANES

## Sandhill Crane
*Grus canadensis*

The Great Plains are a crane-lover's paradise; nowhere else in the world provides so many opportunities for seeing these magnificent birds. Sandhill crane numbers have increased so markedly in recent decades that a species that I was not able observe in the wild until I was in my twenty's can now be fairly easily observed by even pre-school children in many parts of the Great Plains. Cranes are perfect species for stimulating the interest of children in nature; they are large, loud, gregarious, and have other life-history characteristics that are similar in many ways to those of humans. They can effectively be used to teach geography, ecology, conservation, and even bridge the humanities, such as relating cultural attitudes toward nature, art, and even the origins of human dance.

*Winter Distribution.* Root (1988) mapped the area around Muleshoe National Wildlife Refuge in Texas as having by far the highest regional density of sandhill cranes (2,331 birds per party hour). This general pattern of very high Texas panhandle counts still persists, with Muleshoe and Buffalo Lake numbers overshadowing all other regional counts. Mid-continental populations of lesser sandhill cranes increased greatly during the second half of the 20th century, under the influence of abundant grain supplies along major migration routes and on wintering areas. The lesser sandhill crane's mid-continent population has become fairly stable in recent years, with Muleshoe, Buffalo Lake and Bitter Lake (New Mexico) national wildlife refuges providing a primary wintering sites for the Great Plains population. Recent spring surveys of sandhill cranes in the Platte Valley have averaged about 400,000–500,000 birds, about 90 percent of which are considered to be lesser sandhills headed for tundra breeding areas. The rest are regarded as being the greater race that perhaps breed in Minnesota or southern Ontario, although intermediate-sized birds ("Canadian" sandhill cranes) do occur. The marked decade-to-decade number variations apparent in the Texas panhandle probably reflect count sampling variability rather than actual population trends. The may occur because the Bitter Lake population of nearby New Mexico receives varying percentages of the overall mid-continent crane population in different years.

*Seasonality and Migrations.* The sandhill crane is a common to abundant Great Plains seasonal migrant and highly local wetland breeder, wintering southwardly. It was reported three years during Christmas Bird Counts from 1968 to 2007 in North Dakota one year in South Dakota and Nebraska, 18 years in Kansas, 30 years in Oklahoma, and all 40 years in the Texas panhandle. In South Dakota it is extremely rare during winter (Tallman, Swanson and Palmer, 2002). Fifty-three final autumn Nebraska sightings are from October 1 to December 31, with a median of November 5 (Johnsgard, 1980). Several thousand have overwintered in Nebraska a few times in recent years. It is locally common during winter in southern Kansas (Thompson *et al.*, 2011). It is an abundant late fall migrant and a local winter resident in Oklahoma, with continuous monthly records extending from September to May (Woods and Schnell, 1984). It is an abundant winter visitor in the Texas panhandle, and has been seen there in every month except August (Seyffert, 2001). The average number of birds observed per party-hour for the 2013–2014 Christmas Bird Count were: N. Dakota 0, S. Dakota 0, Nebraska 0, Kansas 3.97, Oklahoma 1404, northwest Texas 0. For complete species abundance data, see Appendix 1:48.

*Habitats.* Slowly flowing rivers, with relatively bare bars and islands for roosting, and adjacent wet meadows and croplands for foraging, are used by this species during migration.

*National Population.* Breeding Bird Surveys (of the greater and Florida sandhill crane subspecies) between 1966 and 2012 indicate that the combined races exhibited a population increase (5.1% annually) during that period. No breeding surveys of the arctic-breeding lesser race are performed, but spring counts since the 1960s suggest that the long-term increase in lesser sandhill crane populations that probably began after World War II has ended (Johnsgard, 2003; 2015). Counting all races, the combined Siberian and North American populations of sandhill cranes may exceed 700,000 birds.

*Further Reading.* Bent, 1926; Godfrey, 1986; Johnsgard, 1991, 2011a, 2015; Tacha, Nesbitt and Vohs, 1992 (*The Birds of North America:* No. 32); Tacha and Braun, 1994.

## FAMILY CHARADRIIDAE: PLOVERS

### Killdeer
*Charadrius vociferus*

This most conspicuous and highly abundant of our North American shore-birds is fairly cold-tolerant, often remaining in the northern plains during autumn until ice begins to form on wetlands, and returning about the time the spring waterfowl migration gets underway. Nearly all of its foods are of animal origin, primarily insects, but weed seeds are sometimes eaten and may provide emergency cold-weather food. It is more adaptable to humans than most shorebirds, sometimes nesting on golf courses, in city parks, or even on the roofs of flat-roofed buildings.

*Winter Distribution.* The killdeer is one of the latest autumn and earliest spring shorebird migrants of the Great Plains. Root (1988) mapped the species' winter range as extending north into Kansas, western Nebraska and the Nebraska-South Dakota border region. Our data suggest a nearly stable population in the Great Plains states, with the largest numbers concentrated in Oklahoma. Probably the majority of the Great Plains' killdeers winter farther south in Texas, along the Gulf Coast. Our data do not support a regional population decline, contrary to national Breeding Bird Survey data.

*Seasonality and Migrations.* The killdeer is a common Great Plains seasonal migrant and widespread breeder, wintering southwardly. This species was not reported during Christmas Bird Counts from 1968 to 2007 in North Dakota, but was seen 29 years in South Dakota, 34 years in Nebraska, 39 in Kansas, 40 in Oklahoma, and 33 in the Texas panhandle. In South Dakota it is rare during winter (Tallman, Swanson and Palmer, 2002). The range of 110 final autumn Nebraska records is from August 18 to December 31, with a median of October 19 (Johnsgard, 1980). It is rare during winter in Kansas (Thompson *et al.*, 2011). It is considered a permanent resident in Oklahoma (Woods and Schnell, 1984), and is a rare to fairly common non-breeding resident in the Texas panhandle, recorded every month of the year (Seyffert, 2001). The average number of birds observed per party-hour for the 2013–2014 Christmas Bird Count were: N. Dakota 0, S. Dakota 0.1, Nebraska 0.35, Kansas 0.05, Oklahoma 0.33, northwest Texas 0. For complete species abundance data, see Appendix 1:49.

*Habitats.* This highly adaptable species often occurs on open fields during migration, but typically breeds near wetlands where there is exposed ground nearby. The birds seem to prefer gravely, stony or sandy areas for nesting, such as along the sides of gravel roads, but also nests in a wide variety of locations, sometimes even in suburban gardens.

*National Population.* Breeding Bird Surveys between 1966 and 2012 indicate that this species exhibited a survey-wide population decrease (1.1% annually) during that period. Morrison *et al.* (2001b) estimated the species' total population at about three million birds. The species breeds south to Costa Rica.

*Further Reading.* Bent, 1929; Johnsgard, 1981; Jackson and Jackson, 2000 (*The Birds of North America:* No. 527).

# FAMILY SCOLOPACIDAE:
# SANDPIPERS, SNIPES AND PHALAROPES

## Spotted Sandpiper
*Actitis macularius*

Spotted sandpipers are adept at walking or running along muddy, sandy or rocky shorelines in search of diverse prey. They consume a wide variety of crawling and burrowing insects, and sometimes even snatch flying insects out of the air, or small fry from shallow streams. They also leave the water's edge at times to forage in meadows and gardens. This foraging adaptability perhaps helps explain how the birds are able to temporarily survive subfreezing temperatures.

*Winter Distribution.* This species is marginal in occurrence in the Great Plains by late fall, and it avoids wintering in areas having a mean January temperatures below freezing (Root, 1988). As a result, Christmas Count occurrences in our region can best be regarded as rarities, even in the Texas panhandle.

*Seasonality and Migrations.* The spotted sandpiper is a common Great Plains seasonal migrant and local wetland breeder. It was not reported during Christmas Bird Counts from 1968 to 2007 in North Dakota or South Dakota, but was seen one year in Nebraska, two years in Kansas, 12 years in Oklahoma, and two years in the Texas panhandle. The latest South Dakota records are for mid-November and the earliest spring dates are for early April (Tallman, Swanson and Palmer, 2002). Sixty-two-final autumn Nebraska records are from July 26 to October 26, with a median of September 9 (Johnsgard, 1980), or well before freeze-up. It has been reported a few times during winter in Kansas (Thompson *et al.*, 2011). Records exist through December in Oklahoma (Woods and Schnell, 1984). It has not been reported after October in the Texas panhandle (Seyffert, 2001).

*Habitats.* While on migration and in wintering areas this species is associated with wetlands having exposed or sparsely vegetation shorelines or islands, and ranging from fairly rapidly flowing streams to still-water *Habitats.*

*National Population.* Breeding Bird Surveys between 1966 and 2012 indicate that this species exhibited a survey-wide population decrease (1.5% annually) during that period Morrison *et al.* (2001b) estimated the North American population at 150,000 birds.

*Further Reading.* Bent, 1927; Johnsgard, 1981; Oring, Gray and Reed, 1997 (*The Birds of North America:* No. 289).

## Greater Yellowlegs
*Tringa melanoleuca*

One of the largest of our sandpipers, the greater yellowlegs is mainly a wading forager, and is unlikely to remain in an area long after freezing conditions develop. Thus, winter records anywhere north of Texas are the result of stragglers.

*Winter Distribution.* Like the spotted sandpiper, this species is marginal in occurrence in the Great Plains by late fall, and at least in the East it avoids wintering in regions having minimum subfreezing January temperatures (Root, 1988). Christmas Count numbers in our region are generally small, although Oklahoma regularly supports small numbers through the Christmas period. The Texas coast is probably the final winter destination for the Great Plains population of greater yellowlegs. No population trends are apparent from our limited data.

*Seasonality and Migrations.* The greater yellowlegs is a common seasonal migrant, wintering southwardly. It was not reported during Christmas Bird Counts from 1968 to 2007 in North Dakota or South Dakota, but was reported two years in Nebraska, 11 years in Kansas, 25 years in Oklahoma and 27 years in the Texas panhandle. South Dakota records extend from early March 11 to November 23 (Tallman, Swanson and Palmer, 2002). Thirty-eight final autumn Nebraska sightings are from August 14 to November 16, with a median of October 7 (Johnsgard, 1980). It is occasional during winter in Kansas (Thompson *et al.*, 2011). There are also some winter records in Oklahoma (Woods and Schnell, 1984), and it is a rare winter visitor in the Texas panhandle, but has been seen every month (Seyffert, 2001).

*Habitats.* Migrants and wintering birds use ponds, marshes, creeks, mud flats and flooded meadows.

*National Population.* Breeding Bird Surveys between 1966 and 2012 indicate that this species exhibited a survey-wide population increase (3.4% annually) during that period. Other survey information suggests mixed population trends (Morrison *et al.,* 2001a). Morrison *et al.* (2001b) estimated the North American population at 100,000 birds.

*Further Reading.* Bent, 1927; Johnsgard, 1981; Cramp, 1983; Godfrey, 1986; Elphick and Tibbitts, 1998 (*The Birds of North America:* No. 356).

## Least Sandpiper
*Calidris pusilla*

This tiny sandpiper would seem ill-adapted to handle cold weather, but it is slightly more prone to tarry later in autumn than are other small "peeps" such as semipalmated (*Calidris pusilla*) and Baird's (*C. bairdii*) sandpipers. All of these species are able to forage on land rather than only by wading, but all are closely associated with foraging near the water's edge.

*Winter Distribution.* It is surprising that this species, the smallest of our sand-pipers, has appeared regularly on Christmas Counts in the southern Great Plains, especially Oklahoma. Root (1988) noted that, like most shorebirds, it avoids areas of subfreezing temperatures, and mapped only a small area north of Oklahoma in central Kansas where birds are likely to occur as late as the end of December. Our data show small and possibly increasing numbers in Oklahoma during the Christmas Count period. Obviously, the main wintering grounds are still farther south, and these small numbers are of no real value in judging possible long-term population trends.

*Seasonality and Migrations.* The least sandpiper is a common seasonal migrant, wintering southwardly. This species was not reported during Christmas Bird Counts from 1968 to 2007 in North Dakota or South Dakota, but was seen one year in Nebraska, 16 years in Kansas, 34 years in Oklahoma, and 12 years in the Texas panhandle. Twenty-three final autumn sightings in Nebraska are from July 28 to November 11, with a median of September 18 (Johnsgard, 2007a). Early winter records in Kansas extend through January 25 (Thompson *et al.*, 2011). Monthly Oklahoma records are continuous throughout the year (Woods and Schnell, 1984). It is a rare winter visitor in the Texas panhandle, and has been reported in both December and February (Seyffert, 2001). The average number of birds observed per party-hour for the 2013–2014 Christmas Bird Count were: N. Dakota 0, S. Dakota 0, Nebraska 0, Kansas 0, Oklahoma 0.32, northwest Texas 0. For complete species abundance data, see Appendix 1:50.

*Habitats.* Mud flats, shallow ponds, marsh edges and flooded meadows are used by migrants, which frequently gather in small groups foraging in shallow puddles or wet grasslands well away from the larger "peeps".

*National Population.* Available population survey data for this species show predominantly negative trends (Morrison *et al.*, 2001a). Morrison *et al.* (2001b) estimated the North American population at 600,000 birds.

*Further Reading.* Bent, 1927; Johnsgard, 1981; Cramp, 1983; Godfrey, 1986; Cooper, 1994 (*The Birds of North America:* No. 115).

## Wilson's Snipe
*Gallinago delicata*

This species is perhaps the most cold-hardy of all the Great Plains shorebirds, being reported every year on Christmas Counts from South Dakota southward over the past four decades. Its diet is correspondingly highly adaptive, ranging from aquatic and terrestrial insects through crustaceans, mollusks, and earthworms to vertebrates such as salamanders, frogs and lizards, as well as some seeds. The species is quite territorial, and perhaps the males remain late in the autumn season to patrol and announce their territories, at times remaining even into winter.

*Winter Distribution.* To a greater degree than most shorebirds, the Wilson's snipe seems able to survive in temperatures too cold for typical shorebird species, Root (1988) indicated that it avoids areas with minimum January temperatures below 20° F. This ability allows it to now remain in South Dakota, and rarely even North Dakota, into the Christmas season, although Root's (1988) map shows it barely reaching westernmost South Dakota. By and large, the Great Plains numbers in our sample appear to be small but stable, rather than showing a population decline, as is suggested by national Breeding Bird Survey data.

*Seasonality and Migrations.* The Wilson's snipe is a common but inconspicuous Great Plains seasonal migrant and local wetland breeder, wintering southwardly. It was reported two years in North Dakota Christmas Bird Counts between 1968 and 2007, and all 40 years from South Dakota to Oklahoma and the Texas panhandle. In South Dakota it winters locally and uncommonly (Tallman, Swanson and Palmer, 2002). Eighty-one initial spring Nebraska sightings dating from the mid-1930s to the late 1970s (Johnsgard, 1980) range from January 1 to May 29, with a median of April 13. The data suggest that wintering is rather rare in Nebraska. It is rare during winter in Kansas (Thompson *et al.*, 2011), and is rare during winter in Oklahoma, with continuous monthly records extending from July to May (Woods and Schnell, 1984). It is a rare winter visitor in the Texas panhandle, and has been reported in every month of the year (Seyffert, 2001). The average number of birds observed per party-hour for the 2013–2014 Christmas Bird Count were: N. Dakota 0, S. Dakota 0, Nebraska 0, Kansas 0.05, Oklahoma 2.16, northwest Texas 0. For complete species abundance data, see Appendix 1:51.

*Habitats.* Migrating and wintering birds are associated with marshes, sloughs and other wetlands that support areas of mudflats or mucky organic soil where foraging by probing is readily performed. Marshes rich in shoreline and emergent vegetation and are preferred over more open ones.

*National Population.* Breeding Bird Surveys between 1966 and 2012 indicate
that this species exhibited a survey-wide population increase (0.3% annually) during that period. Morrison *et al.* (2001b) estimated the North American population at about two million birds, but admitted that this number
represents little more than an informed guess. The species also breeds south
to South America.

*Further Reading.* Bent, 1927; Johnsgard, 1981; Mueller, 1999 (*The Birds of
North America:* No. 427).

### American Woodcock
*Scolopax minor*

Woodcocks are almost exclusively probe-feeders under normal conditions,
but during dry, or perhaps frozen, conditions they may pick up foods from
the ground surface, such as eating crickets, centipedes, millipedes, spiders,
snails, frogs, salamanders and even berries or seeds. In recent decades the
species has moved west along prairie rivers as the associated riparian forests
have matured, locally reaching as far upstream along the Platte River as central Nebraska

*Winter Distribution.* Root's (1988) map for this species shows only south-central Oklahoma as having a measurable concentration of winter woodcocks,
and southeastern Texas as having the greatest numbers. Our data show
small but consistent numbers remaining as far north as South Dakota into
the Christmas season, which is quite surprising in view of this species' dependence on probing for earthworms in moist, unfrozen soils. Our sample
sizes are far too small to make an judgments about possible long-term population trends.

*Seasonality and Migrations.* The American woodcock is an uncommon and
inconspicuous Great Plains seasonal migrant and local breeder, wintering southwardly. It was not reported Christmas Bird Counts from 1968
to 2007 in North Dakota, South Dakota or Nebraska, but was seen one
year in Kansas, 22 years in Oklahoma, and none in the Texas panhandle.
South Dakota records extend from February 23 to November 14 (Tallman,
Swanson and Palmer, 2002). Thirteen autumn Nebraska sightings dating
from the mid-1930s to the late 1970s are from September 12 to November
14, with a median of October 15 (Johnsgard, 1980). It has been reported
several times during winter in Kansas (Thompson *et al.*, 2011), and is a

permanent resident in Oklahoma (Woods and Schnell, 1984). It is a rare winter visitor in the Texas panhandle, and has been reported from October to December, and also during February (Seyffert, 2001).

*Habitats.* Migrating and wintering woodcocks are generally associated with floodplain forests, where the trees are rather scattered and the land is poorly drained, so that earthworms can be readily obtained by probing in moist soil.

*National Population.* Breeding Bird Surveys between 1966 and 2006 indicate that this species exhibited a significant national population decline (0.3% annually) during that period. Morrison *et al.* (2001b) estimated the North American population at three million birds, but admitted that, like their estimate for Wilson' snipe, this number represents little more than an informed guess.

*Further Reading.* Bent, 1927; Johnsgard, 1981; Keppie and Whiting, 1994 (*The Birds of North America:* No. 100).

FAMILY LARIDAE:
GULLS AND TERNS

### Black-legged Kittiwake
*Rissa tridactyla*

Kittiwakes are coastal cliff-nesting gulls with loud, multi-note calls that are the basis for their common English name. They wander widely outside the breeding season, and vagrants sometimes visit the northern interior states between the Pacific and Atlantic coasts. They are so adapted to drinking salt water that captives reportedly refuse to drink fresh water.

*Winter Distribution.* This is another arctic-breeding gull that usually winters pelagically along both coasts, but at times has strayed to the Great Plains. It has never attained abundance levels during Christmas Counts above the minimum "present" category.

*Seasonality and Migrations.* This kittiwake is a rare Great Plains seasonal migrant or winter vagrant. It was reported seven years during Christmas Bird Counts from 1968 to 2007 in North Dakota and South Dakota, none in Nebraska, two years in Kansas, one year in Oklahoma, and none in the Texas panhandle. In South Dakota this species does not overwinter (Tallman, Swanson and Palmer, 2002). There are a few Nebraska sightings for November and December (Johnsgard, 2007a). Occurrence records in Kansas are mostly for autumn and spring (Thompson and Ely, 1992). It is considered a rare winter visitor in Oklahoma, with continuous monthly records extending from October to April (Woods and Schnell, 1984). It is an extremely rare winter visitor in the Texas panhandle, and has been reported in November, December and March (Seyffert, 2001).

*Habitats.* Rivers, lakes and coastal shorelines are normally used by migrants; wintering us usually done at sea.

*National Population.* Breeding Bird Survey trend and population data are not available for this arctic-breeding and circumpolar species. An estimate of the world population in the 1990s was 2.6 million individuals (Baird, 1994).

*Further Reading.* Bent, 1921; Cramp, 1983; Baird, 1994 (*The Birds of North America:* No. 92); Olsen and Larson, 2004; Howell and Dunn, 2007.

## Bonaparte's Gull
### *Chroicocephalus philadelphia*

This sub-arctic gull is unique among North American gulls in that, instead of nesting on the ground, it builds a stick nest on low branches of spruce trees in northern Canada and Alaska. The young begin leaving their nests when only about a week old. Recent studies indicate that the species is not a part of the large gull genus *Larus*, but instead is a near relative of the European black-headed gull (*C. ridibundus*).

*Winter Distribution.* Root (1988) noted that the Bonaparte's gull had three areas of concentration in Oklahoma, namely along the Texas panhandle border, in northeastern Oklahoma, and along the Texas border in eastern and central Oklahoma, all of which she associated with the presence of fairly large bodies of water. Our data also show an Oklahoma concentration, and with increasing numbers also occurring in recent years in Kansas, Nebraska and even South Dakota, These changes perhaps reflect later times of freezing for the large impoundments that are now present in these states.

*Seasonality and Migrations.* The Bonaparte's gull is an uncommon seasonal migrant, rarely wintering southwardly. It was not reported during Christmas Bird Counts from 1968 to 2007 in North Dakota, but occurred ten years in South Dakota, six years in Nebraska, 20 years in Kansas, all 40 years in Oklahoma, and four years in the Texas panhandle. In South Dakota there are a few January and February records (Tallman, Swanson and Palmer, 2002). Twenty autumn Nebraska sightings are from August 18 to November 21, with a median of October 26 (Johnsgard, 1980). It is rare during winter in eastern Kansas, with reports extending through February (Thompson *et al.*, 2011), and is also rare in Oklahoma, with continuous monthly records extending from September to May (Woods and Schnell, 1984). It is a rare winter visitor in the Texas panhandle, and has been reported from October to May (Seyffert, 2001). The average number of birds observed per party-hour for the 2013–2014 Christmas Bird Count were: N. Dakota 0, S. Dakota 0, Nebraska 0, Kansas 2.53, Oklahoma 2.16, northwest Texas 0. For complete species abundance data, see Appendix 1:52.

*Habitats.* Migrants and wintering birds are associated with rivers, lakes and marshes, especially large lakes.

*National Population.* Breeding Bird Survey trend data are not available for this unique North American species. An estimate of its total population in the 1990s was 85,000–175,000 pairs (Burger and Gochfeld, 2002).

*Further Reading.* Bent, 1921; Cramp, 1983; Godfrey, 1986; Burger and Gochfeld, 2002 (*The Birds of North America:* No. 634); Olsen and Larson, 2004; Howell and Dunn, 2007.

## Ring-billed Gull
### *Larus delawarensis*

The ring-billed gull is the most typical "seagull" of the Great Plains, and one that is often seen during late autumn and winter as far north as Nebraska. Like many gulls it is a scavenger, and can survive on a wide variety of foods. In some city parks it has become as tame as domestic waterfowl, and barely bothers to get out of the way of pedestrians.

*Winter Distribution.* This species is by far the most abundant of all Great Plains gulls, and is nearly ubiquitous throughout the Plains in its non-breeding distribution. It remains fairly far north during winter, often wintering on partly frozen bodies of water. Root (1988) judged its northern limits to be limited to areas where the average January temperature exceeds 20° F. Our data indicate that the birds now remain north into the Christmas season substantially north of Root's estimated limits, and their numbers have been fairly steadily increasing from Kansas northward. This increase follows trends indicated by the national Breeding Bird Survey data.

*Seasonality and Migrations.* The ring-billed gull is an abundant Great Plains seasonal migrant and local wetland breeder, wintering southwardly. It was reported 14 years during Christmas Bird Counts from 1968 to 2007 in North Dakota, 35 years in South Dakota, 29 years in Nebraska, all 40 years in Kansas and Oklahoma, and 38 years in the Texas panhandle. In South Dakota it is uncommon to common during winter (Tallman, Swanson and Palmer, 2002). Fifty-seven final autumn Nebraska sightings are from August 25 to December 21, with a median of November 28. Eighty initial spring sightings range from January 3 to May 15, with a median of March 16 (Johnsgard, 1980). It is uncommon during winter in Kansas (Thompson *et al.*, 2011), and is a winter resident in Oklahoma, with continuous monthly records extending throughout the year (Woods and Schnell, 1984). It is a common to abundant winter visitor in the Texas panhandle, and has been reported in every month (Seyffert, 2001). The average number of birds observed per party-hour for the 2013–2014 Christmas Bird Count were: N. Dakota 0, S. Dakota 0.1, Nebraska 1.16, Kansas 5.78, Oklahoma 1.15, northwest Texas 0, For complete species abundance data, see Appendix 1:53.

*Habitats.* Migrants and wintering birds use a wide variety of lakes, reservoirs, rivers, marshes and other water areas.

*National Population.* Breeding Bird Surveys between 1966 and 2012 indicate that this species exhibited a survey-wide population increase (2.4% annually) during that period Its total world population has been estimated at 3–4 million birds, most of which nest in Canada (Olsen and Larson, 2004; Alderfer, 2006).

*Further Reading.* Bent, 1921; Cramp, 1983; Ryder, 1993 (*The Birds of North America:* No. 33); Olsen and Larson, 2004; Howell and Dunn, 2007.

## California Gull
### *Larus californicus*

The California gull is famous for being the gull that saved the early Mormon settlers in Utah by consuming a Mormon cricket infestation in 1848. It has expanded its Great Plains nesting range eastward somewhat in recent decades, and has become increasingly frequent as a late autumn migrant in western parts of the Great Plains states.

*Winter Distribution.* Root (1988) showed no wintering of California gulls in the Great Plains. However, in the last few decades the breeding range of this species has expanded east into eastern South Dakota, adding to colonies already breeding in North Dakota and Manitoba. Perhaps these sources are responsible for the small but consistent numbers now seen in the southern Great Plains and the Texas coast during late autumn and winter.

*Seasonality and Migrations.* The California gull is a rare seasonal migrant or winter vagrant, mainly in the central plains, and a local summer resident (South and North Dakota). This species was reported eight years during Christmas Bird Counts from 1968 to 2007 in North Dakota, two years in South Dakota, 16 years in Nebraska, 11 years in Kansas, three years in Oklahoma, and none in the Texas panhandle. It has been reported consistently on western Nebraska Christmas Counts since 1991. Nine late autumn and winter Nebraska sightings extend from December 13 to February 15 (Johnsgard, 2007a), but few birds appear to remain in Nebraska into January. There are also autumn and winter records for Kansas, extending to January 30 (Thompson *et al.*, 2011), and at least one autumn record for Oklahoma (Woods and Schnell, 1984). There are three non-winter sightings for the Texas panhandle (Seyffert, 2001).

*Habitats.* Migrants and wintering birds use reservoirs, lakes, large marshes and similar open-water *Habitats.*

*National Population.* Breeding Bird Surveys between 1966 and 2012 indicate that this species exhibited a survey-wide population increase (0.9% annually) during that period. Its total North American population has been estimated at 50,000–100,000 birds (Olsen and Larson, 2004; Alderfer, 2006).

*Further Reading.* Bent, 1921; Winkler, 1996 (*The Birds of North America:* No. 259); Olsen and Larson, 2004; Howell and Dunn, 2007.


## Herring Gull
*Larus argentatus*


The herring gull is likely to occur anywhere the Great Plains region, although it does not breed here. Four years are needed for this gull to reach maturity, so many of the herring gulls seen in the Plains region are in various plumage stages of immaturity that have not yet established breeding locations.

*Winter Distribution.* This is the largest of our more common gulls in the Great Plains, and non-breeding birds are likely to be seen almost anywhere in the region, although in much smaller numbers than the ring-billed gull. Root (1988) mapped the winter distribution of the herring gull as extending north in the Plains states to central South Dakota, and south to Oklahoma and the Texas panhandle. She identified the Missouri, Arkansas and Platte Rivers as being apparent attractions in this region. Our data show a gradual long-term increase in numbers in the five-state Plains region, with the largest and most rapidly growing population in Kansas. North Dakota and Nebraska have also shown marked increases in count numbers.

*Seasonality and Migrations.* The herring gull is an uncommon seasonal migrant, wintering southwardly. It was reported 24 years during Christmas Bird Counts from 1968 to 2007 in North Dakota, 36 years in South Dakota, 26 years in Nebraska, 39 years in Kansas, all 40 years in Oklahoma, and 15 years in the Texas panhandle. In South Dakota, wintering is uncommon along the Missouri River (Tallman, Swanson and Palmer, 2002). Eighteen final autumn Nebraska sightings are from August 29 to December 21, with a median of November 28. Forty-seven initial spring Nebraska sightings range from January 13 to May 13, with a median of March 18 (Johnsgard, 2007a). It is uncommon during winter in Kansas (Thompson *et al.*, 2011), and a winter resident in Oklahoma, with continuous monthly records extending throughout the year (Woods and Schnell, 1984). It is a rare to uncommon winter visitor in the Texas panhandle, and has been reported there in every month (Seyffert, 2001). For complete species abundance data, see Appendix 1:54.

*Habitats.* Migrants and wintering birds are widely distributed over rivers, lakes, reservoirs and other water areas.

*National Population.* Breeding Bird Surveys between 1966 and 2012 indicate that this species exhibited a survey-wide population decrease (3.2% annually) during that period. Its Atlantic Coast population was estimated at 500,000 pairs in the 1980s (Olsen and Larson, 2004; Alderfer, 2006). Its total breeding range extends across all of Eurasia, where it has been estimated to number over a million pairs.

*Further Reading.* Bent, 1921; Cramp, 1983; Pierotti and Good, 1994 (*The Birds of North America:* No. 124); Olsen and Larson, 2004; Howell and Dunn, 2007.

### Thayer's Gull
*Larus thayeri*

Visually separating the Thayer's gull from the glaucous gull and Iceland gull is a task best left to experts. It is believed that the western-wintering Thayer's gull is more common in the Great Plains than the more easterly-oriented Iceland gull, but both are less common than the glaucous gull.

*Winter Distribution.* This is a high-latitude gull that breeds in east-central arctic Canada and Greenland to the west of the Iceland gull, and is a vagrant in the Great Plains. It has had varied taxonomic treatment, and has at times considered a variant of the herring gull or Iceland gull, but is now generally given species recognition (Alderfer, 2006; Howell and Dunn, 2007). It was not discussed by Root (1988), and has not appeared often enough during Christmas Counts to warrant detailed analysis. It never reached exceeded abundance levels above the minimum "present" category during our study.

*Seasonality and Migrations.* The Thayer's gull is a rare to uncommon Great Plains seasonal migrant or winter vagrant. This species was reported 20 years during Christmas Bird Counts from 1968 to 2007 in North Dakota, six years in South Dakota, ten years in Nebraska, 17 years in Kansas, nine years in Oklahoma, and none in the Texas panhandle. There are more than 20 Nebraska sightings for this species, with most occurring between November and January, and over 100 Kansas sightings, all between September and April (Thompson *et al.*, 2011). It is an extremely rare winter visitor in the Texas panhandle, and has been reported once in February (Seyffert, 2001).

*Habitats.* Rivers, reservoirs and coastal areas are used by wintering birds and migrants.

*National Population.* Breeding Bird Survey trend and population data are not available for this taxonomically problematic arctic-breeding species, which is difficult to distinguish from the closely related Iceland gull. Its total population has been estimated as 6,300 pairs (Olsen and Larson, 2004; Alderfer, 2006), all of which breed in Canada's high arctic and northern Greenland. This species was yellow-listed in the National Audubon Society's 2007 WatchList of rare and declining birds (Butcher *et* al., 2007).

*Further Reading.* Bent, 1921; Godfrey, 1986; Snell, 2003 (*The Birds of North America:* No. 699); Olsen and Larson, 2004; Howell and Dunn, 2007.

### Iceland (Kumlein's) Gull
*Larus glaucoides kumlieni*

The Kumlein's race of this species is a vagrant to the Great Plains, occurring mostly on large lakes and reservoirs.

*Winter Distribution.* This is a pale-colored gull of eastern arctic Canada and Greenland that appears very rarely in the Great Plains, and primarily winters in the northern Atlantic Coast of the U.S. and adjacent Canada. Root (1988) did not map it as being present in the Great Plains, and it has not since appeared sufficiently often during Christmas Counts in the Great Plains region to warrant detailed consideration. It never exceeded the minimum "present" category during our period of study.

*Seasonality and Migrations.* The Iceland gull is a rare to uncommon Great Plains seasonal migrant or winter vagrant. This species was reported five years during Christmas Bird Counts from 1968 to 2007 in North Dakota, one year in South Dakota, three years in Nebraska, one year in Kansas, and none in Oklahoma or the Texas panhandle. There are at least three December records for South Dakota (Tallman, Swanson and Palmer, 2002), and several of the 20-plus Nebraska sightings fall between December and March (Johnsgard, 2007a). It is apparently very rare in Kansas, with no specimen records (Thompson *et al.*, 2011). There is at least one Oklahoma record (Woods and Schnell, 1984), but apparently no Texas panhandle records.

*Habitats.* Rivers, reservoirs and coastal areas are used by wintering birds and migrants.

*National Population.* Breeding Bird Survey trend and population data are not available for this rare arctic-breeding species, which is very easily confused with both the more common glaucous gull and the Thayer's gull. The Iceland gull's total world population has been estimated at 45,000 pairs. Most of these breed in Greenland, but about 5,000 consist of the continental North American Kumlein's race (Olsen and Larson, 2004; Alderfer, 2006). This species was yellow-listed in the National Audubon Society's 2007 WatchList of rare and declining birds (Butcher *et al.*, 2007).

*Further Reading.* Cramp, 1983; Godfrey, 1986; Snell, 2003 (*The Birds of North America:* No. 699); Olsen and Larson, 2004; Howell and Dunn, 2007.

## Lesser Black-backed Gull
### *Larus fuscus*

Lesser black-backed gulls that appear in the Great Plains are likely vagrants from the Great Lakes, region, where the birds are now regular visitors as far west as Lake Superior, and strays have wandered as far west as eastern Colorado.

*Winter Distribution.* This gull is a Eurasian species that regularly appears on the Atlantic and Gulf coasts, as well as increasingly around the Great Lakes, but rarely straggles into the Great Plains region. Like the other large and extra-limital gulls, it has never reached levels during Christmas Counts above the minimum "present" category.

*Seasonality and Migrations.* The lesser black-backed gull is a rare Great Plains seasonal migrant or winter vagrant. This species was reported one year during Christmas Bird Counts from 1968 to 2007 in North Dakota and South Dakota, six years in Nebraska, three years in Kansas, six years in Oklahoma, and none in the Texas panhandle. Many of the 50-plus total Nebraska sightings are for February or March (Johnsgard, 2007a). It was reported once from South Dakota during late February (Tallman, Swanson and Palmer, 2002), and is very rare in Kansas (Thompson *et al.*, 2011). Oklahoma and Texas panhandle records are apparently lacking.

*Habitats.* Rivers, reservoirs and coastal shorelines are used by wintering birds and migrants.

*National Population.* Breeding Bird Survey trend and population data are not available for this rare and extra-limital species, which breeds in Iceland and northern Europe, where it is thought to number in the hundreds of thousands of pairs.

*Further Reading.* Bent, 1921; Cramp, 1983; Olsen and Larson, 2004; Howell and Dunn, 2007.

## Glaucous Gull
*Larus hyperboreus*

This is yet another of the large, pale-backed gulls from the arctic that sometimes strays to the Great Plains, especially during autumn and winter.

*Winter Distribution.* This large arctic-breeding gull of Alaska, northern Canada and Greenland is largely coastally distributed, and has appeared only infrequently in the Great Plains. It has never reached abundance levels during Christmas Counts above the minimum "present" category.

*Seasonality and Migrations.* The glaucous gull is a rare to uncommon Great Plains seasonal migrant or winter vagrant. This species was reported 15 years during Christmas Bird Counts from 1968 to 2007 in North Dakota, 21 years in South Dakota, eight years in Nebraska, 18 years in Kansas, 19 years in Oklahoma, and none in the Texas panhandle. In South Dakota this species is considered an uncommon migrant and winter visitor along the Missouri River (Tallman, Swanson and Palmer, 2002). Five autumn or early winter Nebraska sightings are from December 3 to 27. Ten late winter and spring Nebraska sightings range from January 24 to April 29, with a median of March 24 (Johnsgard, 2007a). Kansas records are mostly from November to March (Thompson and Ely, 1992), and are mainly from eastern counties (Thompson *et al.*, 2011). It is considered a rare winter visitor in Oklahoma, with continuous monthly records extending from November to March (Woods and Schnell, 1984). It is a very rare winter visitor in the Texas panhandle, and has been reported from November to March (Seyffert, 2001).

*Habitats.* Rivers, reservoirs and coastal shorelines are used by wintering birds and migrants.

*National Population.* Breeding Bird Survey trend and population data are not available for this circumpolar species. Its Alaska population has been estimated at 100,000 birds (Olsen and Larson. 2004; Alderfer, 2006), but it breeds very widely across the high arctic of North America and Eurasia, where it is thought to number over 100,000 pairs.

*Further Reading.* Bent, 1921; Cramp, 1983; Godfrey, 1986; Gilchrist, 2001 (*The Birds of North America:* No. 573); Olsen and Larson, 2004; Howell and Dunn, 2007.

## Great Black-backed Gull
*Larus marinus*

This is the largest of all the gulls likely to appear in the Great Plains, and one of the rarest. It breeds along the New England coast, and is slowly increasing its range southward.

*Winter Distribution.* This coastally breeding gull of eastern North America fairly often winters on the eastern Great Lakes, and stragglers may appear west to the Rocky Mountain states. It has never reached abundance levels during Great Plains Christmas Counts above the minimum "present" category.

*Seasonality and Migrations.* The great black-backed gull is a rare to occasional Great Plains seasonal migrant or winter vagrant. This species was reported two years during Christmas Bird Counts from 1968 to 2007 in Nebraska, three years in Kansas, one year in Oklahoma, and none in the Texas panhandle. Several of the 13 Nebraska sightings are from December to February (Johnsgard, 2007a). There are a few November to April records for Kansas (Thompson *et al.*, 2011), and a single November record for the Texas panhandle (Seyffert, 2001).

*Habitats.* Rivers, reservoirs and coastal shorelines are used by wintering birds and migrants.

*National Population.* Breeding Bird Surveys between 1966 and 2012 indicate that this species exhibited a survey-wide population decrease (2.7% annually) during that period. The North American population probably exceeded 50,000 pairs in the 1990s (Good, 1998). It also breeds in Greenland, Iceland and northern Europe, where is it thought to number over 100,000 pairs.

*Further Reading.* Bent, 1921; Cramp, 1983; Godfrey, 1986; Good, 1998 (*The Birds of North America:* No. 330); Olsen and Larson, 2004; Howell and Dunn, 2007.

FAMILY COLUMBIDAE:
PIGEONS AND DOVES

### Rock Pigeon
*Columba livia*

This familiar species is known to nearly everyone, especially city-dwellers, often being called the barnyard pigeon.

*Winter Distribution.* The rock pigeon, an introduced species whose abundance and range are closely associated with humans, occurs throughout the United States, Its winter distribution was not analyzed by Root (1988). Our data for the Great Plains generally suggest a slowly increasing population, in spite of apparent national trends in the opposite direction.

*Seasonality and Migrations.* This familiar urban and farmyard species is a permanent resident across its range.

*Habitats.* The rock pigeon (previously the rock dove) is an abundant resident throughout the entire Great Plains. Mostly associated with human habitations in cities, villages and farms, but also occurs to a limited extent as feral populations around bluffs and cliffs, which was the species' original breeding habitat in Europe.

*National Population.* Breeding Bird Surveys between 1966 and 2012 indicate that this species exhibited a survey-wide population decrease (1.0% annually) during that period. Breeding Bird Surveys between 1966 and 2006 indicate that this species exhibited a statistically nonsignificant national population decline (0.1% annually) during that period. A recent North American population estimate is 26 million birds (Rich *et al.,* 2004). The species breeds south locally to Argentina and Chile. It was not included in Christmas Count summaries until 2003, so no overall winter population trends are apparent yet. The average number of birds observed per party-hour for the 2013–2014 Christmas Bird Count were: N. Dakota 10.58, S. Dakota 9.72, Nebraska 10.7, Kansas 5.4, Oklahoma 3.7, northwest Texas 1.85. For complete species abundance data, see Appendix 1:55.

*Further Reading.* Bent, 1932; Johnston, 1992 (*The Birds of North America:* No. 13); Dunn and Tessaglia-Hymes, 1999.

## Eurasian Collared-Dove
*Streptopelia decaocto*

This species offers a good example of how rapidly a bird species can colonize an entire continent after gaining a small foothold. Unlike such invaders as the European starling and house sparrow, it has not proven ecologically or economically disastrous, and also has not resulted in any apparent adverse competition with any of the Native American doves or pigeons.

*Winter Distribution.* This species was identified by Root (1988) as the very similar and domesticated ringed turtle-dove (*S. risoria*). The collared-dove underwent a population explosion after offspring of a feral population in the Bahamas arrived in Florida during the late 1970s or early 1980s. While expanding from this foothold population in southern Florida, it also was released in a few other states such as Colorado and Missouri. It reached South Dakota by 1996, Montana by 1997, Colorado and Minnesota by 1998, and North Dakota by 2003 (Romagosa, 2002; Baughman, 2003). By 2013 it had been reported west to Baja California and southern Alaska.

*Seasonality and Migrations.* This species is an increasingly common permanent resident throughout its range in the central and southern Great Plains. The average number of birds observed per party-hour for the 2013–2014 Christmas Bird Count were: N. Dakota 0.87, S. Dakota 2.17, Nebraska 4.19, Kansas 8.0, Oklahoma 0.3, northwest Texas 0.5. For complete species abundance data, see Appendix 1:56.

*Habitats.* Closely associated with humans, this range-expanding species is most often seen in and around small towns of the Plains States, where grain elevators are likely to provide a source of waste corn and other grains.

*National Population.* Breeding Bird Surveys between 1966 and 2012 indicate that this species exhibited an astronomically high rate of survey-wide population increase (34.0% annually) during that period. This species did not arrive in the U.S. from the Bahamas until the mid-1970s or later. It first appeared on an Oklahoma Christmas Count in 1995, in the Texas panhandle and Nebraska in 1997, in Kansas and South Dakota in 1998, and in North Dakota in 1999. By 2007 it had been seen on Great Plains Christmas counts as far north as Manitoba and as far south as northern Mexico. No recent total estimates of the still-expanding North American population are available, and any published estimates are likely to be soon outdated.

*Further Reading.* Bent, 1932; Romagosa, 2002 (*The Birds of North America:* No. 630).

## White-winged Dove
*Zenaida asiatica*

Like several other southern species such as the Inca dove, northern mocking-bird, scissor-tailed flycatcher and great-tailed grackle, this dove began a north-ward movement through the central Great Plains during the second half of the 20th century.

*Winter Distribution.* Root (1988) mapped this species as wintering no farther north than southern Texas, along the northern edge of the Chihuahuan Desert. However, its range has gradually expanded northward in recent decades (Baughman, 2003), First noted in Kansas in 1969, it had nested there by 1993. It was first seen during 1988 in Nebraska, and at least one breeding record has since occurred. By 2008, there had been sightings as far north as North Dakota, Montana and Minnesota. Some of these birds linger well into late autumn, and a few may attempt to winter as far north as southwestern Kansas, although these individuals sometimes lose toes to frostbite.

*Seasonality and Migrations.* The white-winged dove is a locally common breeder and migrant in the southern Great Plains, wintering southwardly. It was not reported during Christmas Bird Counts from 1968 to 2007 from North Dakota to Nebraska, but was reported one year in Kansas, six years (start-ing in 2001) in Oklahoma, and two years (starting in 2005) in the Texas panhandle. It is a year-around winter visitor in the Texas panhandle, and has been an uncommon local resident since the 1990s (Seyffert, 2001).

*Habitats.* This species is more arid-adapted than the mourning dove, and of-ten occurs in semiarid woodlands having densely foliaged trees of medium height, such as mesquite (*Prosposis*), acacia (*Acacia*) and even citrus groves. It often visits bird feeders during cold weather.

*National Population.* Breeding Bird Surveys between 1966 and 2012 indicate that this species exhibited a survey-wide population increase (0.8% annu-ally) during that period. The estimated 1990s continental population north of Mexico was about 110.5 million birds (Rich *et al.,* 2004). The species breeds south to Costa Rica.

*Further Reading.* Bent, 1932; Tacha and Braun, 1994; Schwertner *et al.,* 2003 (*The Birds of North America:* No. 710).

## Mourning Dove
*Zenaida macroura*

For many Great Plains dwellers, the call of the mourning doves is one of the most soothing and characteristic sounds of summer. Even non-birders who know the species only as the "turtle dove" are likely to name it as one of their favorite birds. Nebraska lie on the cusp of the species' winter distribution, and during most winters a few mourning doves are likely to stick out the worst weather by remaining close to bird feeders. This may expose the birds to frostbite and possible loss of one or more toes. At these times the birds are also exposed to predation risks from accipiter hawks that regularly visit winter bird feeders.

*Winter Distribution.* According to Root (1988), mourning doves mainly winter in regions where the minimum January temperatures average above 10° F. She mapped their northern limit as extending from southwestern Nebraska to northeastern South Dakota. Our data suggest that some wintering occurs north to North Dakota, but that numbers are highest in Kansas and especially in northwestern Texas. No clear population trend is apparent from our data, and likewise there is no significant national population apparent from Breeding Bird Surveys.

*Seasonality and Migrations.* The mourning dove is an abundant migrant and widespread breeder throughout the region, wintering southwardly. It was reported 36 years during Christmas Bird Counts from 1968 to 2007 in North Dakota, 37 years in South Dakota, and all 40 years in Nebraska, Kansas, Oklahoma, and the Texas panhandle. It is a Great Plains seasonal migrant and local breeder, wintering southwardly. In South Dakota this species winters only rarely (Tallman, Swanson and Palmer, 2002). Sixty-two initial spring Nebraska sightings dating from the mid-1930s to the late 1970s (Johnsgard, 1980) range from January 1 to May 29, with a median of March 26. Ninety final autumn sightings in Nebraska range from August 30 to December 31, with a median of November 1, but the species frequently overwinters in Nebraska. It is local during winter in Kansas (Thompson *et al.*, 2011), but is a permanent resident in Oklahoma (Woods and Schnell, 1984), and is a common to abundant resident in the Texas panhandle (Seyffert, 2001). The average number of birds observed per party-hour for the 2013–2014 Christmas Bird Count were: N. Dakota 0.05, S. Dakota 0.12, Nebraska 0.53, Kansas 1.64, Oklahoma 1.18, northwest Texas 2.21. For complete species abundance data, see Appendix 1:57.

*Habitats.* This is a widely adaptable species, occurring in open woods and edge areas, in parks and cities, on grasslands far from trees, and in cultivated fields.

*National Population.* Breeding Bird Surveys between 1966 and 2012 indicate that this species exhibited a survey-wide population decrease (0.5% annually) during that period. The estimated 1990s continental population north of Mexico was about 110 million birds (Rich *et al.,* 2004). The species breeds south to Panama.

*Further Reading.* Bent, 1932; Tacha and Braun, 1994; Mirarchi and Baskett, 1994 (*The Birds of North America:* No. 117); Dunn and Tessaglia-Hymes, 1999.

## Inca Dove
*Columbina inca*

Barely a third the weight of mourning doves, Inca doves are probably much more cold-sensitive, and are subject to freezing at temperatures more than about ten degrees below freezing. At that time the birds become increasingly prone to share body heat by clumping into heaps of up to twelve birds, stacked on top of one another in a pyramid-like manner. Periodically the birds re-distribute themselves, with those on the bottom moving to the top of the pile. Such close body contact keeps the birds warm but increases the chances of transmission of external parasites or diseases such as salmonellosis.

*Winter Distribution.* Originally native to the Sonoran and Chihuahuan deserts of Mexico and, the southwestern U.S., the Inca dove was still confined to that region during the period of Christmas Counts studied by Root (1988), with a small disjunctive population in east-central Texas. The Texas influx began in 1889, but the first Oklahoma record did not occur until the 1950s. By 1980, the birds had occupied much of eastern Texas. They were first reported in Oklahoma in 1992, and by 2000 had spread to southeastern Oklahoma and adjoining Arkansas (Baughman, 2003). They were first seen in the Texas panhandle in 1954, and the first regional breeding record was in 1992. Northward movements of birds during autumn and early winter gradually resulted in expansions into western Oklahoma and western Kansas. The first nesting record for Kansas occurred in 1993 (Busby and Zimmerman, 2001), and it nested in southeastern Colorado in 2000. As of 2007, there have been at least five Nebraska sightings, including a wintering bird in 1988 (Johnsgard, 2007a). Our Christmas Count data indicate a scattering of late December records extending north to Kansas in recent decades, but only in the Texas panhandle have the numbers risen above minimal "present" status.

*Seasonality and Migrations.* This species is a permanent resident within its still-expanding southern range. Expanding north from Mexico, it now is a rare to uncommon local resident in the Texas panhandle, having been first reported in 1954 and observed in at least 18 counties by 1996 (Seyffert, 2001). In Oklahoma is it a rare visitor, reported from November through May (Woods and Schnell, 1984). It is a rare local resident in southwestern Kansas (Thompson *et al.*, 2011). Nesting begins as early as February in the Texas panhandle.

*Habitats.* The Inca dove, once largely limited to the Rio Grande Valley, has moved north with settlements and agricultural developments, becoming increasingly associated with irrigated yards and gardens. The tiny birds are highly sensitive to cold, and resort to nocturnal hypothermia and multi-layer clumping in cold weather to conserve heat.

*National Population.* Breeding Bird Surveys between 1966 and 2012 indicate that this species exhibited a survey-wide population increase (2.0% annually) during that period. The estimated 1990s continental population north of Mexico was about 475,000 birds (Rich *et al.*, 2004). The species breeds south to Costa Rica.

*Further Reading.* Bent, 1932; Mueller, 1992 (*The Birds of North America:* No. 28); Dunn and Tessaglia-Hymes, 1999.

## FAMILY CUCULIDAE:
## CUCKOOS AND ROADRUNNERS

### Greater Roadrunner
*Geococcyx californianus*

An iconic species of the American Southwest, it seems strange to imagine roadrunners existing in the central Great Plains. Its diet is highly variable, allowing it to survive in varied habitats, and include insects, spiders, centipedes, millipedes, scorpions and snails, plus such vertebrates as birds' eggs and young, lizards and snakes. Various fruits and seeds are sometimes also eaten. Its need for a nearly bare substrate for running escapes and chasing prey limits its primary habitats to desert and semi-desert situations, but the grasslands of southern Kansas apparently allow for its marginal survival. To what extent winter temperatures may be range-limiting is uncertain, but its U.S. range seems to correspond roughly to regions where at least 140 days of sunshine exists. On cold days, the birds can raise their body temperatures by about five degrees Fahrenheit by orienting their body laterally toward the sun and raising their feathers as to expose their black skin, which absorbs the sun's rays and warms the body.

*Winter Distribution.* Root (1988) mapped the distribution of this desert-adapted bird as extending north to north-central Kansas, which corresponds fairly well with the known distribution of the species during the 1980s (Thompson and Ely, 1989). The birds expanded east from Kansas into southwestern Missouri in the 1950s, and from there had moved north to about Jefferson City by the late 1990s (Jacobs, 2001). There has been no comparable northward movement in Kansas beyond the Arkansas River. Our data suggest a probably increasing population in northwestern Texas, and a small but seemingly stable population in Oklahoma. The tiny Kansas population has also shown no sign of increasing measurably. The average number of birds observed per party-hour for the 2013–2014 Christmas Bird Count were: N. Dakota 0, S. Dakota 0, Nebraska 0, Kansas 0.09, Oklahoma 0.11, northwest Texas 0.05. For complete species abundance data, see Appendix 1:58.

*Seasonality and Migrations.* The greater roadrunner is an uncommon to rare permanent resident in the central and southern Great Plains.

*Habitats.* Preferred habitats include arid shrublands and pinyon-juniper woodlands. Occasionally roadrunners may also be found in cholla (*Opuntia imbricata*) grasslands. Open areas for foraging and taller vegetation for nesting and roosting are basic habitat needs.

*National Population.* This species was reported eight years during Christmas Bird Counts from 1968 to 2007 in Kansas, all 40 years in Oklahoma, and 39 years in the Texas panhandle. Breeding Bird Surveys between 1966 and 2012 indicate that this species exhibited a survey-wide population stability (0.0% annually) during that period. The estimated 1990s continental population north of Mexico was about 550,000 birds (Rich *et al.,* 2004). The species breeds south to central Mexico.

*Further Reading.* Bent, 1940; Geluso, 1970; Ohmart and Lasiewski, 1971; Hughes, 1996 (*The Birds of North America:* No. 244).

# FAMILY TYTONIDAE:
# BARN OWLS

## Barn Owl
*Tyto alba*

Barn owls have a nearly worldwide distribution and occur widely in North America, but generally at quite low densities. Although they are relatively common nesters over much of western Nebraska they are evidently cold sensitive, and some Nebraska barn owls that were banded and released after rehabilitation from injuries have been found to have subsequently moved substantially, as far south as Texas. Root (1988) suggested that the species is usually found during winter where the January temperature is from 10° to 20° F. Barn owls in Nebraska that are maintained in captivity lay their eggs at all seasons, suggesting that they are not limited by photoperiod changes or some other intrinsic timing factor.

*Winter Distribution.* Over most of the Great Plains, the winter population of barn owls is very small, with at least part of the population moving southward. Root's (1988) map indicates a possible concentration of barn owls in the Texas panhandle and adjacent Oklahoma, but she doubted the reliability of her data in defining the species' actual distribution. Our data show small numbers of barn owls throughout the southern parts of the region, but very few if any from Nebraska northward.

*Seasonality and Migrations.* The barn owl is a local permanent resident in the southern Plains, and a variable migrant northwardly. It was not reported during Christmas Bird Counts from 1968 to 2007 in North Dakota, but was reported twice in South Dakota, once in Nebraska, 19 times each in Kansas and Oklahoma and six times in the Texas panhandle.

*Habitats.* Open to semi-open habitats, where small rodents are abundant and where hollow trees, old buildings or caves, are available to provide roosting and nesting sites are favored by this species. Rats are frequently chosen as prey, but many other rodents are also consumed.

*National Population.* Breeding Bird Surveys between 1966 and 2012 indicate that this species exhibited a survey-wide population increase (1.5% annually) during that period. The estimated 1990s continental population north of Mexico was about 343,000 birds (Rich *et al.,* 2004). The species breeds south to southernmost South America (Tierra del Fuego). There are many other barn owl populations around the world; barn owls and peregrines are the most widely distributed of all raptors (del Hoyo, Elliott and Sargatal, 1999).

*Further Reading.* Marti, 1992 (*The Birds of North America:* No. 1); del Hoyo, Elliott and Sargatal, 1999; Johnsgard, 2002b.

# FAMILY STRIGIDAE:
## TYPICAL OWLS

### Western Screech-Owl
*Otus kennicottii*

Western and eastern screech-owls have generally non-overlapping ranges, but there are a few places where both seem to occur in the same general region, such as in the Oklahoma panhandle, where they might breed as close as in adjacent counties (Reinking, 2004). In any areas of actual contact their quite different vocalizations probably would serve to prevent hybridization. Root (1988) did not attempt to separately map their winter distributions.

*Winter Distribution.* This species has not yet been proven to breed within the five-state Plains region or the Texas panhandle. Reinking (2004) reported one possible nesting record for Cimarron County in northwestern Oklahoma during a five-year (1997–2001) atlasing project, and noted that the best evidence of nesting within the state is based on a female with a fully formed egg that was collected north of Boise City in 1966. Seyffert (2001) considered the species to be a vagrant autumn visitor to the Texas panhandle. Although it did not appear in any Texas panhandle Christmas Counts during the study period, it has sometimes appeared on Kansas (3 years) and Oklahoma (12 years) counts, but only at minimal "present" levels.

*Seasonality and Migrations.* The western screech-owl is a permanent resident throughout its range, which only barely reaches the Great Plains states.

*Habitats.* Habitats used by this species are similar to those of the eastern screech-owl.

*National Population.* Breeding Bird Surveys between 1966 and 2012 indicate that this species exhibited a survey-wide population decrease (1.7% annually) during that period. The estimated 1990s continental population north of Mexico was about 740,000 birds (Rich *et al.*, 2004), including about 1,000–2,000 pairs in Canada (Lynch, 2007). The species breeds south to central Mexico.

*Further Reading.* Bent, 1938; del Hoyo, Elliott and Sargatal, 1999; Cannings and Angell, 2001 *(The Birds of North America:* No. 597); Johnsgard, 2002b; Lynch, 2007.

## Eastern Screech-Owl
*Megascops asio*

For being such a small owl, the eastern screech-owl is quite cold-tolerant, being residential as far north as northernmost North Dakota and adjacent Manitoba and Saskatchewan. It probably is able to survive cold winters by caching prey items such as small rodents in winter roosting cavities, and shifting its food sources from insects and other invertebrates to vertebrates as temperatures decline during fall.

*Winter Distribution.* Root (1988) mapped the eastern and western screech-owls as a single combined population, but since the western screech-owl is essentially extra-limital in our region, only the eastern species is relevant here. Root's map indicates a concentration of birds in the Texas panhandle and adjacent Oklahoma, and generally small or absent populations elsewhere in the region. Our data show no panhandle concentration, but a larger and possibly increasing population in Kansas.

*Seasonality and Migrations.* The eastern screech-owl is an uncommon permanent resident throughout the Plains states.

*Habitats.* This widespread species occurs in a variety of wooded habitats, including farmyards, cities, orchards, and other human-made habitats, as well as in forests and woodlands. It is probably more common in cities than in heavy woodlands, where it is preyed upon by larger owls.

*National Population.* This species was reported all 25 years during Christmas Bird Counts between 1982 and 2007 from North Dakota to Oklahoma, and 19 years in the Texas panhandle. Breeding Bird Surveys between 1966 and 2012 indicate that this species exhibited a survey-wide population decrease (1.5% annually) during that period. The estimated 1990s continental population north of Mexico was about 740,000 birds (Rich *et al.,* 2004), including about 10,000–15,000 pairs in Canada (Lynch, 2007).

*Further Reading.* Bent, 1938; Gehlbach, 1994, 1995 (*The Birds of North America:* No. 165); del Hoyo, Elliott and Sargatal, 1999; Johnsgard, 2002b; Lynch, 2007.

## Great Horned Owl
*Bubo virginianus*

This large, common and fierce owl dominates all others in the Great Plains. Smaller owls such as barn owls and barred owls often fall prey to it, and its near-dark hunting periods force these other owls to be actively hunting at other times, to avoid being killed themselves. It is an owl of forest-edge situations rather than a deep-forest species like the barred owl, and it often reaches high populations where nearby grassland areas are available for hunting open-country mammals such as rabbits and ground squirrels.

*Winter Distribution.* As an abundant and fairly conspicuous owl, the great horned can probably be mapped more effectively than most of the other Great Plains owls. Root (1988) indicated a substantial peak in density in the vicinity of the South Platte River of eastern Colorado and adjacent western Nebraska, continuing south into Kansas and Oklahoma. Lesser concentrations were mapped in the eastern Dakotas. Our data suggest a broadly distributed population across the entire Plains region, with no clear-cut indications of any long-term population trends. The average number of birds observed per party-hour for the 2013–2014 Christmas Bird Count were: N. Dakota 0, S. Dakota 0.12, Nebraska 0.13, Kansas 0.13, Oklahoma 0.12, northwest Texas 0.19. For complete species abundance data, see Appendix 1:59.

*Seasonality and Migrations.* The great horned owl is an uncommon but widespread permanent resident throughout the Plains states.

*Habitats.* This highly adaptable species occurs in a variety of habitat types ranging from dense forests to city parks and farm woodlands, and extends into non-wooded environments in rocky canyons and gullies. Raptor rehabilitation records from Nebraska suggest it is the commonest of all owls in that state, followed closely by the eastern screech-owl.

*National Population.* Breeding Bird Surveys between 1966 and 2012 indicate that this species exhibited a survey-wide population decrease (0.89% annually) during that period. This species was reported all 40 years during Christmas Bird Counts from 1968 to 2007 from North Dakota to Oklahoma, and the Texas panhandle. Breeding Bird Surveys between 1966 and 2006 indicate that this species exhibited a statistically nonsignificant national population decline (0.1% annually) during that period. The estimated 1990s continental population north of Mexico was about 2,280,000 birds (Rich *et al.,* 2004). The species breeds south to southernmost South America (Tierra del Fuego).

*Further Reading.* Bent, 1938; Houston, Smith and Rohner, 1998 (*The Birds of North America:* No. 372); del Hoyo, Elliott and Sargatal, 1999; Johnsgard, 2002b; Lynch, 2007; Mikkola, 2012.

## Snowy Owl
*Bubo scandiacus*

One of the few pleasant memories of childhood winters in North Dakota during the Great Depression was the frequent sight of snowy owls, which often suddenly appeared on haystacks in snow-covered fields, only to disappear again a few days later. These silent visitors made me realize that they are places that must be even colder during winter than North Dakota, although that was hard to believe at the time. Then, snowy owls were ignored or, worse yet, used for target practice, but now the news that a snowy owl has arrived as far south as Kansas might attract bird-watchers from far away. Most of the snowy owls that appear as far south as Nebraska are young birds, often in poor condition. The adult females, which are heavier than males, normally winter the farthest north, and immature males the farthest south.

*Winter Distribution.* Root's (1988) map indicates the highest Plains concentration of snowy owls as being in western Nebraska, with smaller numbers in the eastern Dakotas. That winter pattern has evidently shifted northward, as our data indicate that only North and South Dakota have had snowy owls in numbers averaging above the minimal "present" category. There is no evidence of any long-term population trend in our data.

*Seasonality and Migrations.* The snowy owl is a regular but usually rare winter visitor in the northern plains. This species was reported on all Christmas Bird Counts from 1968 to 2007 in North Dakota, as compared with 26 in South Dakota, four in Nebraska, eight in Kansas, four in Oklahoma, and none in the Texas panhandle. There is a range in 18 initial autumn Nebraska sightings dating from the mid-1930s to the late 1970s (Johnsgard, 1980) from November 6 to December 29, with a median of December 4. Twenty-three final spring Nebraska sightings are from January 3 to April 30, with a median of February 5. Among a total of 2,571 owls handled by Raptor Recovery Nebraska over a 25-year period, there were 14 snowy owls (Johnsgard, 2002c). The snowy owl is a rare and irregular winter visitor in Kansas, with records from November 1 to April 15 (Thompson and Ely, 1989). Oklahoma records extend from November to February Woods and Schnell, 1984). There are a few unproven sightings for the Texas panhandle (Seyffert, 2001). Periodic incursions into the Great Plains occur when lemming populations have surged, resulting in high reproductive success and corresponding high owl reproduction. competitions for food the following winter drives young, inexperienced owls varying distances southward, often to Nebraska and sometimes as far as Oklahoma (Johnsgard, 2012c).

*Habitats.* Wintering birds are usually associated with open fields, plains, marshes, and grassy lowlands, often perching on haystacks or other somewhat elevated sites.

*National Population.* Breeding Bird Survey trend data are not available for this arctic-breeding breeder. The estimated 1990s continental population was about 290,000 birds (Rich *et al.,* 2004). The Canadian breeding population has been estimated at 10,000–30,000 pairs (Lynch, 2007). There is also Greenland, Iceland and Eurasian populations of unknown total sizes.

*Further Reading.* Bent, 1938; Kerlinger, Lein and Sevick, 1985; Godfrey, 1986; Parmelee, 1992 (*The Birds of North America:* No. 10); del Hoyo, Elliott and Sargatal, 1999; Johnsgard, 2002b, 2012c; Lynch, 2007.

## Burrowing Owl
*Athene cunicularia*

The burrowing owl weighs almost exactly as much as an eastern screech-owl, but is much less cold-tolerant, and it migrates south fairly early in the fall. This post-breeding movement occurs as the insect food supply declines, and although burrowing owls do at times prey on small mammals, they are apparently less able to achieve this autumn dietary shift than are screech-owls.

*Winter Distribution.* Root (1988) identified the Texas panhandle and adjoining parts of Texas and Oklahoma as the primary wintering area of burrowing owls in the Great Plains, especially the area around Plainview and Lubbock, where prairie dogs at the time of her study were still quite abundant. Since then prairie dog populations have crashed throughout the Great Plains, owing to uncontrolled poisoning, hunting, and habitat destruction (Johnsgard, 2002). This results from an abject failure of both state and federal agencies to fulfill their obligations to monitor and protect not only prairie dogs, but also the several threatened or endangered species of birds and mammals that are part of the associated shortgrass ecosystem (Johnsgard, 2005). The winter distribution of burrowing owls in the Great Plains is now so small that their numbers scarcely register, even in their one-time Texas stronghold.

*Seasonality and Migrations.* The burrowing owl is a Great Plains seasonal migrant and local breeder, mainly in western regions, wintering southwardly. This species was not reported during Christmas Bird Counts from 1968 to 2007 in North Dakota, South Dakota or Nebraska, but was reported eight years in Kansas, ten years in Oklahoma and 25 years in the Texas panhandle. In South Dakota wintering is considered extremely rare (Tallman,

Swanson and Palmer, 2002). The range of 119 initial spring Nebraska sightings dating from the mid-1930s to the late 1970s (Johnsgard, 1980) is from
March 10 to June 10, with a median of April 24. Forty-three final autumn
sightings are from July 21 to November 9, with a median of September 16.
It is a very rare winter visitor in southwestern Kansas (Thompson and Ely,
1989; Thompson *et al.*, 2011). Although a permanent resident in Oklahoma
by Woods and Schnell (1984), the number estimated to winter is the state
is only about one percent (Reiking, 2004). It is a rare winter visitor in the
Texas panhandle, but has been reported in every month (Seyffert, 2001).

*Habitats.* This species is normally associated with heavily grazed grasslands,
especially those supporting colonies of large rodents such as prairie dogs.
Normally colonial, scattered nestings may also occur by individual pairs
where suitable cavities are available.

*National Population.* Breeding Bird Surveys between 1966 and 2012 indicate
that this species exhibited a survey-wide population decrease (1.1% annually) during that period. Most state surveys have also indicated marked declines in recent years as prairie dog populations have also declined. The estimated 1990s continental population north of Mexico was about 620,000
birds (Rich *et al.*, 2004). The Canadian population is nationally endangered,
with only about 400-500 remaining pairs (Lynch, 2007). There are also
West Indian and South American (south to Tierra del Fuego) populations
of unknown sizes (del Hoyo, Elliott and Sargatal, 1999).

*Further Reading.* Bent, 1938; James and Ethier, 1989; Haug, Millsap and Martell, 1993 (*The Birds of North America:* No. 61); Dechant *et al.*, 1999a; del
Hoyo, Elliott and Sargatal, 1999; Johnsgard, 2001b, 2002b, 2005; Lynch,
2007.

## Barred Owl
### *Strix varia*

This forest-adapted owl seems to be highly residential, with the same territories typically occupied for year after year, over periods for as long as three
decades. A few long-distance movements of banded birds have been documented, but there is no indication of any winter movements to warmer climates. Barred owls are highly territorial, in contrast to great horned owls, and
defend their entire home range throughout the year. As a result, they often respond to the imitation or playback of their calls at any season, which aids in
population surveys.

*Winter Distribution.* Root (1968) indicated the western edge of the barred owl's range as extending south from northeast Nebraska though central Kansai and Oklahoma to parts of the Texas panhandle, especially where riparian forests extend out into the grasslands of the Great Plains. Our data are in general agreement, with Kansas and Oklahoma supporting the largest numbers. There is no apparent long-term trend in populations, so the substantial upward trend indicated by national Breeding Bird Surveys perhaps largely reflects range expansion rather than increased densities in the historic range.

*Seasonality and Migrations.* The barred owl is an uncommon permanent resident in densely wooded areas of the Plains states, although the northernmost populations may be somewhat mobile seasonally, depending on prey availability.

*Habitats.* Throughout the year, this species is found in dense river-bottom hardwood forests of the Great Plains. However, coniferous forests seem to be preferred habitats, and are used when they are available.

*National Population.* Breeding Bird Surveys between 1966 and 2012 indicate that this species exhibited a survey-wide population increase (1.6% annually) during that period, much of which is related to extensive range expansion in the Pacific Northwest during the past several decades. The estimated 1990s continental population was about 560,000 birds (Rich *et al.,* 2004), including about 50,000 pairs in Canada (Lynch, 2007). The species breeds south to southern Mexico.

*Further Reading.* Bent, 1938; del Hoyo, Elliott and Sargatal, 1999; Maser and James, 2000 (*The Birds of North America:* No. 508); Johnsgard, 2002b; Lynch, 2007.

## Long-eared Owl
*Asio otus*

This is a large but very shy owl, sometimes living for years in a well-traveled area without ever being noticed. One year I found two nearly fledged long-eared owls being reared within a few hundred yards of our western Nebraska field station, none of the faculty or students had been aware of the nesting activity that was occurring almost under our noses. In 2008 I also learned from a rancher of an owl roost in a dense grove of junipers only a few miles from the field station. The roost had evidently been there for many years without the knowledge of station biologists. As a result of such near-invisibility, it is very difficult to map or survey long-eared owls accurately.

*Winter Distribution.* The long-eared owl is a nocturnal and inconspicuous owl, so Christmas Count data are unlikely to offer much information in terms of distribution and abundance. Root (1988) suggested that the South Platte River valley of eastern Colorado and western Nebraska may represent an area of high density, but no other parts of the Plains region showed notable concentrations on her map. Our data for the species also do not allow delineating the species' distribution or population.

*Seasonality and Migrations.* The long-eared owl is an uncommon to occasional Great Plains seasonal migrant or permanent resident throughout the region. These owls often aggregate and roost in dense woods during winter within the breeding range, but may also migrate considerable distance southward, as far as central Mexico. An individual banded in Saskatchewan was later recovered in Oaxaca, Mexico. Another, also banded in Saskatchewan, was recovered five years later in Mississippi. Long-eared owls were reported on 15 Christmas Bird Counts from 1968 to 2007 in North Dakota, 17 in South Dakota, 35 in Nebraska, 38 in Kansas, 14 in Oklahoma and 17 in the Texas panhandle. In South Dakota this species is an irregular winter visitor, but it apparently does not breed there (Tallman, Swanson and Palmer, 2002). Nineteen autumn sightings are from July 21 to December 31, with a median of November 24 (Johnsgard, 1980). Twenty-four spring Nebraska sightings dating from the mid-1930s to the late 1970s range from January 2 to May 14, with a median of March 9th. These limited data suggest that this species is a summer resident and a late autumn and early spring migrant in Nebraska, with frequent wintering. It is a permanent resident in Kansas (Thompson *et al.*, 2011), and a permanent resident (Woods and Schnell, 1984) or a regular winter visitor (Reinking, 2004) in Oklahoma. It is an uncommon winter visitor in the Texas panhandle, and has been reported in every month but July and August (Seyffert, 2001).

*Habitats.* Throughout the year, this species is associated with wooded areas, including river bottom forests, parks, orchards and woodlots. Coniferous as well as hardwood forests are utilized, with the former apparently preferred.

*National Population.* Breeding Bird Survey trend data are not available for this widespread but inconspicuous species. The estimated 1990s continental population north of Mexico was about 36,000 birds (Rich *et al.*, 2004), including about 10,000 to 20,000 pairs in Canada (Lynch, 2007). The species breeds south to central Mexico, and there is also a Eurasian population that might number in the hundreds of thousands (del Hoyo, Elliott and Sargatal, 1999).

*Further Reading.* Bent, 1938; Marks, Evans and Holt, 1994 (*The Birds of North America:* No. 134); del Hoyo, Elliott and Sargatal, 1999; Johnsgard, 2002b; Lynch, 2007.

## Short-eared Owl
*Asio flammeus*

Short-eared owls are grassland-adapted owls, sharing their habitat with and to a degree competing with harriers as low-altitude predators of small rodents. They tend to replace harriers in late afternoon, starting their hunting activities as daylight begins to fade, but probably have better directional hearing and near-dark vision than do harriers.

*Winter Distribution.* This is an owl that is often seen foraging during daylight hours, so it should be easily detected during Christmas Counts. Root's (1988) map indicates relatively high populations around several grass-dominated North Dakota wildlife refuges, as well as one in eastern Oklahoma. However, our data indicate very low and widely scattered numbers of birds across the Plains states throughout the entire four-decade period.

*Seasonality and Migrations.* The short-eared owl is an uncommon seasonal or permanent resident throughout, being more common in the summer in natural grasslands, and moving variably southward in winter, probably in relation to relative prey abundance. This species was reported on 33 Christmas Bird Counts from 1968 to 2007 in North Dakota, 31 in South Dakota, 18 in Nebraska, 40 in Kansas, 38 in Oklahoma, and 12 in the Texas panhandle. In South Dakota it is considered a summer resident and irruptive migrant (Tallman, Swanson and Palmer, 2002). During winter the Nebraska population is apparently supplemented by migrants from farther north. Twenty-nine autumn Nebraska sightings dating from the mid-1930s to the late 1970s are from July 20 to December 31, while 35 spring sightings range from January 8 to June 6, (Johnsgard, 1980). The data suggest that this specie is a late autumn and early spring migrant in Nebraska, with frequent wintering. It is local during winter in Kansas (Thompson *et al.*, 2011), and a winter resident in Oklahoma, with continuous monthly records extending from September to May (Woods and Schnell, 1984). It is an uncommon to fairly common winter visitor in the Texas panhandle, having been reported in every month but July (Seyffert, 2001).

*Habitats.* Throughout the year, this species is found in open, grass-dominated environments, and in Nebraska the Sandhills prairie and other natural grasslands are favored habitats. Nesting usually occurs in grassy cover, with several pairs often nesting fairly close to one another in a loose colonial situation.

*National Population.* Breeding Bird Surveys between 1966 and 2012 indicate that this species exhibited a survey-wide population decrease (2.48% annually) during that period. The estimated 1990s continental population was about 700,000 birds (Rich *et al.,* 2004). It has been designated a species of continental conservation importance, with the Prairie Avifaunal Biome supporting an estimated 12 percent of the continental winter population (Rich *et al.,* 2004). The species breeds south to central Mexico and the West Indies. There is also a Eurasian population that might number in the hundreds of thousands (del Hoyo, Elliott and Sargatal, 1999). This species was yellow-listed in the National Audubon Society's 2007 WatchList of rare and declining birds (Butcher *et* al., 2007).

*Further Reading.* Bent, 1938; Colvin and Spaulding, 1983; Holt and Leasure, 1993 (*The Birds of North America:* No. 62); Dechant *et al.,* 1999i; del Hoyo, Elliott and Sargatal, 1999; Johnsgard, 2001b, 2002b; Lynch, 2007.

## Northern Saw-whet Owl
*Aegolius acadicus*

This is the smallest of our Great Plains owls, weighing around three ounces, roughly the same body mass as an American robin, or half the weight of a screech-owl. As a result, it should not be surprising that the species is strongly migratory, at least toward the northern end of its Great Plains range. The species' huge facial disk provides a clue that its hearing abilities are extraordinary, and it typically hunts in total darkness. Small rodents such as voles (*Microtis*) are the species' primary prey, with most of the animals taken weighing less than two grams. However, mammals as large as pocket gophers (*Thomomys, Geomys*) have rarely been reported as prey, as well as birds as large as rock pigeons.

*Winter Distribution.* As one of the most secretive of Great Plains owls, Christmas Counts are unlikely to be of much value in mapping the winter distribution of this species. Root (1988) noted that her data for this species were too limited to be reliable, and our numbers likewise were very small and not indicative of possible regional differences in its distribution or populations.

*Seasonality and Migrations.* The northern saw-whet owl is a rare to uncommon seasonal migrant, wintering widely. This species was reported on 16 (40 percent) of the North Dakota Christmas Bird Counts between 1968 and 2007. In South Dakota and Nebraska, this species occurred on nine years each of the counts, in Kansas on six, in Oklahoma on four, and on none in the Texas panhandle. It is a permanent resident in the Black Hills, but a migrant or winter visitor elsewhere (Tallman, Swanson and Palmer, 2002). Ten autumn Nebraska sightings dating from the mid-1930s to the late 1970s are from July 29 to December 22, with a median of November 8. Seven spring records are from January 1 to May 16, with a median of February

20. These and other data suggest that this species is primarily a winter visitor in Nebraska (Johnsgard, 1980, 2007a). It is a winter visitor in Kansas (Thompson *et al.*, 2011), reported from October 14 to April 30 (Thompson and Ely, 1992). It has been reported from November through January in Oklahoma (Woods and Schnell, 1984), but there are no winter records from the Texas panhandle (Seyffert, 2001).

*Habitats.* In winter these inconspicuous owls roost during the day in dense coniferous trees and similar dense vegetation, such as grapevine tangles.

*National Population.* Breeding Bird Survey trend data are not available for this elusive species. The estimated 1990s continental population north of Mexico was about 1,950,000 birds (Rich *et al.*, 2004), including about 100,000–300,000 pairs in Canada (Lynch, 2007). The species breeds south to southern Mexico.

*Further Reading.* Bent, 1938; Cannings, 1993 (*The Birds of North America:* No. 42); del Hoyo, Elliott and Sargatal, 1999; Johnsgard, 2002b; Lynch, 2007.

# FAMILY ALCEDINIDAE:
## KINGFISHERS

### Belted Kingfisher
*Ceryle alcyon*

This familiar species has an extremely broad winter range in North America, the northern edge of which probably fluctuates from year to year with the varied boundaries of open water. Certainly in Nebraska it is one of the last water-dependent birds to head south as ice closes in around it, and is one of the first to re-appear as thawing conditions allow.

*Winter Distribution.* Kingfishers require open water to survive the winter, so their northern winter limits are clearly defined by ice-free conditions. As such, they should respond quickly in their wintering behavior to a warming climatic trend. Root (1988) mapped the northern winter limits of the species near the South Dakota, North Dakota border, and the greatest densities to be found in eastern Oklahoma, where they are associated with smaller rivers. In our study, Kansas had substantially higher densities than Oklahoma, but both states showed apparent long-term population declines, in accordance with national trends indicated by Breeding Bird Surveys.

*Seasonality and Migrations.* The belted kingfisher is an uncommon Great Plains seasonal migrant and local wetland breeder, wintering southwardly. This species was reported 11 years during Christmas Bird Counts from 1968 to 2007 in North Dakota. It occurred every year in South Dakota, Nebraska, Kansas and Oklahoma counts, and 39 years in the Texas panhandle. In South Dakota this species rarely winters (Tallman, Swanson and Palmer, 2002). Forty-three initial spring Nebraska sightings dating from the mid-1930s to the late 1970s (Johnsgard, 1980) range from January 2 to May 10, with a median of March 20. Forty-seven final autumn sightings are from July 26 to December 31, with a median of November 15. The frequency of late December and January records suggests that this species winters occasionally as far north as Nebraska. It is uncommon to rare during winter in Kansas (Thompson *et al.*, 2011), and is a permanent resident in Oklahoma (Woods and Schnell, 1984) and the Texas panhandle (Seyffert, 2001). The average number of birds observed per party-hour for the 2013–2014 Christmas Bird Count were: N. Dakota 0, S. Dakota 0, Nebraska 0.27, Kansas 0.36, Oklahoma 0.44, northwest Texas 1.17. For complete species abundance data, see Appendix 1:60.

*Habitats.* Throughout the year this species occurs near water areas support-ing populations of fish, amphibians and similar aquatic life. Nests are ex-cavated from nearly vertical earth exposures in bluffs, road cuts, eroded stream banks, and the like.

*National Population.* Breeding Bird Surveys between 1966 and 2012 indicate that this species exhibited a survey-wide population decrease (1.45% annu-ally) during that period. The estimated 1990s continental population was about 2.2 million birds (Rich *et al.,* 2004).

*Further Reading.* Bent, 1940; Hamas, 1994 (*The Birds of North America:* No. 84).

# FAMILY PICIDAE:
# WOODPECKERS

## Red-headed Woodpecker
*Melanerpes erythrocephalus*

This familiar and colorful woodpecker often strays out well away from the generally open woods where it breeds, sometimes fly-catching over open grasslands, or extracting insects from cracks in wooden telephone poles or fence posts. The species also regularly nests in cities where mature trees occur, such as parks. It tends to prefer somewhat more open habitats, with more widely scattered trees and less undergrowth, than does the red-bellied woodpecker. To a greater degree than with other Plains woodpeckers, it is not rare to see the carcasses of red-headed woodpeckers along the sides of highways, where the birds had been feeding on insects in the roadways. Competition for nesting sites with starlings has evidently reduced this woodpecker's population over much of eastern North America, and perhaps this same factor has partly caused the population reduction that is increasingly apparent in the Great Plains.

*Winter Distribution.* According to Root (1988) this woodpecker reaches its highest winter abundance in eastern Oklahoma and southeastern Kansas, but with some birds remaining as far north as southern South Dakota. Our data are similar, with Oklahoma having easily the highest numbers in the region, and Kansas substantially fewer, and very few being found north of Nebraska as late as Christmas. The significant national population decline indicated by national Breeding Bird Surveys is supported by declines in our five-state average numbers.

*Seasonality and Migrations.* The red-headed woodpecker is an uncommon Great Plains seasonal migrant and widespread breeder, wintering southwardly. This species was reported ten years during Christmas Bird Counts from 1968 to 2007 in North Dakota, 19 years in South Dakota, 38 years in Nebraska, all 40 years in Kansas and Oklahoma, and four years in the Texas panhandle. In South Dakota this species is extremely rare after December (Tallman, Swanson and Palmer, 2002). The range of 106 final Nebraska sightings dating from the mid-1930s to the late 1970s (Johnsgard, 1980) is from August 8 to December 31. Less than ten percent of the autumn records are for December, suggesting that this species only rarely winters in Nebraska. Presumably its relatively high dependence on aerial insects accounts for this species' migration tendencies, as compared with most other Great Plains woodpeckers. It is a local and irregular wintering bird in Kansas (Thompson *et al.*, 2011). It is considered a permanent resident in Oklahoma (Woods and Schnell, 1984), and is a rare winter visitor in the Texas panhandle, having been reported in every month (Seyffert, 2001). The average number of birds observed per party-hour for the 2013–2014 Christmas Bird Count were: N. Dakota 0, S. Dakota 0.23, Nebraska 0.62, Kansas 0.92, Oklahoma 0.73, northwest Texas 0. For complete species abundance data, see Appendix 1:61.

*Habitats.* This species occurs in fairly open forests, woodlots, urban parks, and wooded housing areas. It occupies somewhat more open areas than does the red-bellied woodpecker, and is more widespread than is that species.

*National Population.* Breeding Bird Surveys between 1966 and 2012 indicate that this species exhibited a survey-wide population decrease (2.6% annually) during that period. It has been designated a species of continental conservation importance, with the Prairie Avifaunal Biome supporting an estimated 39 percent of the continental winter population (Rich *et al.,* 2004). The estimated 1990s continental population was about 2,5 million birds (Rich *et al.,* 2004). This species was yellow-listed in the National Audubon Society's 2007 WatchList of rare and declining birds (Butcher *et* al., 2007).

*Further Reading.* Bent, 1938; Short, 1982; Winkler, Christie and Nurney, 1995; Smith, Withgott and Rodewald, 2000 (*The Birds of North America:* No. 518).

## Lewis' Woodpecker
### *Melanerpes lewis*

It is easy to think of this species as a red-headed woodpecker that has been sprinkled with soot, and it seems appropriate that, like the blackish three-toed woodpecker, it is especially typical of charred forests that are recovering from forest fires. In flight it lacks the contrasting and distinctive white wing-markings of the red-headed woodpecker, and is so uniformly blackish in flight that the first time I saw the species as a teenager I thought I was seeing a miniature crow.

*Winter Distribution.* Root (1988) indicated no winter populations of this woodpecker in the Great Plains except for extreme western Kansas and the western edge of the Texas panhandle. This western species exhibits considerable wandering outside the breeding season, and has been reported east to Lake Superior and Arkansas (Baughman, 2003). Our data show moderate Christmas Count numbers in South Dakota, and some in Oklahoma, no doubt representing birds persisting around breeding areas in the western parts of these states.

*Seasonality and Migrations.* Lewis' woodpecker is an occasional to rare seasonal migrant, or very local resident and breeder, often wintering locally. In South Dakota this species occurred at least once on seven percent of Christmas Bird Count locations between 1949 and 1998, and is a permanent resident of the Black Hills (Tallman, Swanson and Palmer, 2002). It was not reported during counts in North Dakota, Nebraska or Texas, but appeared two years on Kansas counts, and nine years on Oklahoma counts. There are too few records to judge this rare species' migration in Nebraska, but 15 total records range from January 20 to September 23, with the largest number in May. Like the red-headed woodpecker, it is somewhat dependent on aerial insects, and so is more migration-prone than most woodpeckers. It has been observed in Kansas as late as November 7 (Thompson *et al.*, 2011). It is considered a permanent resident in extreme western Oklahoma (Woods and Schnell, 1984), with nesting apparently limited to Cimarron County (Reinking, 2004). It is a very rare visitor in the Texas panhandle, and has been reported as late as December (Seyffert, 2001). The average number of birds observed per party-hour for the 2013–2014 Christmas Bird Count were: N. Dakota 0, S. Dakota 0.02, Nebraska 0, Kansas 0, Oklahoma 0, northwest Texas 0. For complete species abundance data, see Appendix 1:62.

*Habitats.* The edges of pine forests and streamside cottonwood groves having considerable dead growth are favored Black Hills habitats, and probably similar habitats are used in Nebraska's nearby Pine Ridge region.

*National Population.* Breeding Bird Surveys between 1966 and 2012 indicate that this species exhibited a survey-wide population decrease (3.5% annually) during that period. The estimated 1990s continental population was about 130,000 birds (Rich *et al.*, 2004). This species was red-listed in the National Audubon Society's 2007 WatchList of rare and declining birds (Butcher *et* al., 2007).

*Further Reading.* Bent, 1938; Short, 1982; Winkler, Christie and Nurney, 1995; Tobalske, 1997 (*The Birds of North America:* No. 284).

## Golden-fronted Woodpecker
*Melanerpes aurifrons*

The golden-fronted woodpecker neatly replaces the red-bellied in southwestern Oklahoma and the southern half of the Texas panhandle, although in the narrow zone of contact there is some interaction and occasional hybrids.

*Winter Distribution.* The very limited Texas and southwestern Oklahoma distribution of this resident species shown by Root (1988) has not significantly changed, judging from our data, although the Oklahoma population is barely indicated. The Oklahoma population is limited to three southwestern counties, where it first appeared in 1954 and began nesting in 1958. There it has contacts the red-bellied woodpecker, which had expanded west into that region by 1959, and the two have since hybridized (Reinking, 2004). Numbers for the Texas panhandle show no clear population trend to compare with Breeding Bird Survey results. Texas Breeding Bird Survey data suggest an essentially stable population (Benson and Arnold, 2001). The average number of birds observed per party-hour for the 2013–2014 Christmas Bird Count were: N. Dakota 0, S. Dakota 0, Nebraska 0, Kansas 0, Oklahoma 0, northwest Texas 0.46. For complete species abundance data, see Appendix 1:63

*Seasonality and Migrations.* The golden-fronted woodpecker is a permanent resident throughout it range.

*Habitats.* Riparian woodlands of cottonwood, willow and cypress, and mesquite-oak-juniper brush lands are typical habitats of the golden-fronted woodpecker.

*National Population.* This species was reported six years during Christmas Bird Counts from 1968 to 2007 in Oklahoma, and 39 years in the Texas panhandle. Breeding Bird Surveys between 1966 and 2012 indicate that this species exhibited a survey-wide population decrease (1.04% annually) during that period. The estimated 1990s continental population was about 850,000 birds (Rich *et al.,* 2004).

*Further Reading.* Bent, 1938; Short, 1982; Winkler, Christie and Nurney, 1995; Husak and Maxwell, 1998 (*The Birds of North America:* No. 373).

### Red-bellied Woodpecker
*Melanerpes carolinus*

In contrast to the red-headed woodpecker, this is a sedentary species, and is just as likely to turn up at winter feeding stations as the downy or hairy woodpecker. Like other woodpeckers, suet is a favorite food, as are acorns, but sunflower seeds are also readily taken. These items are carried away for caching in well-hidden sites, often probably far from the feeding station, judging from the time elapsed between trips. Red-bellies often return to a feeder repeatedly during these caching trips, each time carrying the last item it picked up in its beak as it flies away.

*Winter Distribution.* Root (1988) mapped the Great Plains distribution of the red-bellied woodpecker as extending out to the western limits of the southern and central Plains, and northward to eastern parts of the Dakotas in the northern Plains. In the central and northern plains it is closely associated with gallery forests that extend west varying distances. In Nebraska, for example, such riparian corridors have allowed the species to extend west into Colorado and Wyoming during the past half-century. It has also expanded northwest from southern Minnesota into North Dakota since the 1960s (Baughman, 2003). These Great Plains incursions are part of a much broader range-expansion trend (Jackson and Davis, 1998). Judging from our data, similar range expansions have resulted in population increases in states from North Dakota through Oklahoma, contrary to the national population decline suggested by Breeding Birds Survey results. Jackson and Jackson (1987) attribute its overall range expansion to be the result of climate change and increased urban tree-planting.

*Seasonality and Migrations.* The red-bellied woodpecker is a permanent resident throughout it range. The average number of birds observed per party-hour for the 2013–2014 Christmas Bird Count were: N. Dakota 0.07, S. Dakota 0.22, Nebraska 0.7, Kansas 1.09, Oklahoma 0.92, northwest Texas 0.09. For complete species abundance data, see Appendix 1:64.

*Habitats.* Throughout the year the red-bellied woodpecker occupies somewhat open stands of coniferous or hardwood forests, often river bottom forests. It also frequents orchards and gardens in urban or suburban locations.

*National Population.* This is a Great Plains seasonal migrant and widespread breeder, wintering southwardly. It was reported 27 years during Christmas Bird Counts from 1968 to 2007 in North Dakota, as compared with 39 years in South Dakota, all 40 years in Nebraska, Kansas, Oklahoma, and 27 years in the Texas panhandle. Breeding Bird Surveys between 1966 and 2012 indicate that this species exhibited a survey-wide population increase (1.05% annually) during that period The estimated 1990s continental population was about ten million birds (Rich *et al.,* 2004).

*Further Reading.* Bent, 1938; Kilham, 1963; Short, 1982; Winkler, Christie and Nurney, 1995; Dunn and Tessaglia-Hymes, 1999; Shackelford, Brown and Conner, 2000 (*The Birds of North America:* No. 500).

## Yellow-bellied Sapsucker
*Sphyrapicus varius*

Root (1988) mentioned that sapsuckers are likely to winter in areas where the usual temperature is high enough so that sap will flow, or where the average minimum January temperature is above 15° F. They are more common where the temperature rarely falls below freezing. They also tend to occur in regions where there is enough annual precipitation to support adequate growth of sap-bearing trees. Many other species use the sap that sapsuckers make available through their drilling activities, ranging from insects to hummingbirds and flying squirrels, and including at least 35 species of insect- or sap-eating birds.

*Winter Distribution.* Root (1988) mapped the winter distribution of this species as extending north to northwestern Nebraska and northeastern South Dakota, with larger numbers in southeastern Oklahoma. She noted that the dependence of sapsuckers on sap means that the birds must avoid subfreezing winter climates. Our data show a progressive increase in numbers southwardly, with Oklahoma and the Texas panhandle having the highest numbers, The long-term average numbers appear to be stable, rather than showing any increase, as is suggested by national Breeding Bird Surveys.

*Seasonality and Migrations.* The yellow-bellied sapsucker is an uncommon seasonal migrant, and a very local breeder in western South Dakota (*nuchalis*) and northeastern North Dakota (*varius*), wintering southwardly. This species was reported one year during Christmas Bird Counts from 1968 to 2007 in North Dakota, seven years in South Dakota, 37 years in Nebraska, all 40 years in Kansas and Oklahoma, and 24 years in the Texas panhandle. Thirty-four initial autumn Nebraska sightings dating from the mid-1930s to the late 1970s are from September 1 to December 30, with a median of October 3. Twenty-five final autumn sightings in Nebraska are from October 9 to December 31, with a median of December 18. Sixteen initial spring Nebraska sightings are from January 1 to May 28, with a median of March 14 (Johnsgard, 1980). These data would suggest that this species is a very late autumn migrant in Nebraska, that it frequently winters in the state, and remains for a rather variable period in spring. It is an uncommon winter resident in Kansas (Thompson *et al.*, 2011). and is a winter resident in Oklahoma, with continuous monthly records extending from September to May (Woods and Schnell, 1984). It is a rare to uncommon winter visitor in the Texas panhandle, with records from September to May (Seyffert, 2001). The average number of birds observed per party-hour for the 2013–2014 Christmas Bird Count were: N. Dakota 0, S. Dakota 0, Nebraska 0.04, Kansas 0.11, Oklahoma 0.14, northwest Texas 0. For complete species abundance data, see Appendix 1:65.

*Habitats.* While in Nebraska sapsuckers are associated with various woodlands, especially those having poplars or aspens, which are favored foraging trees. However, they also drill in birches (*Betula*), maples (*Acer*), cottonwoods (*Populus*), apple trees (*Pyrus*) and junipers (*Juniperus*), but only infrequently in such hardwoods as oaks (*Quercus*) and hackberries (*Celtis*).

*National Population.* Breeding Bird Surveys between 1966 and 2012 indicate that this species exhibited a survey-wide population increase (0.56% annually) during that period. The estimated 1990s continental population of *varius* was about 9.2 million birds (Rich *et al.,* 2004).

*Further Reading.* Bent, 1938; Short, 1982; Winkler, Christie and Nurney, 1995; Dunn and Tessaglia-Hymes, 1999; Walters, Miller and Lowther, 2002 (*The Birds of North America:* No. 662).

## Red-naped Sapsucker
### *Sphyrapicus nuchalis*

This species is a close western relative of the yellow-bellied sapsucker (at times the two have been considered subspecies), and the two are known to hybridize where they come into contact. Many western trees, including both conifers and hardwoods, are used by these birds for their sap, or for the soft cambium layer just below the bark.

*Winter Distribution.* This woodpecker was only recently recognized as a distinct species, and only one record of a single bird has so far been noted on Christmas Counts in the Great Plains since 1968. Because of this single record, no tabular summary of its Plains status is presented.

*Seasonality and Migrations.* The red-naped sapsucker is a rare Great Plains seasonal migrant and very local breeder in western South Dakota (Black Hills), wintering peripherally and probably also southwardly. There are a few specimen records extending from western Nebraska and western Kansas, but the winter distribution of this species farther south is still very uncertain. It is a rare to uncommon winter visitor in the Texas panhandle, the few reports mostly occurring during autumn (Seyffert, 2001).

*Habitats.* All sapsuckers are associated with various woodlands, especially those having aspens (*Populus*), which are notable sap-producing trees.

*National Population.* Breeding Bird Surveys between 1966 and 2012 indicate that this species exhibited a survey-wide population increase (1.4% annually) during that period.

*Further Reading.* Bent, 1938; Short, 1982; Winkler, Christie and Nurney, 1995; Walters, Miller and Lowther, 2002 (*The Birds of North America:* No. 663).

## Ladder-backed Woodpecker
*Picoides scalaris*

The distribution of ladder-backed woodpeckers in Oklahoma and northwestern Texas is probably strongly influenced by the distribution of mesquite (*Prosopsis glandulosa*), which is a favored breeding habitat. Its distribution is thus similar to that of the golden-fronted woodpecker, which is also strongly attracted to mesquite woodlands. However, the ladder-backed has a broader ecological range, and farther west it is also common in oak woodlands, bringing it into contact and local competition with the closely related Nuttall's woodpecker (*P. nutallii*) of the Southwest.

*Winter Distribution.* Root (1988) mapped the distribution of this southwestern woodpecker as extending north to northwestern Kansas and east to eastern Oklahoma. Our data suggest only low populations in Oklahoma, still fewer birds in northwestern Texas, and almost none in Kansas. This would seem to be a better description of its current status in Kansas and Oklahoma, but might under-estimate of its status in the Texas panhandle, where it is considered fairly common (Seyffert, 2002).

*Seasonality and Migrations.* The ladder-backed woodpecker is an uncommon to rare permanent resident in the southern Plains states.

*Habitats.* During winter this species may be found foraging in cottonwoods (*Populus*), willows (*Salix*) and hackberry (*Celtis*) trees, but its limited range in Oklahoma generally coincides with that of mesquite (Sutton, 1967). Desert scrub dominated by mesquite is the species' favored breeding habitat in Texas (Benson and Arnold, 2001).

*National Population.* This species was reported 23 years during Christmas Bird Counts from 1968 to 2007 in Kansas, 36 years in Oklahoma, and all 40 years in the Texas panhandle. Breeding Bird Surveys between 1966 and 2012 indicate that this species exhibited a survey-wide population decrease (0.1% annually) during that period. Texas populations have shown corresponding probable population declines (Benson and Arnold, 2001).

The estimated 1990s continental population north of Mexico was about 693,000 birds (Rich *et al.*, 2004). The species breeds south to Nicaragua.

*Further Reading.* Bent, 1938; Short, 1982; Winkler, Christie and Nurney, 1995; Lowther, 2001 (*The Birds of North America:* No. 565).

## Red-cockaded Woodpecker
### *Picoides borealis*

This is a seriously declining species that is nationally endangered and has been nearly eliminated from several eastern states, including Virginia, Kentucky and Tennessee. This decline has been attributed to habitat fragmentation and loss resulting from fire-restriction, short-term forestry rotation, and forest-clearing. Nesting is done in old living pines that have rotten heartwood affected by red-heart fungus. In Oklahoma the species is limited to McCurtain County Wilderness Area, where in recent years fewer than a dozen nests have been found and monitored.

*Winter Distribution.* Within the study region, this woodpecker is limited to Oklahoma. It is only found in the southeastern quarter of the state, in mature pine and pine-oak forests, and specifically where diseased trees have weakened trunks that allow for nest-hole excavation. In our study, the species was too rare to appear in Oklahoma's Christmas Counts at any level higher than the minimal (under 0.01 bird/party-hour) "present" category.

*Seasonality and Migrations.* The red-cockaded Woodpecker is an uncommon to rare permanent resident throughout its highly limited Great Plains range in southeastern Oklahoma.

*Habitats.* In its confined Oklahoma range this species is closely associated with large shortleaf pines (*Pinus echinata*) (Sutton, 1967), but longleaf pines (*P. palustris*) seem to be generally favored elsewhere for nesting sites.

*National Population.* Breeding Bird Surveys between 1966 and 2012 indicate that this species exhibited a survey-wide population decrease (3.47% annually) during that period. The estimated 1990s continental population was about 20,000 birds (Rich *et al.*, 2004).

*Further Reading.* Bent, 1938; Morse, 1972; Kilham, 1976; Short, 1982; Jackson, 1994 (*The Birds of North America:* No. 85); Winkler, Christie and Nurney, 1995.

## American Three-toed Woodpecker
*Picoides dorsalis*

Four toes on woodpeckers seem to be so universal that Roger Tory Peterson once inadvertently included four toes on a three-toed woodpecker that he painted for a book on the birds of Nova Scotia (it was corrected in later printings). In any case, one wonders what the selective significance of having only three toes might be for a woodpecker; the first, or hind inner toe, is lacking. Apparently small woodpeckers can maintain their climbing and clinging abilities with only three toes. In contrast, larger and more arboreal woodpeckers have an unusually lengthened fourth toe (the outer front toe) that is evidently important in facilitating their climbing, clinging and excavation behavior.

*Winter Distribution.* Within the Great Plains this species occurs only in South Dakota's Black Hills, where it forages on wood-boring insects from both living and dead trees, especially burnt conifers. Because of its specialized niche, the species is very rare, and its presence has barely registered during the four decades of Christmas Counts analyzed.

*Seasonality and Migrations.* The American three-toed woodpecker is a rare permanent resident in its limited Black Hills range. In South Dakota it occurred only once on Christmas Bird Counts between 1968 and 2007.

*Habitats.* This species occupies spruce forests and recently burned forests, especially of western conifers.

*National Population.* Breeding Bird Surveys between 1966 and 2012 indicate that this species exhibited a survey-wide population increase (3.15% annually) during that period. The great increase in forest fires during recent years associated with global warming and droughts has resulted in many charred forests and heavy infestations of bark beetles, the primary food of three-toed and black-backed woodpeckers. The estimated 1990s continental population was about 830,000 birds (Rich *et al.,* 2004).

*Further Reading.* Bent, 1938; Short, 1982; Winkler, Christie and Nurney, 1995; Leonard, 2001 (*The Birds of North America:* No. 588).

## Black-backed Woodpecker
*Picoides arcticus*

Like the previous species, this woodpecker is very rare in the Great Plains, because of the rarity of extensive forests. In both species the larvae of wood-boring beetles that infect dead conifers are their primary foods. Scaling off the bark exposes the larvae. At times the inner bark of living trees may also be consumed, as well as various nuts, fruits and acorns.

*Winter Distribution.* Like the previous species, the year-around range of this woodpecker is confined in the Great Plains to South Dakota's Black Hills. It has been reported slightly more frequently than has that species, but both have occurred at very low abundance levels.

*Seasonality and Migrations.* The black-backed woodpecker is a rare permanent resident in its limited Black Hills range. In South Dakota it was reported six years on Christmas Bird Counts between 1968 and 2007, but not elsewhere in the Plains states.

*Habitats.* This woodpecker occupies coniferous forests, especially recently burned ones.

*National Population.* Breeding Bird Surveys between 1966 and 2012 indicate that this species exhibited a survey-wide population increase (1.5% annually) during that period. The estimated 1990s continental population was about 1.3 million birds (Rich *et al.,* 2004).

*Further Reading.* Bent, 1938; Short, 1982; Winkler, Christie and Nurney, 1995; Dixon and Saab, 2000 (*The Birds of North America:* No. 509).

## Downy Woodpecker
*Picoides pubescens*

One of the most widespread and common of American woodpeckers, the downy is tough enough to withstand the coldest of Great Plains winters in spite of weighing only about an ounce. Downy woodpeckers maintain their territories through the winter, with pairs sometimes feeding close to one another but often some distance apart, since the slightly larger males are dominant, and a female is unlikely to approach her mate where he is actively feeding. Downies accept a wide variety of foods at feeders, but suet is their favorite feeder food. Otherwise, they spend a good deal of time searching goldenrod stems for galls in which fly larvae may be found and extracted.

*Winter Distribution.* Root (1988) mapped the downy woodpecker as extending west to western parts of all the Great Plains states, as well as to the Texas panhandle, with a few areas of slightly higher densities. Our data also shows a fairly uniform Great Plains distribution, with the highest averages in Kansas, followed closely by Nebraska, but very low numbers in the Texas panhandle. There is no clear evidence of either a long-term population increase or decrease. The average number of birds observed per party-hour for the 2013–2014 Christmas Bird Count were: N. Dakota 0,72, S. Dakota 0.52, Nebraska 0.9, Kansas 0.7, Oklahoma 0.55, north-west Texas 0.38. For complete species abundance data, see Appendix 1:66.

*Seasonality and Migrations.* The downy woodpecker is a common permanent resident throughout its range.

*Habitats.* Throughout the year this species is found in dense or open forests, but also extends into cities to visit parks, gardens, and the like. Besides foraging in smaller trees and the smaller branches of large trees, it also sometimes visits shrubs and tall weeds, which hairy woodpecker rarely do. Christmas Count records indicate a downy-to-hairy woodpecker ratio of 1.1:1 in North Dakota, 2.4:1 in South Dakota, 4.2:1 in Nebraska, 4.8:1 in Kansas, and 6.1:1 in Oklahoma. These ratios suggest that the smaller downy woodpecker can survive and compete more effectively with hairy woodpeckers in southern regions, where winter temperatures are less severe.

*National Population.* Breeding Bird Surveys between 1966 and 2012 indicate that this species exhibited a survey-wide population increase (0.29% annually) during that period The estimated 1990s continental population was about 13 million birds (Rich *et al.,* 2004).

*Further Reading.* Bent, 1938; Short, 1982; Winkler, Christie and Nurney, 1995; Dunn and Tessaglia-Hymes, 1999; Jackson and Ouellet, 2002 (*The Birds of North America:* No. 613).

## Hairy Woodpecker
*Picoides villosus*

Hairy woodpeckers are more than twice as heavy as downies, and their beaks are proportionately longer and more massive. This allows them to feed on larger branches and dig deeper for food than do downies. Like the downy, the hairy woodpecker exhibits some sexual dimorphism in both body size and behavior. Thus, foraging males are more likely to excavate wood, while females are more likely to scale off bark while in search of prey. The two sexes remain together all year and sometimes pairs forage together. They only occasionally visit bird feeders, and tend to remain at feeders for shorter periods than do downies (Dunn and Tessaglia-Hymes, 1999).

*Winter Distribution.* Root (1988) mapped the winter distribution of this species as extending throughout the Great Plains states, with a definite peak in North Dakota, and relatively low numbers in the Texas panhandle. Our data show a progressive decrease in Christmas Count numbers from North Dakota to the Texas panhandle, and a seemingly stable long-term population. The average number of birds observed per party-hour for the 2013–2014 Christmas Bird Count were: N. Dakota 0.21, S. Dakota 0.2, Nebraska 0.25, Kansas 0.15, Oklahoma 0.12, northwest Texas 0.06. For complete species abundance data, see Appendix 1:67

*Seasonality and Migrations.* The hairy woodpecker is an uncommon permanent resident throughout its range.

*Habitats.* Throughout the year this species prefers fairly extensive areas of coniferous or deciduous forest, or streamside groves of trees. Although sometimes seen in urban areas, this species more commonly remains in mature forests, especially hardwood forests, where it typically forages on tree trunks and larger branches. It breeds virtually throughout the entire Plains States region.

*National Population.* Breeding Bird Surveys between 1966 and 2012 indicate that this species exhibited a survey-wide population increase (0.88% annually) during that period. The estimated 1990s continental population north of Mexico was about 7.5 million birds (Rich *et al.,* 2004). The species breeds south to Panama.

*Further Reading.* Bent, 1938; Short, 1982; Winkler, Christie and Nurney, 1995; Dunn and Tessaglia-Hymes, 1999; Jackson, Ouellet and Jackson, 2003 (*The Birds of North America:* No. 702).

## Northern Flicker
*Colaptes auratus*

Probably the most common of the Great Plains woodpeckers, this is a species well adapted to foraging on open grasslands, especially for ants. Although it often forages on the ground, it is a competent excavator, and its nesting cavities are regularly used secondarily by many other bird species. Compared with downy woodpeckers, flickers are infrequent visitors to feeders in eastern Nebraska, but they also are likely to have left the state by the time winter weathers begins to be a serious issue. Flickers become semi-gregarious during migration, forming loose flocks. The eastern yellow-shafted race seems to be more migration-prone than the western red-shafted form.

*Winter Distribution.* Root (1988) mapped the distributions of the yellow-shafted and red-shafted races of this species separately, with the yellow-shafted having the highest concentration of birds in eastern Oklahoma, and the red-shafted's highest numbers in Oklahoma and the Texas panhandle. Our data show a progressive increase in average numbers from North Dakota to Oklahoma, and still higher average numbers in the Texas panhandle. Over the entire Plains region there appears to be a gradually increasing population, which is counter to national Breeding Bird Survey trend results.

*Seasonality and Migrations.* The northern flicker is a common permanent resident throughout nearly all the five Great Plains states, with some southward movements out of the Dakotas during winter, and some eastward winter movements by the western (red-shafted) population. This species was reported 25 years during Christmas Bird Counts from 1968 to 2007 in North Dakota, 33 years each in South Dakota and Nebraska, 35 years in Kansas, 34 years in Oklahoma and 33 years in the Texas panhandle. In South Dakota, the species is uncommon during winter (Tallman, Swanson and Palmer, 2002). In Kansas, the red-shafted race is a winter resident only, and many of the wintering birds appear to be racial hybrids (Thompson and Ely, 1989; (Thompson *et al.*, 2011). It is considered a permanent resident in Oklahoma, but more common during winter (Woods and Schnell, 1984), and is a fairly common to common resident in the Texas panhandle, with winter populations much higher than summer numbers, and with red-shafted phenotypes outnumbering yellow-shafted by a five-to-one ratio (Seyffert, 2001). The average number of birds observed per party-hour for the 2013–2014 Christmas Bird Count were: N. Dakota 0.22, S. Dakota 0.3, Nebraska 0.37, Kansas 0.6, Oklahoma 0.61, northwest Texas 0.2. For complete species abundance data, see Appendix 1:67.

*Habitats.* Throughout the year this species occupies diverse habitats, including relatively open woodlands, orchards, woodlots, and urban environments. Dense forests are apparently avoided, and foraging is often done by probing in moist ground.

*National Population.* Breeding Bird Surveys between 1966 and 2012 indicate that this species exhibited a survey-wide population increase (1.36% annually) during that period. The estimated 1990s continental population north of Mexico was about 14,550,000 birds (Rich *et al.*, 2004). The species breeds south to Nicaragua.

*Further Reading.* Bent, 1938; Kilham, 1959; Short, 1982; Winkler, Christie and Nurney, 1995; Moore, 1995 (*The Birds of North America:* No. 166); Dunn and Tessaglia-Hymes, 1999.

### Pileated Woodpecker
*Dryocopus pileatus*

One of the things I miss about not living near eastern deciduous woods is the chance to easily see pileated woodpeckers. Although one or two pairs have nested recently in eastern Nebraska, that is a far cry from being able to walk out into a nearby forest and have a chance of at least hearing, if not seeing, one of these spectacular birds. These woodpeckers mate for life and hold year-around territories. Thus, they are not hard to locate if one happens to be visiting their preferred habitat of mature forests with many dead trees, snags and fallen logs.

*Winter Distribution.* Root's (1988) map shows the western boundary of this woodpecker's range to include the eastern parts of the Dakotas, the southeastern quarter of Kansas, and the eastern half of Oklahoma. Mature deciduous forests are its primary habitat in the Plains, and the highest numbers were found by Root to occur in southeastern Oklahoma. Our data show extremely small population in North Dakota and Kansas, but a moderate and seemingly stable population in Oklahoma.

*Seasonality and Migrations.* The pileated woodpecker is an uncommon to occasional permanent resident throughout its range. The average number of birds observed per party-hour for the 2013–2014 Christmas Bird Count were: N. Dakota 0.06, S. Dakota 0, Nebraska 0.05, Kansas 0.13, Oklahoma 0.2, northwest Texas 0. For complete species abundance data, see Appendix 1:69.

*Habitats.* This species is generally limited to mature forests, often river bottom forests, having a mixture of tall living trees and dead stubs.

*National Population.* This species was reported 39 years during Christmas Bird Counts from 1968 to 2007 in North Dakota, as compared with none in South Dakota, two years in Nebraska. 39 years in Kansas, all 40 years in Oklahoma, and none in the Texas panhandle. Breeding Bird Surveys between 1966 and 2012 indicate that this species exhibited a survey-wide population increase (1.35% annually) during that period. The estimated 1990s continental population north of Mexico was about 930,000 birds (Rich *et al.,* 2004).

*Further Reading.* Bent, 1938; Kilham, 1959; Short, 1982; Winkler, Christie and Nurney, 1995; Bull and Jackson, 1995 (*The Birds of North America:* No. 148); Dunn and Tessaglia-Hymes, 1999.

# FAMILY FALCONIDAE: FALCONS

## American Kestrel
*Falco sparverius*

The smallest and easily most numerous of the Great Plains falcons, kestrels are the only hawks likely to be seen perching on telephone wires, since their toes are small enough to clutch the wires. They often hunt from such perches, swooping down and out to catch large insects or, increasingly in cold weather, small rodents. Their breeding distribution and abundance is strongly influenced by the presence of medium-sized woodpeckers such as the red-headed and red-bellied, whose abandoned nesting cavities provide convenient nest sites for them.

*Winter Distribution.* The map generated by Root (1988) showed the Oklahoma and Texas panhandles as the center of winter distribution for the American kestrel, with numbers declining northward. Our data indicate that maximum numbers in the Great Plains states still occur in northwest Texas, but with Kansas also supporting significant numbers of kestrels. There is an indication of gradual increases in numbers across the five-state Plains region and especially in northwest Texas, where the population appears to have trebled during the study period.

*Seasonality and Migrations.* The American kestrel is a common seasonal migrant and breeder in the northern Great Plains, wintering southwardly. It was reported 38 years during Christmas Bird Counts from 1968 to 2007 in North Dakota, 39 years in South Dakota, and all 40 years in Nebraska, Kansas, Oklahoma and the Texas panhandle. In South Dakota it is uncommon during winter (Tallman, Swanson and Palmer, 2002). Twenty-nine spring Nebraska sightings dating from the mid-1930s to the late 1970s and 22 autumn records are widely scattered, suggesting that this species is largely residential (Johnsgard, 2007a). It is present year-around in Kansas, but is rarer westward during winter (Thompson and Ely, 1989, Thompson *et al.*, 2011). It is considered a permanent resident in Oklahoma (Woods and Schnell, 1984), and is a fairly common to common resident in the Texas panhandle (Seyffert, 2001). For complete species abundance data, see Appendix 1:70.

Habitats, Open country with elevated perching sites such as telephone lines or scattered trees are used throughout the year.

*National Population.* Breeding Bird Surveys between 1966 and 2012 indicate that this species exhibited a survey-wide population decrease (1.6%

annually) during that period. The estimated 1990s continental population
north of Mexico was about 4,350,000 birds (Rich *et al.*, 2004). The species
breeds south to southern South America.

*Further Reading.* Bent, 1937; Cade, 1955, 1982; Palmer, 1988; Johnsgard,
1990; del Hoyo, Elliott and Sargatal, 1994; Smallwood and Bird, 2002 (*The
Birds of North America:* No. 602).

## Merlin
*Falco columbarius*

Only slightly larger than a kestrel, the merlin is much more a bird-hunter, with
small to medium-sized songbirds is favorite prey. Open-country birds such as
horned larks are especially favored. When it enters cities, house sparrows are
often favored prey, but unlike accipiters it rarely if ever visits bird feeders when
hunting. While hunting, it seems quite oblivious to humans. I have seen merlins flying full-bore in parallel with me while driving interstate highways in
large cities, or cross the road directly in front of my car while in close pursuit
of a fleeing bird. In some northern cities, merlins have established local resident populations, taking advantage of the close supply of house sparrows and
other small urban-adapted birds.

*Winter Distribution.* Root (1988) mapped southwestern Nebraska and northwestern Kansas as important wintering areas. The numbers reported in
more recent Christmas Counts have been too low to suggest any areas of
significant density, although Nebraska, Kansas and northwest Texas all
have slightly higher counts than other parts of the Great Plains states. The
national population increase indicated by Breeding Bird Surveys is not apparent from our data.

*Seasonality and Migrations.* The merlin is an uncommon Great Plains seasonal
migrant and local breeder. It was reported 32 years during Christmas Bird
Counts from 1968 to 2007 in North Dakota, 36 years in South Dakota, 34
years in Nebraska, 38 in Kansas, 34 in Oklahoma, and 24 in the Texas panhandle. In South Dakota it is uncommon during winter (Tallman, Swanson
and Palmer, 2002). Forty-eight autumn Nebraska sightings dating from the
mid-1930s to the late 1970s extend from August 16 to December 31, with
a median of October 23. The largest number (21) of autumn records are
for December, followed by September (15) and October (7). Ninety-nine
spring sightings range from January 1 to June 6, with a median of March
19. Half of the records fall within the two periods January 1–20 and March
30–April 24, suggesting that this species is primarily a winter visitor and

spring migrant in Nebraska (Johnsgard, 2007a). It is rare during winter in Kansas (Thompson *et al.*, 2011), with extreme dates of August 14 and June 10 (Thompson and Ely, 1989). It is a rare winter visitor in Oklahoma, with continuous monthly records extending from September to April (Woods and Schnell, 1984), and is a rare to uncommon winter visitor in the Texas panhandle, and has been reported every month, although the summer records are unreliable (Seyffert, 2001).

*Habitats.* Open country with elevated perches such as telephone lines or scattered trees are used throughout the year. Nesting season habitat typically consists of scattered trees or groves near large areas of grasslands, croplands or badlands.

*National Population.* Breeding Bird Surveys between 1966 and 2012 indicate that this species exhibited a survey-wide population increase (3.0% annually) during that period. Niven *et al.* (2004) estimated a national annual increase rate of 3.3 percent, based on Christmas Counts from 1965–67 to 2002–2003. The estimated 1990s continental population was about 650,000 birds (Rich *et al.,* 2004). There is also a Eurasian population that may number in the tens of thousands of pairs (Ferguson-Lees and Christie, 2001; del Hoyo, Elliott and Sargatal, 1994).

*Further Reading.* Bent, 1937; Palmer, 1988; Cade, 1982; Johnsgard, 1990; Sodhi *et al.,* 1993 (*The Birds of North America:* No. 44); del Hoyo, Elliott and Sargatal, 1994.

### Peregrine Falcon
*Falco peregrinus*

The peregrine is the classic raptor species, just common enough that any birder in the Great Plains might hope to see one or two in the course of a year, but rare enough so that every sighting is likely to be remembered with pleasure for the rest of one's life. Following the disastrous population decline brought on by pesticide poisoning following the second World War, peregrines began a slow comeback in the 1970s, which continues to the present time. As a result of intensive re-introduction efforts, peregrines now nest in several Great Plains cities, usually on very tall buildings. Thus, in Lincoln, Nebraska, peregrines have recently nested successfully at the base of the dome of the State Capitol. There they can look over out the entire city, and select from an almost unlimited supply of rock pigeons or songbirds. Using the lights illuminating the building they can even hunt well into near-darkness, taking such unlikely prey as nighthawks.

*Winter Distribution.* The species is too rare to get a clear sense of its regional distribution; Root (1988) mapped slightly higher numbers in the eastern parts of the region, from North Dakota south to Oklahoma. Numbers seen during the four-decade study period were too small to warrant much interpretation, but no upward trend in numbers was evident over the four-decade period, as is indicated by national Breeding Bird Surveys.

*Seasonality and Migrations.* The peregrine is an uncommon to rare Great Plains seasonal migrant and very local breeder. It was reported nine years during Christmas Bird Counts from 1968 to 2007 in North Dakota, three years in South Dakota, 12 years in Nebraska, 28 years in Kansas, 20 years in Oklahoma, and three years in the Texas panhandle. In South Dakota ,it is rare during winter (Tallman, Swanson and Palmer, 2002). Twenty total autumn Nebraska records dating from the mid-1930s to the late 1970s (Johnsgard, 1980) extend from July 26 to December 26, with a median of September 22. The largest number of autumn records (eight) is for September. A total of 97 initial spring Nebraska sightings range from January 1 to May 17, with a median of March 20. Half of the records fall within the two periods January 1–20 and April 21–May 11, suggesting that this species is a winter visitor and spring migrant. It is occasional during winter in Kansas (Thompson *et al.*, 2011), reported September 5 to May 31 (Thompson and Ely, 1989), and rare in Oklahoma, with nearly all records between September and May (Woods and Schnell, 1984). It is a very rare winter visitor in the Texas panhandle, and has been seen from December 19 to February 11 (Seyffert, 2001).

*Habitats.* During migration or wintering this species is most likely found in open, grassland habitats, but it sometimes enters cities while hunting rock pigeons or other urban birds.

*National Population.* Breeding Bird Surveys between 1966 and 2012 indicate that this once-endangered species exhibited a survey-wide population increase (2.4% annually) during that period. The estimated 1990s continental population north of Mexico was about 340,000 birds (Rich *et al.*, 2004). The species breeds south to southern South America. There are also many widely scattered peregrine populations elsewhere in the world that may number in the tens of thousands of pairs (Ferguson-Lees and Christie, 2001; del Hoyo, Elliott and Sargatal, 1994).

*Further Reading.* Bent, 1937; Ratcliffe, 1980; Cade, 1982; Palmer, 1988; Johnsgard, 1990; del Hoyo, Elliott and Sargatal, 1994; White *et al.*, 2002 (*The Birds of North America:* No. 660).

## Gyrfalcon
*Falco rusticolus*

For most birders the gyrfalcon is a near-mythic creature that they will probably never see in the lifetimes, unless perhaps they are lucky enough to visit arctic tundra. Even there, gyrfalcons are rare; during five extended visits to the
arctic I have yet to observe one. However, I did happen to see one briefly in
North Dakota while a youngster more than a half-century ago. The moment
has stuck in memory as clearly as if it has just happened. Such is the power
and attraction of nature; almost everything else that happened that year has
gradually faded from memory.

*Winter Distribution.* This rare arctic species usually only reaches as far south as
the Dakotas during winter, and is rare even there. All 20 records that Root
(1988) found in her early survey were from sites north of the 40th parallel,
and all the records that have occurred on more recent Christmas Counts
have been limited to the Dakotas. In central South Dakota the birds are
said to prey mostly on waterfowl prior to winter freeze-up, and on sharp-
tailed grouse thereafter (Tallman, Swanson and Palmer, 2002).

*Seasonality and Migrations.* The gyrfalcon is a very rare winter visitor in the
northern Great Plains. In Nebraska, only one gyrfalcon was received by
Raptor Recovery Nebraska in its first 25 years, as compared with 12 peregrines, 37 merlins, 52 prairie falcons and over 800 American kestrels (Johnsgard, 2002c). This species was reported 11 years during Christmas Bird
Counts from 1968 to 2007 in North Dakota, six years in South Dakota
(with maximum numbers of 0.02 bird/party-hour during two years), and
none in Nebraska, Kansas, Oklahoma, or the Texas panhandle. In South
Dakota, autumn arrival of this rarely seen species occurs from October to
mid-November, and spring departure ranges from February to early April.
The earliest reported autumn date is September 8, and the latest spring
date is April 23. Eight total Nebraska sightings dating from the mid-1930s
to the late 1970s range from November to March. Kansas records extend
from September 7 though April 23 (Thompson *et al.*, 2011), There are only
a few Oklahoma records, and no verified records for the Texas panhandle.

*Habitats.* Open plains and prairies are used during migration and while wintering, especially where grouse and similar-sized birds are available as prey.

*National Population.* Breeding Bird Survey trend data are not available for this
arctic-breeding species. The estimated 1990s continental population was
about 55,000 birds (Rich *et al.*, 2004). There is also a Eurasian population
that probably numbers in the thousands of pairs (Ferguson-Lees and Christie, 2001; del Hoyo, Elliott and Sargatal, 1994).

*Further Reading.* Bent, 1937; Godfrey, 1986; Cade, 1982; Palmer, 1988; Johns-gard, 1990; Clum and Cade, 1994 (*The Birds of North America:* No. 114); del Hoyo, Elliott and Sargatal, 1994.

### Prairie Falcon
*Falco mexicanus*

Great Plains birders are lucky that five of the six American falcons are spe-cies that represent possible regional sightings. For people living or birding in western parts of the region, seeing a prairie falcon is always a possibility, but because of its speedy flight the opportunity can easily be missed. The falcon's brown upperparts plumage is the color of dead vegetation, adding to the diffi-culties of noticing the bird. Once, when looking directly down from the top of a steep cliff in western Nebraska (Scott's Bluff) to the dead-grass landscape be-low, I saw the moving shadow of a flying falcon. I was able to locate the prai-rie falcon only by judging where it should be, based on the angle of the sun in the sky, since its dorsal plumage almost perfectly matched the color of the landscape it was flying over.

*Winter Distribution.* Root (198) reported a center of Great Plains winter dis-tribution along the South Dakota, Nebraska border, in the general vicin-ity of Fort Niobrara and Valentine national wildlife refuges, with a sec-ondary peak in Kansas along the Arkansas River. Our data suggest that the Texas panhandle now supports the largest concentration of wintering prai-rie falcons in the southern Great Plains, with Kansas a secondary area of concentration. Small mammals and songbirds such as meadowlarks and smaller shorebirds are commonly chosen for winter prey, and the playas and open grasslands of the southern Great Plains provide ideal hunting habitat. Our data, although limited, supports the likelihood of a recent pop-ulation increase, as is also suggested by national Breeding Bird Survey data.

*Seasonality and Migrations.* The prairie falcon is a uncommon to rare local per-manent resident in western areas, elsewhere it is a occasional to rare mi-grant and winter visitor. This species was reported 30 years during Christ-mas Bird Counts from 1968 to 2007 in North Dakota, 37 years in South Dakota, 32 years in Nebraska, all 40 years in Kansas and Oklahoma, and 36 years in the Texas panhandle. In South Dakota, some of the state's win-tering birds may originate in the Rocky Mountain region (Tallman, Swan-son and Palmer, 2002). Forty-five autumn Nebraska records extend from July 21 to December 31, with a median of November 13 and no obvious autumn peak in records. A total of 135 initial spring Nebraska sightings range from January 1 to May 22, with a median of January 30, suggesting that this species is primarily a resident and winter visitor, with no obvious peak of spring migration (Johnsgard, 1980). It is uncommon during winter in western Kansas (Thompson *et al.*, 2011), from early September to late March (Thompson and Ely, 1989), and a permanent resident in Oklahoma (Woods and Schnell, 1984), and is a fairly common winter visitor in the Texas panhandle, and has been seen in every month (Seyffert, 2001).

*Habitats.* This species is associated with large expanses of open grasslands or sagebrush scrub, with nearby cliffs, bluffs or rocky outcrops for roosting and nest sites.

*National Population.* Breeding Bird Surveys between 1966 and 2012 indicate that this species exhibited a survey-wide population increase (1.1% annually) during that period. The estimated 1990s continental population north of Mexico was about 35,000 birds (Rich *et al.*, 2004). The species breeds south to northern Mexico.

*Further Reading.* Bent, 1937; Cade, 1982; Palmer, 1988; Anderson and Squires, 1997; del Hoyo, Elliott and Sargatal, 1994; Steenhof, 1998 (*The Birds of North America:* No. 346); Johnsgard, 1990, 2001b.

# FAMILY TYRANNIDAE:
# TYRANT FLYCATCHERS

## Eastern Phoebe
*Sayornis phoebe*

Phoebes are usually the first of the insectivorous passerines to return to Nebraska and Kansas in the spring, and one of the last to leave in the fall, so their rare appearance on Christmas Counts in these states should not be too surprising. Besides a very wide variety of insects, the birds eat many other invertebrates, and occasionally even catch tiny fish.

*Winter Distribution.* The winter distribution of this insectivorous and Neotropical migrant is largely sound of the Mexican border, but some birds concentrate along the Gulf Coast of Texas, and a few occur north to the southeastern parts of Oklahoma (Root, 1988). Our data indicate that the numbers in recent decades are very few, but perhaps suggest an increase over the last 20 years of the study. There have even been a few recent Kansas and Nebraska sightings dating from the mid-1930s to the late 1970s, supporting the possibility of a gradual winter movement northward or a long-term population increase, as suggested by national Breeding Bird Survey results.

*Seasonality and Migrations.* The eastern phoebe is a common Great Plains seasonal migrant and local breeder, wintering southwardly. This species was reported on no Christmas Bird Counts from 1968 to 2007 in North Dakota or South Dakota, once in Nebraska, seven years in Kansas, 35 years in Oklahoma, and seven years in the Texas panhandle. The latest record for South Dakota is December 1 (Tallman, Swanson and Palmer, 2002). Sixty-seven final autumn Nebraska sightings range from August 4 to October 25, with a median of September 26 (Johnsgard, 1980). It is not known to winter in Kansas, with records to January 4 (Thompson *et al.*, 2011), but is considered a permanent resident in Oklahoma (Woods and Schnell, 1984), and is a breeder and an extremely rare winter visitor in the Texas panhandle (Seyffert, 2001). The average number of birds observed per party-hour for the 2013–2014 Christmas Bird Count were: N. Dakota 0, S. Dakota 0, Nebraska 0, Kansas 0.07, Oklahoma 0.07, northwest Texas 0.02. For complete species abundance data, see Appendix 1:71

*Habitats.* This species is usually found near water in woodlands or partially wooded areas, including farmsteads.

*National Population.* Breeding Bird Surveys between 1966 and 2012 indicate that this species exhibited a survey-wide population increase (0.45% annually) during that period. The estimated 1990s continental population was about 16 million birds (Rich *et al.,* 2004).

*Further Reading.* Bent, 1942; Martin. Zim and Nelson, 1951; Weeks, 1994 (*The Birds of North America:* No. 94).

## Say's Phoebe
### *Sayornis saya*

This slightly larger western relative of the eastern phoebe appears to be about as cold tolerant as the eastern, but it occurs only at the western edge of the Plains States region, and is unlikely to be seen in winter. One pair of Say's phoebes nested for several years in a highway culvert near the University of Nebraska's Cedar Point field station in western Nebraska. While the female was incubating, the male would spend the hot summer days sitting in the shadow of a telephone pole, one of the few places where any shade was to be found in the vicinity.

*Winter Distribution.* Root (1988) mapped the winter range of this phoebe as barely reaching the southwestern corner of the Texas panhandle, and the small numbers reported during our period of study suggests that the situation is still much the same. As the largest of the phoebes, the Say's should be the most cold-tolerant, so insect food availability is likely to set its northern winter limits.

*Seasonality and Migrations.* Say's phoebe is an uncommon Great Plains seasonal migrant and local breeder, wintering southwardly. This species was reported on no Christmas Bird Counts from 1968 to 2007 in North Dakota, South Dakota, or Nebraska, during four years in Kansas, six years in Oklahoma, and 16 years in the Texas panhandle. South Dakota records extend only to late October (Tallman, Swanson and Palmer, 2002). Fifty-two final autumn Nebraska sightings range from July 29 to October 29, with a median of September 14 (Johnsgard, 1980). Records in Kansas extend to December 31 (Thompson *et al.*, 2011), and Oklahoma records to mid-December (Woods and Schnell, 1984), and is a breeder and an extremely rare winter visitor in the Texas panhandle (Seyffert, 2001).

*Habitats.* This species is typically found in fairly open and dry habitats, includes rocky canyons, badlands and ranchlands.

*National Population.* Breeding Bird Surveys between 1966 and 2012 indicate that this species exhibited a survey-wide population increase (0.75% annually) during that period. The estimated 1990s continental population north of Mexico was about 3,370,000 birds (Rich *et al.,* 2004). The species breeds south to central Mexico.

*Further Reading.* Bent, 1942; Schukman and Wolf, 1998 (*The Birds of North America:* No. 374).

# FAMILY LANIIDAE: SHRIKES

## Loggerhead Shrike
*Lanius ludovicianus*

Although the loggerhead shrike is declining seriously over nearly all of its national range, it is still a regular nesting species in the Nebraska Sandhills, where the open grassland habitat and widely scattered clumps of mostly low trees seem to provide optimum conditions. Most of the loggerhead shrikes have left Nebraska by the time the first northern shrikes arrive, usually in November, and the northerns in turn are mostly gone by the time the loggerheads return in the spring.

*Winter Distribution.* Root (1988) was unable find any specific environmental factor coinciding with the northern winter limits of this shrike. but it is then largely replaced by the larger northern shrike, and perhaps these competitive interactions might play a role in influencing the loggerhead's *Winter Distribution.* Root mapped its northern limits as reaching southeastern Nebraska, and its highest abundance in southeastern Oklahoma. Our data show very small numbers occurring north to North Dakota, but no indication of any general winter shift north over the past four decades. Instead, there as been a consistent downward trend in all states over this time period, coinciding with a pronounced national decline in loggerhead shrike populations.

*Seasonality and Migrations.* The loggerhead shrike is an uncommon to rare Great Plains seasonal migrant and local breeder, wintering southwardly. This species was reported six years during Christmas Bird Counts from 1968 to 2007 in North Dakota, 20 years in South Dakota, 31 years in Nebraska and all 40 years in Kansas, Oklahoma and the Texas panhandle. In South Dakota the frequency of wintering is uncertain (Tallman, Swanson and Palmer, 2002). The range of 95 initial spring Nebraska sightings dating from the mid-1930s to the late 1970' (Johnsgard, 1980) is from January 2 to May 28, with a median of April 4. Ninety-eight final autumn sightings are from July 26 to December 30, with a median of September 19, suggesting that wintering in Nebraska is probably rare. It is rare during winter in Kansas, the birds usually departing by November (Thompson and Ely, 1992), but is considered a permanent resident in Oklahoma (Woods and Schnell, 1984), and is a breeder and uncommon winter visitor in the Texas panhandle, and has been reported every month (Seyffert, 2001). The average number of birds observed per party-hour for the 2013–2014 Christmas Bird

Count were: N. Dakota 0, S. Dakota 0, Nebraska 0, Kansas 0.29, Oklahoma 0, northwest Texas 0.4. For complete species abundance data, see Appendix 1:72

*Habitats.* Outside the breeding season these birds occupy the same open country that northern shrikes utilize, and during the nesting period they are also associated with open habitat with scattered or clustered shrubs or small trees.

*National Population.* Breeding Bird Surveys between 1966 and 2012 indicate that this species exhibited a survey-wide population decline (3.18% annually) during that period, making it one of America's many seriously declining grassland species. The estimated 1990s continental population north of Mexico was about 3.7 million birds (Rich *et al.,* 2004). The species breeds south to central Mexico.

*Further Reading.* Bent, 1950; Yosef, 1996 (*The Birds of North America:* No. 230); Lefrane, 1997.

## Northern Shrike
*Lanius excubitor*

Northern shrikes weigh almost twice as much as do loggerheads, and are about the mass and size of a red-headed woodpecker. They are nevertheless fierce predators, sometimes taking prey as large as themselves, such as blue jays, robins, and mourning doves, but in turn may be preyed upon by accipiter hawks. During winter they may be attracted to rural feeding stations if suct or hamburger is provided.

*Winter Distribution.* Root's (1988) map of the northern shrike's distribution in the 1960s indicated a peak in numbers in South Dakota, and some birds wintering south to northern Oklahoma, with periodic irruptions in some years. She also noted that the northern shrike and American kestrel exhibit similar winter distribution patterns. Our data show North Dakota as having the highest numbers throughout the entire four-decade period, with far fewer in South Dakota and Nebraska, and very few reaching Kansas. Apparent population irruptions occurred in the mid-1980s and mid-1990s, but no clear long-term population trend is apparent.

*Seasonality and Migrations.* The northern shrike is an uncommon to rare winter visitor throughout the region. This species was reported all 40 years during Christmas Bird Counts from 1968 to 2007 in North Dakota, South Dakota, and Nebraska, 31 years in Kansas, 20 years in Oklahoma and 17 years in the Texas panhandle. In South Dakota wintering is regular (Tallman,

Swanson and Palmer, 2002). The range of 44 initial autumn Nebraska sightings dating from the mid-1930s to the late 1970s (Johnsgard, 1980) is from August 28 to December 26, with a median of November 9. Twenty-four final spring Nebraska sightings range from January 7 to April 24, with a median of March 11. It is rare and irregular during winter in Kansas, with extreme reports from October to April (Thompson *et al.*, 2011). and a rare winter visitor in Oklahoma, with continuous monthly records extending from October to February (Woods and Schnell, 1984). It is a rare winter visitor in the Texas panhandle, and has been reported from October to March (Seyffert, 2001). The average number of birds observed per party-hour for the 2013–2014 Christmas Bird Count were: N. Dakota 0.1, S. Dakota 0.18, Nebraska 0.1, Kansas 0.81, Oklahoma 0.07, northwest Texas 0. For complete species abundance data, see Appendix 1:73.

*Habitats.* Migrants and wintering birds are found on open plains or prairies having scattered trees, shrubs or posts for perches.

*National Population.* Breeding Bird Survey trend data are not available. Niven *et al.* (2004) estimated a national annual decline rate of 1.3 percent, based on Christmas Counts from 1965–67 to 2002–2003. The estimated 1990s continental population was about 210,000 birds (Rich *et al.,* 2004). There is also a Eurasian ("great grey shrike") population.

*Further Reading.* Bent, 1950; Godfrey, 1986; Lefrane, 1997; Caddie and Atkinson, 2002 (*The Birds of North America:* No. 671).

# FAMILY CORVIDAE:
## JAYS, MAGPIES AND CROWS

### Blue Jay
*Cyanocitta cristata*

One of the most familiar and most conspicuous of winter birds, blue jays add a touch of color and interest to the coldest and darkest of winter days. They usually dominate bird feeders, giving ground only at the approach of woodpeckers or crows. Often arriving in small parties of four or five birds, probably family groups, one bird at a time will visit the feeder, probably on the basis of well-established dominance relationships. Within a few minutes it will stuff its throat with so many food items (more than 100 sunflower seeds have been reported) that the bird's swollen throat would make it appear to have a severe case of goiter. They are constantly on the alert for danger; their warning notes will put most birds to flight, and put squirrels instantly on the alert. In spite of their bullying nature, these other species probably benefit from the presence of jays at a bird feeder because of their extreme watchfulness for predators.

*Winter Distribution.* The blue jay is one of the best-known winter birds of the Great Plains, since it is large, loud, colorful and often dominates other species at bird feeders. Root (1988) reported its highest numbers in the Plains as occurring in southeastern Kansas and eastern Oklahoma, grading off westwardly to almost none in the Texas panhandle, westernmost Oklahoma, western Kansas and western North Dakota. Our data indicate high populations in Oklahoma, tapering off gradually to the north, much as found by Root. Our data also suggests a gradual population increase through the 1980s and 1990s, but with a decline starting in the early 2000s, coinciding with the onset of West Nile virus. This disease, first reported in New York City in August, 1999, spread to the Rocky Mountains within a month. Jays, crows and other corvids seem to be especially susceptible to the virus, as well as some owls, especially great horned owls (Pelikan, 2002).

*Seasonality and Migrations.* The blue jay is a common to abundant permanent resident throughout its range. The average number of birds observed per party-hour for the 2013–2014 Christmas Bird Count were: N. Dakota 1.07, S. Dakota 1.67, Nebraska 1.42, Kansas 0, Oklahoma 0, northwest Texas 0. For complete species abundance data, see Appendix 1:74.

*Habitats.* This species was reported on all 40 Christmas Bird Counts from 1968 to 2007 from North Dakota to Oklahoma, and 35 years in the Texas panhandle. Throughout the year this species is widely distributed in forests, parks suburbs, cities, and almost anywhere a combination of trees and grasslands occurs. It is seemingly somewhat better adapted to city life than is the Steller's jay (*Cyanocitta stelleri*).

*National Population.* Breeding Bird Surveys between 1966 and 2012 indicate that this species exhibited a survey-wide population decrease (0.66% annually) during that period The estimated 1990s continental population was about 22 million birds (Rich *et al.,* 2004).

*Further Reading.* Bent, 1946; Bock and Lepthien, 1976; Goodwin, 1976; Dunn and Tessaglia-Hymes, 1999; Tarvin and Woolfenden, 1999 (*The Birds of North America:* No. 469); Madge and Burn, 2001.

## Gray Jay
### *Perisoreus canadensis*

Rare in the Great Plains, gray jays share some of the same behavioral attributes as blue jays, and if anything are tamer and more adapted at surviving cold weather. Although smaller than blue jays, they have more northerly ranges, and their plumage is unusually dense, and highly effective as insulation. When roosting the feathers are fully spread, nearly hiding the bird's normal outline, and body temperatures are also reduced a few degrees to help conserve energy loss. Caching of food items occurs both during summer and winter, and like several other jays the birds have a remarkable capacity for locating items that might have been cached months previously.

*Winter Distribution.* The only residential population of gray jays in the Great Plains is in western South Dakota. There is some winter scattering of records outside the breeding areas, and perhaps the North Dakota winter records are the result of birds moving in temporarily from Manitoba or northern Minnesota. The numbers seen in both states are too small to speculate on possible population trends.

Seasonally and Migrations. The gray jay is a common to occasional permanent resident in the Black Hills of South Dakota. This species has appeared 26 out of 40 years of Christmas Counts in South Dakota between 1968 and 2007. They occurred in all four decades, but only twice reached yearly mean abundance levels above 0.01 bird per party-hour, It also appeared during 12 North Dakota counts, but not on any Nebraska counts during the four-decade period under study.

*Habitats.* This species is associated with cool coniferous forests throughout its range.

*National Population.* Breeding Bird Surveys between 1966 and 2012 indicate that this species exhibited a survey-wide population decrease (0.59% annually) during that period. The estimated 1990s continental population was about 16 million birds (Rich *et al.,* 2004).

*Further Reading.* Bent, 1946; Goodwin, 1976; Strickland and Ouellet, 1993 (*The Birds of North America:* No. 40); Madge and Burn, 2001.

## Western Scrub-Jay
### *Aphelocoma californicus*

Western scrub-jays differ some from blue jays in their winter behavior. For example, young birds leave their parents to form winter flocks that may wander some distance from the places where they were raised. Resident adults like other jays, habitually cache foods and have remarkable abilities as remembering the locations of hundreds of such caching sites. Acorns are also collected and cached by wild birds. Sunflower seeds and various nuts such as peanuts are favorite foods at feeding stations.

*Winter Distribution.* Root (1988) showed this species (then including the Florida population, which is now considered a separate species) distribution as limited to southwestern Kansas, the Oklahoma panhandle, and the northern panhandle of Texas, terminating eastwardly where the vegetation changes from dense brush land to grassland. Our data indicate a very small Oklahoma population that might be increasing, and a northwestern Texas population that is probably increasing. This increasing trend is in accordance with national trends indicated by Breeding Bird Survey data. For complete species abundance data, see Appendix 1:75.

*Seasonality and Migrations.* The western scrub-jay is an uncommon to occasional permanent resident throughout its range.

*Habitats.* In Colorado scrub-jays primarily occupy Gambel's oak (*Quercus gambelli*) woodlands, especially those with interspersed openings. They also occur in mountain-mahogany (*Cercocarpus*) shrublands and pinyon-juniper woodlands (Andrews and Righter, 1992), and is also often found brushy ravines or wooded creek bottoms.

*National Population.* This species was not reported during Christmas Bird Counts from 1968 to 2007 in North Dakota, South Dakota, or Nebraska, but was seen one year in Kansas, 11 years in Oklahoma, and eight years in the Texas panhandle. Breeding Bird Surveys between 1966 and 2012 indicate that this species exhibited a survey-wide population decrease (0.36% annually) during that period The estimated 1990s continental population north of Mexico was about 2,720,000 birds (Rich *et al.,* 2004). The species breeds south to central Mexico.

*Further Reading.* Bent, 1946; Goodwin, 1976; Godfrey, 1986; Dunn and Tessaglia-Hymes, 1999; Madge and Burn, 2001; Curry, Peterson and Langen, 2003 (*The Birds of North America:* No. 712).


### Pinyon Jay
*Gymnorhinus cyanocephalus*

More than any other Americans jay, this species is closely associated with a single food source, the seeds of pinyon pine (*Pinus edulis).* Up to 90 percent of the species' winter foods are comprised of this single source, which is gathered assiduously and cached in areas that may be more than half a mile away from where it was gathered. Like other jays, its throat is highly expandable, and up to nearly 50 pine seeds can be ingested before the bird takes flight to hide its stash. Availability of the annual pine crop determines the timing of breeding, and whether there will be more than one breeding cycle in a single year. Post-breeding flocks of wandering birds can be very large, and their movements might cover substantial distances, especially when food supplies are low.

*Winter Distribution.* Root (1988) indicated this jay's winter distribution as barely touching North Dakota's southwestern border, but including the western half of South Dakota, northwestern Nebraska, the western third of Kansas, the Oklahoma panhandle, and the northern panhandle of Texas. Breeding is limited to the Black Hills of South Dakota, the Pine Ridge region of northwestern Nebraska, and the Black Mesa region of the Oklahoma panhandle. Because of a fluctuating food supply, the birds are somewhat nomadic outside the breeding season, accounting for winter records in North Dakota, Kansas and Texas. Our data indicate a declining population in both South Dakota and Nebraska, which is in accordance with national trends as indicated by Breeding Bird Surveys. For complete species abundance data, see Appendix 1:76

*Seasonality and Migrations.* The pinyon jay is probably a permanent resident in most regions where it occurs regularly. Vagrants often appear in other regions, especially during late winter or spring.

*Habitats.* In the Black Hills and Nebraska's Pine Ridge region this species is found in ponderosa pine(*Pinus ponderosa*) forests where the soil is fairly dry and the trees are small and scattered. Farther west it is mainly associated with the pinyon pine-juniper woodland, but during the non-breeding season it extends into mountain-mahogany (*Cercocarpus*), sagebrush (*Artemisia* spp.) and desert scrub habitats. The seeds of pinyon pine provide the primary winter food for this jay in the southern Great Plains, although juniper berries are also eaten.

*National Population.* This species was reported one year during Christmas Bird Counts from 1968 to 2007 in North Dakota, as compared with 35 years in South Dakota, 15 years in Nebraska, one year in Kansas, 17 years in Oklahoma, and one year in the Texas panhandle. Breeding Bird Surveys between 1966 and 2012 indicate that this species exhibited a survey-wide population decrease (4.24% annually) during that period. The estimated 1990s continental population was about 4.1 million birds (Rich *et al.,* 2004). This species was yellow-listed in the National Audubon Society's 2007 Watch-List of rare and declining birds (Butcher *et* al., 2007)

*Further Reading.* Bent, 1946; Goodwin, 1976; Ligon, 1978; Dunn and Tessaglia-Hymes, 1999; Madge and Burn, 2001; Balda, 2002 (*The Birds of North America:* No. 605).


## Clark's Nutcracker
### *Nucifraga columbiana*

This species is closely associated with conifer forests, as their seeds are a primary source of its foods. Like many other corvids, food-caching is a typical trait, and the birds are famous for their abilities to remember cache locations.

*Winter Distribution.* Limited to isolated populations in South Dakota's Black Hills, where it is rare. and the Pine Ridge region of Nebraska (Sioux County), where it is very rare.

*Seasonality and Migrations.* The Clark's nutcracker is an uncommon to occasional permanent resident throughout its range. Some fall and winter wandering occurs outside the breeding range.

*Habitats.* Associated with coniferous forests, including spruce, fir and pine forests. Outside of the breeding season it may also occur in aspens, juniper-pinyon woodlands and mountain-mahogany (*Cercocarpus*) scrub. The birds generally move to lower altitudes during fall and winter.

*National Population.* This species was reported five years during Christmas Bird Counts from 1968 to 2007 in South Dakota, but never achieved abundance levels above the "present" category. It was not reported from other Great Plains states. Breeding Bird Surveys between 1966 and 2012 indicate that this species exhibited a survey-wide population decrease (0.22% annually) during that period.The estimated 1990s continental population was about one million birds (Rich *et al.,* 2004).

*Further Reading.* Bent, 1946; Goodwin, 1976; Tomback, 1998 (*The Birds of North America:* No. 331); Madge and Burn, 2001.

## Black-billed Magpie
*Pica hudsonia*

Magpies are such spectacular birds that it is easy to forgive their tendencies to steal the eggs and consume the young of other species. They quickly discover any new source of food, and don't hesitate to chase other birds from a feeder if it is well stocked. A pair will defend its breeding territory year-around, but winter foraging flocks may also form and wander some distances. In very cold weather these flocks may persist through the night, where like crows they may gather to share common roosting sites and thereby probably gain some vigilant protection against nocturnal visits by owls, especially great horned owls.

*Winter Distribution.* Magpies are somewhat nomadic in winter, and Root's distribution map shows the species extending east across North Dakota and into Minnesota, all of South Dakota and Nebraska except for their easternmost regions, most of Kansas, and the panhandles of Oklahoma and Texas. Our data show a fairly uniform winter distribution across the Plains states except for the Texas panhandle, with probable peak numbers in North Dakota and Nebraska. Although hidden by the decade-averaging of the data between 1997–98 and 2006–07, all of the Plains states had a population crash beginning 2002, a few years after the arrival of West Nile virus (Pelikan, 2002). The total numbers of magpies seen in the five-state Plains region dropped 70 percent, from 1,048 in 2000–01 (0.51 per party-hour) to 311 in 2006–07 (0.15 per party-hour). Oklahoma counts in 2006–07 recorded no magpies for the first time in 28 years, and Kansas counts then were the lowest since the start of the four-decade study period in the 1960s.

*Seasonality and Migrations.* The black-billed magpie is an uncommon to occasional permanent resident, with vagrants often wandering east of the breeding range in winter. This species was observed every year on Christmas Counts from 1968 to 2007 in all states from North Dakota to Kansas, 35 years in Oklahoma and none in the Texas panhandle. The average number of birds observed per party-hour for the 2013–2014 Christmas Bird Count were: N. Dakota 0.05, S. Dakota 0.45, Nebraska 0.26, Kansas 0, Oklahoma 1.0, northwest Texas 0. For complete species abundance data, see Appendix 1:77.

*Habitats.* Throughout the year this species normally frequents wooded canyons and river-bottom forests and forest edges, but ranges out into more arid environments wherever there are thickets of shrubs or small trees that provide nest sites.

*National Population.* Breeding Bird Surveys between 1966 and 2012 indicate that this species exhibited a survey-wide population decrease (0.46% annually) during that period The estimated 1990s continental population was about 3.4 million birds (Rich *et al.,* 2004).

*Further Reading.* Bent, 1946; Goodwin, 1976; Birkhead, 1991; Dunn and Tessaglia-Hymes, 1999; Trost, 1999 (*The Birds of North America:* No. 389); Madge and Burn, 2001.

## American Crow
### *Corvus brachyrhynchos*

Crows lack the visual attraction of magpies or blue jays, and perhaps in part for that reason are generally disliked, if not reviled. They are also even more intelligent than either of these species, which makes them more frustrating for humans to try control. For biologists, that feature simply adds to their attraction. During the winter crows in some areas gather to form large winter roosts, which in southern Nebraska and Kansas can occasionally grow to enormous sizes. These groups perhaps mostly represent migrant birds, as resident birds remain on their breeding territories through the winter. After the appearance of West Nile virus such enormous roosts have disappeared, and it may be many years before winter crow populations regain the numbers they once had.

*Winter Distribution.* Root's (1988) distribution map from the 1960s shows a high peak of numbers in central Kansas in the Arkansas River valley, and a secondary peak in southeastern Oklahoma, with the northern limits close to the South Dakota-North Dakota border. At that time enormous winter roosts in Kansas and southern Nebraska were often subjected to dynamiting to try control their numbers. Our data indicate that crow populations in the five-state Plains region continued to increase until the late 1990s, especially in Kansas. However, the arrival of the West Nile virus in 1999 was associated with a prolonged drop in Kansas' crow numbers, from 47.54 birds per party hour in 1999–2000 to 3.51 birds per party-hour in 2005–06. Recent party-hour counts for South Dakota, Nebraska and Kansas suggest that the population may have begun a slight recovery. The average number of birds observed per party-hour for the 2013–2014 Christmas Bird Count were: N. Dakota 1.95, S. Dakota 1.85, Nebraska 2.63, Kansas 6.13, Oklahoma 0.49, northwest Texas 0.12. For complete species abundance data, see Appendix 1:78.

*Seasonality and Migrations.* The American crow is a common to abundant permanent resident in the central and southern Great Plains, and somewhat migratory in the northern plains. This species was observed every year on Christmas Counts from 1968 to 2007 in all states from North Dakota to Oklahoma and the Texas panhandle. Six initial spring Nebraska sightings dating from the mid-1930s to the late 1970s (Johnsgard, 1980) range from January 1 to June 6, with no obvious clustering of dates, suggesting permanent residency.

*Habitats.* Throughout the year this species occurs in a wide variety of forests, wooded river bottoms, suburban areas, orchards, parks and woodlots.

*National Population.* Breeding Bird Surveys between 1966 and 2012 indicate that this species exhibited a survey-wide population decrease (0.25% annually) during that period. The estimated 1990s continental population was about 31 million birds (Rich *et al.,* 2004).

*Further Reading.* Bent, 1946; Goodwin, 1976; Dunn and Tessaglia-Hymes, 1999; Madge and Burn, 2001; Verbeek and Caffrey, 2002 (*The Birds of North America:* No. 647).

### Fish Crow
*Corvus ossifragus*

The fish crow is smaller and more gregarious than the American crow; it often nests in small colonies, and its winter roosts may number in the thousands of individuals. It gets much of its food from the surface of the water, often by snatching it with its beak, or even by catching live minnows with its feet. It will also steal the eggs of birds or turtles, or pirate food from gulls and terns by harassing them until they drop their food.

*Winter Distribution.* Root's (1988) map does not show the fish crow reaching the Great Plains states region in winter, but indicated its northern limits as occurring near Houston, Texas. Sutton (1967) summarized early Oklahoma records for the fish crow, including it's breeding along the Arkansas and Red Rivers, but noted no winter records. Reinking (2004) described the species' later movement into eastern Oklahoma, and noted that it now winters in southeastern Oklahoma. The first Kansas record was obtained in 1984 (Thompson and Ely, 1992), and a range expansion also occurred into southwestern Missouri during the late 1900s, especially along trout streams (Jacobs, 2001). This breeding expansion into the southern Great Plains no doubt accounts for its now increasingly regular occurrence in Oklahoma Christmas Counts.

*Seasonality and Migrations.* The fish crow is generally a permanent resident, but possibly is a limited migrant at the northern edge of its small Great Plains range.

*Habitats.* Over much of its range this species is associated with tidal marshes and river systems.

*National Population.* This species appeared 15 out of 40 years of Oklahoma Christmas Counts between 1968 and 2007, with a maximum frequency (2004 and 2006) of 0.03 birds per party hour. It was not reported on Kansas counts, where it is a very local breeder. Breeding Bird Surveys between 1966 and 2012 indicate that this species exhibited a survey-wide population increase (0.44% annually) during that period. The estimated 1990s continental population was about 790,000 birds (Rich *et al.*, 2004).

*Further Reading.* Bent, 1946; Goodwin, 1976; McGowan, 2001 (*The Birds of North America:* No. 589); Madge and Burn, 2001.

## Chihuahuan Raven
*Corvus cryptoleucus*

This small desert-adapted raven is only slightly larger than the American crow, and about half the weight of the common raven (*Corvus corax*). Like other ravens it is largely a scavenger, and is opportunistic as to what it eats. Normally solitary nesters, the birds may form large flocks when on migration or while roosting.

*Winter Distribution.* Root (1988 provided no map for this uncommon species, which breeds locally north to southwestern Colorado (historically to southern Nebraska), and winters in New Mexico, Texas and western Oklahoma. Our data indicate a few birds may linger in southern Kansas into the Christmas season as well. The small size of these ravens may make them unable to compete effectively with common ravens (*Corvus corax*) during winter, which may be why they tend to winter to the east and south of common ravens.

*Seasonality and Migrations.* The Chihuahuan raven is an uncommon to occasional seasonal migrant and breeder in the southwestern Great Plains, wintering southwardly. This species was observed 14 years on Christmas Counts from 1968 to 2007 in Kansas, 23 years in Oklahoma, and 17 years in the Texas panhandle. It is rare during winter in Kansas (Thompson *et al.*, 2011), and a summer resident in Oklahoma, with continuous monthly records into December (Woods and Schnell, 1984), and is an uncommon to common resident in the Texas panhandle (Seyffert, 2001).

*Habitats.* The usual breeding habitat consists of open and arid grasslands, using scattered trees, telephone poles, or windmills for nest sites. It is generally not associated with river valleys or other wooded areas, and is probably unable to compete with the larger common raven where they are in contact.

*National Population.* Breeding Bird Surveys between 1966 and 2012 indicate that this species exhibited a survey-wide population decrease (0.19% annually) during that period. The estimated 1990s continental population north of Mexico was about 370,000 birds (Rich *et al.,* 2004). The species breeds south to northern Mexico.

*Further Reading.* Bent, 1946; Goodwin, 1976; Madge and Burn, 2001; Bednarz and Raitt, 2002 (*The Birds of North America:* No. 606).

# FAMILY ALAUDIDAE: LARKS

## Horned Lark
### *Eremophila alpestris*

Judging from breeding surveys, the horned lark is probably the most common breeding bird in North Dakota. Besides the races *E. a. pratincola* and *E. a. erthymia,* the respective breeding populations of the eastern and western Great Plains, the wintering population over most of the region is *alpestris,* a tundra-breeding subspecies that is more richly colored than the paler plains breeders. The breeding populations of the northern plains also tend to move variably southward in winter, but those of the central and southern plains are increasingly residential.

*Winter Distribution.* This is one of the most widespread of North American birds, and it is especially prevalent on the Great Plains grasslands, at all seasons. Root (1988) mapped it as occurring throughout the region except for northeastern North Dakota, with the greatest numbers occurring in western Kansas, and a smaller peak in the western panhandle of Texas. Our data also show a maximum density of horned larks in Kansas, but with Nebraska having the second-highest long-term numbers. However, the long-term trends for Kansas and North Dakota are upward, while populations of South Dakota, Oklahoma and the Texas panhandle appear to be stable, and Nebraska's numbers show a gradual decline. There is no obvious explanation for these trends, except that some regional shifting of wintering birds may have occurred. Adding to the complexity of interpretation is that Breeding Bird Surveys indicate a significant long-term national decline in horned lark populations. Examining the data on a yearly basis, Great Plains numbers seem to be highly irruptive, with Nebraska numbers varying by as much as 70 birds per party hour in a single year, and Kansas as much as 143 birds per party-hour. These massive shifts in year-to-year distributions probably account for the seemingly opposing trend figures among different states.

*Seasonality and Migrations.* The horned lark is a common to abundant permanent resident throughout the region, with some Great Plains populations variably migratory, and winter populations supplemented by migrants from farther north. This species was reported every year during Christmas Bird Counts from 1968 to 2007 from North Dakota through Oklahoma and the Texas panhandle. The average number of birds observed per party-hour for the 2013–2014 Christmas Bird Count were: N. Dakota 2.7, S. Dakota 6.16, Nebraska 3.4, Kansas 10.94, Oklahoma 0.6, northwest Texas 3.17. For complete species abundance data, see Appendix 1:79.

*Habitats.* A variety of low-stature open habitats are used by this species throughout the year, but in Nebraska it is mostly found in natural grasslands and cultivated fields. The sparse grasslands of the Sandhills are probably a nearly optimum habitat.

*National Population.* Breeding Bird Surveys between 1966 and 2012 indicate that this species exhibited a survey-wide population decrease (2.22% annually) during that period. The estimated 1990s continental population north of Mexico was about 98 million birds (Rich *et al.,* 2004). The species breeds south to Colombia, and there is also a Eurasian ("shore lark") population.

*Further Reading.* Bent, 1942; Beason, 1995 (*The Birds of North America:* No. 195); Johnsgard, 2001b.

FAMILY PARIDAE: TITMICE

## Carolina Chickadee
*Poecile carolinensis*

The Carolina chickadee's range is directly south of the black-capped, with southeastern Kansas representing the narrow transition zone. The Carolina is very slightly larger than the black-capped, and during times in autumn and winter when the Carolina population expands, the ranges of the black-capped may be pushed farther south temporarily. Both species have *chick-a-dee* calls, but the Carolina's is higher-pitched and faster, and their meanings are apparently slightly different. Carolina chickadees mate for life, but the species' fairly high mortality rate dictates that many of the pairs will be disrupted by the time two breeding seasons have passed.

*Winter Distribution.* This species occurs immediately south of the black-capped chickadee, and may hybridize with it in the zone of contact. Root (1988) mapped it as extending north to central Kansas, and with maximum numbers in eastern Oklahoma. That pattern fits with our more generalized data, which further suggests an essentially stable or slightly increasing long-term population, in contrast to the possible national population decline suggested by Breeding Bird Survey data. There may have been a recent slight decline in numbers associated with the advent in 1999 of West Nile virus, but this trend is not clear-cut.

*Seasonality and Migrations.* The Carolina chickadee is a common permanent resident throughout its range. This species has not been seen on Christmas Counts in the Dakotas or Nebraska, but was reported every year during counts from 1968 to 2007 in Kansas and Oklahoma, and during 33 years of counts in the Texas panhandle. The average number of birds observed per party-hour for the 2013–2014 Christmas Bird Count were: N. Dakota 0, S. Dakota 0, Nebraska 0, Kansas 1.74, Oklahoma 2.09, northwest Texas 0. For complete species abundance data, see Appendix 1:80.

*Habitats.* Like the adaptable black-capped chickadee, the Carolina chickadee is typically found in mature forests, streamside woods, and well-wooded parks where nest cavities are available.

*National Population.* Breeding Bird Surveys between 1966 and 2012 indicate that this species exhibited a survey-wide population decrease (0.17% annually) during that period. The estimated 1990s continental population was about 18 million birds (Rich *et al.,* 2004).

*Further Reading.* Bent, 1946; Harraq and Quin, 1995; Dunn and Tessaglia-Hymes, 1999; Mostrom, Curry, and Lohr, 2002 (*The Birds of North America:* No. 696).

### Black-capped Chickadee
*Poecile atricapillus*

Perhaps the most familiar and most beloved of winter birds for many people in the northern Great Plains, black-capped chickadees have all the attributes of a perfect yard bird. With a disproportionately large head and small beak that produces an innate visual perception of juvenile "cuteness," a chickadee is likely to be one of the most appealing subjects for Christmas cards and similar winter-theme subjects. Unlike jays and woodpeckers, which rapidly stuff their throats with food while at a feeder, chickadees and nuthatches daintily pick up only one or two seeds at a time before flying off to hide them somewhere nearby, and return a minute or two later to repeat the process. Up to about 75 visits may be made in a single day by individual chickadees. During winter, chickadees may consume as much as twenty times more food than during summer, and although daytime body temperatures may be kept at about 108° F, their nighttime roosting temperature under freezing conditions may be reduced as much as 20 degrees.

*Winter Distribution.* Root (1988) indicated the winter range of this familiar species as extending south to northern Oklahoma and the northern Texas panhandle, with peak numbers in northern Kansas and adjacent southern Nebraska. Our data suggest a more northerly distribution, with North Dakota having the greatest numbers in recent years, and virtually none extending south to Oklahoma or the Texas panhandle. While North Dakota shows a gradual increase in birds at least through the 1990s, long-term counts for the five-state Plains region suggest that peak numbered occurred in the late 1980s, followed by a decline since, especially in the early 2000s, when the effects of West Nile virus began to appear. However, the Plains population of chickadees may be starting to recover slightly from this disastrous influence, since numbers seem to have stabilized or even slightly increased by about 2005.

*Seasonality and Migrations.* The black-capped chickadee is a common permanent resident throughout its range. This species was seen every year on Christmas Counts from 1968 to 2007 in the Dakotas, Nebraska and Kansas, but was reported only three years in Oklahoma, and one year in the Texas panhandle. The average number of birds observed per party-hour for the 2013–2014 Christmas Bird Count were: N. Dakota 1.77, S. Dakota 1.12, Nebraska 1.07, Kansas 1.19, Oklahoma 0, northwest Texas 0. For complete species abundance data, see Appendix 1:81

*Habitats.* Throughout the year this species is found in deciduous and coniferous forests, as well as orchards and woodlots. Nesting often occurs in edge situations or forest openings, but during the winter period the birds frequently appear at residential feeding stations, especially where suet is provided.

*National Population.* Breeding Bird Surveys between 1966 and 2012 indicate that this species exhibited a survey-wide population decrease (0.79% annually) during that period. The estimated 1990s continental population was about 34 million birds (Rich *et al.,* 2004).

*Further Reading.* Bent, 1946; Odum, 1941; Chaplin, 1974; Desrochers, Hannon and. Nordin, 1988; Harrison and Harrison, 1990; Hitchcock and Sherry, 1990; Smith, 1991, 1993 (*The Birds of North America:* No. 39); Harraq and Quin, 1995; Dunn and Tessaglia-Hymes, 1999.

## Mountain Chickadee
*Poecile gambeli*

Mountain chickadees are strictly peripheral in the Great Plains states, and are most likely to occur in winter, when the birds move out of coniferous forests to lower altitudes and spread out over the plains. At that time they are still prone to forage in conifers, but will visit bird feeders to a limited extent.

*Winter Distribution.* Mountain chickadees do not breed within the Great Plains states region, but eastward movements from the Rocky Mountains to the western high plains bring small numbers of birds into the region periodically. Root (1988) indicated that during her study period these influxes were most common in North Dakota, Kansas, Oklahoma and the Texas panhandle. Our data show the greatest incidence of occurrences in Oklahoma, but very few in North Dakota or the Texas panhandle. These numbers and influx patterns probably have no significant bearing on the species' population generally.

*Seasonality and Migrations.* The mountain chickadee is a common permanent resident throughout its range. This species was seen two years on Christmas Counts from 1968 to 2007 in North Dakota, none in South Dakota, seven years in Nebraska, eight years in Kansas, 16 years in Oklahoma, and one year in the Texas panhandle.

*Habitats.* In winter this species is found in shrublands, lowland riparian forests, and urban areas. It is associated with coniferous and mixed woodlands in summer.

*National Population.* Breeding Bird Surveys between 1966 and 2012 indicate that this species exhibited a survey-wide population decrease (1.38% annually) during that period. The estimated 1990s continental population was about 12 million birds (Rich *et al.,* 2004).

*Further Reading.* Bent, 1946; Harraq and Quin, 1995; Dunn and Tessaglia-Hymes, 1999; McCallum, Grundel, and Dahlsten, 1999 (*The Birds of North America:* No. 453).

## Juniper Titmouse
### *Baolophus ridgwayi*

This is a bird of the oak-juniper woodlands of the Southwest, and is generally much like chickadees in its life-history characteristics, including its vocalizations. They are quite sedentary, with small home ranges, and have prolonged pair bonds that may last for several years. Titmice have more robust beaks than chickadees, and are better able to crack open seeds or nuts. During cold weather they typically roost in tree cavities.

*Winter Distribution.* This species, once considered part of a larger species, the plain titmouse, has a highly restricted distribution in the Black Mesa region of western Oklahoma. It was seen in such small numbers on the Christmas Counts analyzed that nothing can be said of possible population trends.

*Seasonality and Migrations.* The juniper titmouse is a locally common permanent resident throughout its range.

*Habitats.* In Oklahoma this species occupies pinyon pines, junipers, and arid-adapted oaks located on the sides and tops of mesas (Sutton, 1967).

*National Population.* This species was seen 32 years on Christmas Counts from 1968 to 2007 in Oklahoma, but none elsewhere. Breeding Bird Surveys between 1966 and 2012 indicate that this species exhibited a survey-wide population decrease (0.44% annually) during that period. The estimated 1990s continental population was about 330,000 birds (Rich *et al.,* 2004).

*Further Reading.* Bent, 1946; Harraq and Quin, 1995; Dunn and Tessaglia-Hymes, 1999; Cicero, 2000 (*The Birds of North America:* No. 485).

## Tufted Titmouse
*Baeolophus bicolor*

Tufted titmice are near relatives of chickadees, as indicated by the similarities in the vocalizations, behavior and life-history characteristics. Like chickadees, they form strong pair bonds that persist indefinitely. In one case a pair was known to remain together for more than three years, a surprising length in view of the high mortality rates of most small passerines. A few birds reach old ages; one banded bird survived for at least 12 years. Tufted titmice are strongly residential. This attribute might help account for their occasional long survival, as they are not exposed to the perils of migration. The species' range has expanded north somewhat in recent decades, perhaps in part owing to climate change.

*Winter Distribution.* The western edge of distribution this forest-adapted species was mapped by Root (1988) as extending from southeastern Nebraska south to southwestern Oklahoma, plus the eastern edge o the Texas panhandle. Our data indicate the highest Plains density to occur in Oklahoma, as also indicated by Root, and reduced numbers in both directions from that center, with very few in the Texas panhandle. The numbers appear to be generally stable over the entire period, but may be increasing somewhat in the Texas panhandle. Except for that region, there is no support for a nationally increasing population trend, as indicated by Breeding Bird Survey results.

*Seasonality and Migrations.* The tufted titmouse is a common permanent resident throughout its range. This species was not seen on Christmas Counts from 1968 to 2007 in North Dakota or South Dakota, but was reported every year in Nebraska, Kansas and Oklahoma, and ten years in the Texas panhandle. The average number of birds observed per party-hour for the 2013–2014 Christmas Bird Count were: N. Dakota 0, S. Dakota 0, Nebraska 0.28, Kansas 0.98, Oklahoma 1.25, northwest Texas 0. For complete species abundance data, see Appendix 1:82.

*Habitats.* Throughout most of its range, this species is generally found in coniferous or deciduous forests, orchards, woodlots and suburban areas. At the edge of its range in Nebraska it is confined to bottomland deciduous forest.

*National Population.* Breeding Bird Surveys between 1966 and 2012 indicate that this species (including the black-crested form) exhibited a survey-wide population increase (0.85% annually) during that period. The estimated 1990s continental population north of Mexico was about 12 million birds (Rich *et al.*, 2004). The species breeds south to northeastern Mexico.

*Further Reading.* Bent, 1946; Gillespie, 1930; Condor, 1970; Grubb and Pravo-sudov, 1994 (*The Birds of North America:* No. 86); Harraq and Quin, 1995; Dunn and Tessaglia-Hymes, 1999.

## Black-crested Titmouse
*Baeolophus alticristatus*

This is mainly a scrub oak woodland species that in most ways resembles the tufted titmouse, except for its black crest. The two forms hybridize in a nar-row zone where they overlap, and they exhibit very few behavioral differences. The black-crested's loud *peter, peter* call notes are very similar but shorter than the tufted titmouse's, while its *chick-a-dee* notes are higher-pitched and more slurred. During winter the birds move about in loose flocks, sometimes in the company of other small passerines.

*Winter Distribution.* This close relative of the tufted titmouse replaces it to the west. Within our region it is limited to southwestern Oklahoma and the Texas panhandle. The numbers seen in Oklahoma are too small to be sig-nificant, but in the Texas panhandle the species evidently underwent a long-term increase in average numbers until about the year 2000, but was not re-ported during the last eight years of the study period. The average number of birds observed per party-hour for the 2013–2014 Christmas Bird Count were: N. Dakota 0, S. Dakota 0, Nebraska 0, Kansas 0, Oklahoma 0, north-west Texas 0.4. For complete species abundance data, see Appendix 1:83.

*Seasonality and Migrations.* The black-crested titmouse is a locally common per-manent resident throughout its range

*Habitats.* Associated with scrub oak and bottomland woods during breeding, but it also inhabits city parks and towns.

*National Population.* This species was seen on four years on Christmas Counts from 1968 to 2007 in Oklahoma, and 31 years in the Texas panhandle. No data are available from Breeding Bird Surveys of this bird, which at times has been considered a race of the tufted titmouse. Breeding Bird Surveys between 1966 and 2012 indicate that the collective species exhibited a sur-vey-wide population increase (0.85% annually) during that period. The es-timated 1990s continental population was about 740,000 birds (Rich *et al.,* 2004).

*Further Reading.* Bent, 1946; Grubb and Pravosudov, 1994 (*The Birds of North America:* No. 86); Harraq and Quin, 1995.

2

## Bushtit
*Psaltriparus minimus*

The bushtit is a tiny, inconspicuous, and seemingly "friendly" bird, which moves about in small flocks of up to few dozen birds for much of the year, foraging amiably for insects while uttering soft, twittering contact calls. The birds often associate during winter with other small birds, such as chickadees, wrens, or kinglets, and only occasionally visit bird feeders. At about six grams, these bird vie with kinglets for being the smallest of the Great Plains winter songbirds, and as such they spend a high proportion of their daily food energy on simply maintaining their body temperature. During cold weather they may consume up to 80 percent of their weight per day, and during cold nights will huddle together to help maintain body heat.

*Winter Distribution.* Root's (1988) map indicated that the bushtit has its eastern range limits in western Kansas, the western third of Oklahoma and the Texas panhandle. Our data indicate virtually none in Kansas, few in Oklahoma, and by far the largest number in the Texas panhandle. Even in the Texas panhandle the numbers are too small to indicate any possible long-term population trend.

*Seasonality and Migrations.* The bushtit is a locally common permanent resident throughout its range. This species was seen on six years on Christmas Counts from 1968 to 2007 in Kansas, 30 years in Oklahoma, and 36 years in the Texas panhandle. The average number of birds observed per party-hour for the 2013–2014 Christmas Bird Count were: N. Dakota 0, S. Dakota 0, Nebraska 0, Kansas 0, Oklahoma 0.4, northwest Texas 0. For complete species abundance data, see Appendix 1:84.

*Habitats.* In western Oklahoma this species occurs in scrubby oak thickets, pinyon pines (*Pinus edulis*) and junipers located on the sides and tops of mesas (Sutton, 1967). They may also occur along the margins of ponderosa pine (*Pinus ponderosa*) forests at lower to middle elevations (Reinking, 2004).

*National Population.* Breeding Bird Surveys between 1966 and 2012 indicate that this species exhibited a survey-wide population decrease (0.51% annually) during that period. The estimated 1990s continental population north of Mexico was about 2,970,000 birds (Rich *et al.,* 2004). The species breeds south to Guatemala.

*Further Reading.* Bent, 1946; Ervin, 1977; Chaplin, 1982; Harraq and Quin, 1995; Dunn and Tessaglia-Hymes, 1999; Sloane, 2001 (*The Birds of North America:* No. 598).

# FAMILY SITTIDAE: NUTHATCHES

## Red-breasted Nuthatch
*Sitta canadensis*

Half the weight of white-breasted nuthatches, or about ten grams, red-breasted nuthatches are tiny bundles of energy that will boldly approach a person to within a few feet and wait patiently as a bird-feeder is being replenished. They then take a single seed, quickly fly off to hide it, and return for more. They breed in coniferous forests, but when the crops of conifer seeds are low they will readily move into towns and become regular visitors at feeders. In spite of being short-term visitors to bird-feeders, these birds cache large numbers of seeds. In the winter of 2007–2008, there was a major invasion of red-breasted nuthatches, pine siskins and purple finches into Lincoln, Nebraska. Some of the nuthatches then remained to breed, hundreds of miles their nearest usual breeding areas in the northern part of the state.

*Winter Distribution.* Root (1988) mapped a rather uniform winter distribution of this species throughout the Plains states, with few gaps and no areas of higher concentration. Our data also show a broad distribution, but with a large and apparently increasing population trend in the two Dakotas and Kansas, and a smaller and stable population in Oklahoma and Texas. The Nebraska trend is not clear. The possibility of a general population increase is in accordance with the results of national Breeding Bird Surveys.

*Seasonality and Migrations.* The red-breasted nuthatch is a wintering migrant in much of the Great Plains, excepting western South Dakota and northwestern Nebraska, where it is a permanent resident. This species was observed every year on Christmas Counts from 1968 to 2007 in all states from North Dakota to Oklahoma, and during 16 years in the Texas panhandle. Seventy-two initial autumn Nebraska sightings dating from the mid-1930s to the late 1970s (Johnsgard, 1980) range from August 10 to December 31, with a median of October 9. Thirty-nine final spring sightings are from January 4 to June 8, with a median of April 3. It is common to uncommon during winter in Kansas, usually present from mid-August to mid-May (Thompson and Ely, 1992), and a winter resident in Oklahoma, with continuous monthly records extending from August to June (Woods and Schnell, 1984). It is a rare to uncommon winter visitor in the Texas panhandle (Seyffert, 2001). The average number of birds observed per party-hour for the 2013–2014 Christmas Bird Count were: N. Dakota 0.24, S. Dakota

0.26, Nebraska 0.11, Kansas 0, Oklahoma 0.24, northwest Texas 0. For complete species abundance data, see Appendix 1:85.

*Habitats.* This species typically breeds in coniferous forests, but is likely found in conifer plantations, mixed woodlands, and sometimes also appears at urban bird feeders during winter.

*National Population.* Breeding Bird Surveys between 1966 and 2012 indicate that this species exhibited a survey-wide population increase (1.78% annually) during that period. The estimated 1990s continental population was about 18 million birds (Rich *et al.,* 2004).

*Further Reading.* Bent, 1946; Bock and Lepthien, 1972; Harraq and Quin, 1995; Dunn and Tessaglia-Hymes, 1999; Ghalambor and Martin, 1999 (*The Birds of North America:* No. 459).

## White-breasted Nuthatch
### *Sitta carolinensis*

This is the largest and most widespread of the North Americana nuthatches, and one that shows little migratory tendencies, although some autumn dispersal does occur. As sedentary birds, territories are maintained year-around. A pair's territory during winter may be fairly large, and its edges are not always defended against other neighboring pairs. However, the territory often centers on a feeding station, and this part is defended against intrusion. The species' familiar *ank* call is used as a contact note between pair members when uttered singly. When uttered as a double note, mild excitement is indicated, and the call is rapidly repeated under extreme excitement, such as during inter-pair conflicts.

*Winter Distribution.* This familiar winter species has a broad range across the Great Plains states, with smaller numbers in western Kansas, Oklahoma and the Texas panhandle, and with increasing numbers along the eastern boundaries of all five Plains states (Root, 1988). Our data indicate that the highest average numbers occur in North Dakota and Nebraska, and virtually none winter in the Texas panhandle. Trends in Nebraska, Kansas and Oklahoma appear to be upward, following apparent national population trends, while those farther north seem to be stable.

*Seasonality and Migrations.* The white-breasted nuthatch is a common permanent resident in most parts of the Great Plains. It is a rare to uncommon winter visitor and very rare breeder in the Texas panhandle (Seyffert, 2001). This species was observed every year on Christmas Counts from 1968 to

2007 in all states from North Dakota to Oklahoma, and during 12 years in the Texas panhandle. The average number of birds observed per party-hour for the 2013–2014 Christmas Bird Count were: N. Dakota 0.5, S. Dakota 0.34, Nebraska 1.06, Kansas 0.42, Oklahoma 0.29, northwest Texas 0. For complete species abundance data, see Appendix 1:86.

*Habitats.* This species is generally associated with fairly mature floodplain forests during the breeding season, while during the rest of the year it is more widespread and often visits residential feeding stations, especially where suet is provided.

*National Population.* Breeding Bird Surveys between 1966 and 2012 indicate that this species exhibited a survey-wide population increase (2.08% annually) during that period. The estimated 1990s continental population north of Mexico was about nine million birds (Rich *et al.,* 2004). The species breeds south to southern Mexico.

*Further Reading.* Bent, 1948; Kilham, 1971; Grubb, 1977; Woodrey, 1990, 1991; Pravosudov and Grubb, 1993 (*The Birds of North America:* No. 54); Harraq and Quin, 1995; Dunn and Tessaglia-Hymes, 1999.

## Pygmy Nuthatch
### *Sitta pygmaea*

The pygmy nuthatch is virtually the same size as the red-breasted nuthatch, and like it is adapted to coniferous forests, especially ponderosa pine forests. Unlike the red-breasted, it does not exhibit periodic irruptive invasions into the Great Plains states, but during winter loose flocks of up to about two dozen or more may form. Like bushtits, these winter flocks are marked by constant contact-calling, and the birds may be joined by other species, such as titmice, chickadees, or yellow-rumped warblers. Group roosting in tree cavities may occur in winter, and there is are reports of more than 100 birds being found in a single cavity. The collective body heat of a large group of roosting birds may keep the cavity temperature as much as 18° F. above the outside temperature, and may allow the birds to remain in the roost for as long as 40 hours without having to eat.

*Winter Distribution.* This nuthatch is a local breeder and resident in the Black Hills and western Nebraska (Pine Ridge and Wildcat Hills). It is a rare vagrant in Kansas, Oklahoma (but may be rare breeder in the Black Mesa region) and Texas. Root (1988) mapped the South Dakota and Nebraska populations as very low-density, and the Kansas and Texas populations as marginal. Our data suggest a very low winter population in the Great Plains, at least outside their limited breeding areas of South Dakota and Nebraska.

*Seasonality and Migrations.* The pygmy nuthatch is a local and uncommon permanent resident, with some winter wandering outside the breeding range. It is a rare to very rare winter visitor in western Kansas, Oklahoma and the Texas panhandle. It has been seen in Kansas from September to January (Thompson and Ely, 1992), during December in Oklahoma, and from October to April in the Texas panhandle (Seyffert, 2001).

*Habitats._*This nuthatch is closely associated with western pine forests, especially ponderosa pine (*Pinus ponderosa*).

*National Population.* This species was seen 14 years on Christmas Counts from 1968 to 2007 in South Dakota, 12 years in Nebraska, one year in Kansas, two years in Oklahoma and none in the Texas panhandle. Breeding Bird Surveys between 1966 and 2012 indicate that this species exhibited a survey-wide population decrease (0.94% annually) during that period. The estimated 1990s continental population north of Mexico was about 1.7 million birds (Rich *et al.,* 2004). The species breeds south to central Mexico.

*Further Reading.* Bent, 1948; Sydeman and Guntert, 1983; Guntert, Hay and Balda, 1988; Harraq and Quin, 1995; Dunn and Tessaglia-Hymes, 1999; Kingery and Ghalambor, 2001 (*The Birds of North America:* No. 567).

## Brown-headed Nuthatch
### *Sitta pusilla*

Like some other nuthatches, this species occurs in pairs through the breeding season, but later on family groups coalesce to form larger foraging assemblages. Such winter foraging flocks may include other species, such as pine warblers, kinglets, woodpeckers, titmice and chickadees. Single-species flocks of brown-headed nuthatches may include up to about two-dozen birds. Some sociality is evident even during the breeding season, since about a fifth of the pairs may be helped by another male while rearing their young. These birds perhaps are the offspring from a previous breeding.

*Winter Distribution.* This species has a very limited distribution in Oklahoma. The relatively few Christmas Count records for Oklahoma might suggest an increasing state population, but the sample is too small to offer such a conclusion, and it would be counter to national trends suggested by Breeding Bird Survey data.

*Seasonality and Migrations.* The brown-headed nuthatch is a local permanent resident over its limited breeding range.

*Habitats.* Associated with forests of mature shortleaf pines (*Pinus echinata*) in southeastern Oklahoma, and pine forests of the southeastern U.S. generally.

*National Population.* This species has not been seen on Christmas Counts in the Dakotas or Nebraska, but was reported 19 years in counts from 1968 to 2007 Oklahoma, and during 33 years of counts in the Texas panhandle. Breeding Bird Surveys between 1966 and 2012 indicate that this species exhibited a survey-wide population decrease (0.39% annually) during that period. The estimated 1990s continental population was about 1.5 million birds (Rich *et al.,* 2004).

*Further Reading.* Bent, 1948; Morse, 1967a, b; Harraq and Quin, 1995; Withgott and Smith, 1998 (*The Birds of North America:* No. 349) Dunn and Tessaglia-Hymes, 1999.

# FAMILY CERTHIIDAE: CREEPERS

## Brown Creeper
### Certhia americana

Among the least conspicuous of our wintering birds, brown creepers are almost perfectly camouflaged when clinging to tree bark. It is also a very quiet bird, producing only faint hissing, tinkling and, while courting, high-pitched musical notes. During winter it visits feeders only infrequently, and stays only for very brief periods. Creepers sometimes join the mixed foraging flocks that might include wintering chickadees, titmice, nuthatches and bushtits. Their tiny beaks can handle only the smallest of food items, and have little or no ability to cope with large seeds or nuts.

*Winter Distribution.* Root (1988) mapped this species as widely distributed over the Plains states in winter, with the lowest numbers in North Dakota and local areas of higher densities in Nebraska, Kansas, Oklahoma and the Texas panhandle. Our data suggest fairly high population from Nebraska through Oklahoma, and declining both toward the north and south. The entire Plains population would appear to be stable throughout the entire four-decade period.

*Seasonality and Migrations.* The brown creeper is a local permanent resident in the northern Great Plains (Black Hills, northwestern Nebraska). Elsewhere, it is a widespread winter visitor, as in Kansas (Thompson and Ely, 1992), Oklahoma (Woods and Schnell, 1984) and the Texas panhandle (Seyffert, 2001). This species was observed every year on Christmas Counts from 1968 to 2007 in all states from North Dakota to Oklahoma, and during 32 years in the Texas panhandle. The average number of birds observed per party-hour for the 2013–2014 Christmas Bird Count were: N. Dakota 0.06, S. Dakota 0.09, Nebraska 0.18, Kansas 0.12, Oklahoma 0.07, northwest Texas 0.06. For complete species abundance data, see Appendix 1:87.

*Habitats.* While breeding, these birds are associated with fairly mature deciduous or coniferous forests, but in the winter the birds often move to woodland streams, city parks and suburbs.

*National Population.* Breeding Bird Surveys between 1966 and 2012 indicate that this species exhibited a survey-wide population increase (0.63% annually) during that period. The estimated 1990s continental population north of Mexico was about five million birds (Rich *et al.,* 2004). The species breeds south to Nicaragua.

*Further Reading.* Bent, 1948; Franzreb, 1985; Harraq and Quin, 1995; Dunn and Tessaglia-Hymes, 1999; Hejl *et al.,* 2002 *(The Birds of North America:* No. 669).

# FAMILY TROGLODYTIDAE: WRENS

## Cactus Wren
*Campylorhynchus brunneicapillus*

Easily the largest of North American wrens, the cactus wren is rarely far from cactus. Like all wrens, cactus wrens are highly territorial, and announce their territories with loud and frequent singing. They temporarily cease their territorial behavior in winter, when they may wander somewhat, often moving into somewhat denser vegetation. Both the male and female build winter roost nests that are elaborate domed structures shaped like retorts and usually placed in a secure location, such as in a cholla (*Opuntia imbricata*) cactus. Sometimes a brood nest may be retained for use as a winter roosting nest, and a new one is constructed for the next laying cycle. Over time, several spare roosting nests may be constructed and maintained within a pair's territory.

*Winter Distribution.* Root's (1988) map does not show this wren as quite reaching the Texas panhandle, but in recent years it has moved north, and Seyffert (2001) reported it seen in every month except January, May and November. Evidence of nesting there occurred as early as 1977. It has appeared on a least 19 panhandle Christmas counts since 1968, with a maximum density of 0.17 bird/party hour, but has usually achieved only "present" levels of abundance, especially in recent decades.

*Seasonality and Migrations.* The cactus wren is a local permanent resident in the western panhandle of Texas, and a rare visitor elsewhere.

*Habitats.* Associated with cactus-covered country sides, especially in areas of hilly topography.

*National Population.* The tiny Texas panhandle population would appear to be in decline, judging from our very limited data. Breeding Bird Surveys between 1966 and 2012 indicate that this species exhibited a national population decline (2.3% annually) during that period. The estimated 1990s continental population north of Mexico was about 4.1 million birds (Rich *et al.,* 2004). The species breeds south to central Mexico.

*Further Reading.* Bent, 1948; Anderson and Anderson, 1973; Dunn and Tessaglia-Hymes, 1999; Proudfoot, Sherry and Johnson, 2000 (*The Birds of North America:* No. 558).

## Rock Wren
*Salpinctes obsoletus*

This is a loud, rather raucous wren that is always found near cliffs, canyons, talus slopes and similar rocky substrates. It nests in crevices within such sites, and although the nest is thus well hidden from view, its entrance often has an entrance pathway of tiny pebbles, revealing its location. Perhaps for this reason, rock wren nests are parasitized fairly frequently by brown-headed cowbirds. Rock wrens breeding in the central and northern plains are migratory, while those in the southern plains are apparently sedentary.

*Winter Distribution.* According to Root (1988), this species extends during winter to western Kansas, southwestern Oklahoma, and the Texas panhandle. Our data indicate that winter populations are very low in all these areas, even in Oklahoma and the Texas panhandle, where it is regarded as a permanent resident. Probably its highly localized, topography-dependent distribution results in very low overall state-wide densities. No population trends can be inferred from these low numbers.

*Seasonality and Migrations.* The rock wren is a Great Plains seasonal migrant and local breeder in western parts of the region, north to southwestern North Dakota, wintering southwardly. It is a common summer resident in South Dakota; the latest records are for mid-October (Tallman, Swanson and Palmer, 2002). This species was observed 12 years on Christmas Counts from 1968 to 2007 in Kansas, 34 years in Oklahoma and 39 years in the Texas panhandle. Thirty–three final autumn Nebraska sightings are from August 18 to October 29, with a median of October 27. Eighty-three initial spring sightings dating from the mid-1930s to the late 1970s range from April 2 to June 9, with a median of May 2 (Johnsgard, 1980). It is a rare during winter in southwestern Kansas (Thompson *et al.*, 2011) but is considered a permanent resident in Oklahoma (Woods and Schnell, 1984), and is also an uncommon to common resident in the Texas panhandle (Seyffert, 2001).

*Habitats.* This species occurs on eroded slopes and badlands, rocky outcrops, cliff walls, talus slopes and similar environments.

*National Population.* Breeding Bird Surveys between 1966 and 2012 indicate that this species exhibited a survey-wide population decrease (1.24% annually) during that period.

The estimated 1990s continental population north of Mexico was about 3,360,000 birds (Rich *et al.*, 2004). The species breeds south to Nicaragua.

*Further Reading.* Bent, 1948; Lowther *et al.*, 2000 (*The Birds of North America:* No. 486).

## Canyon Wren
*Catherpes mexicanus*

The canyon wren has well residential populations in the Oklahoma and Texas panhandles, and an isolated breeding population in the Black Hills. It is believed that at least some of that population migrates seasonally out of the Black Hills (Tallman, Swanson and Palmer, 2002).

*Winter Distribution.* Like the rock wren, the distribution of this species is closely associated with rocky topography, and so statewide densities are likely to be very low. This was true in all three states where the species was seen on Christmas Counts, but the numbers in Texas were substantially higher than in South Dakota or Oklahoma. The Texas numbers may indicate a population decline, but the sample size is too small to make such an inference.

*Seasonality and Migrations.* The canyon wren is a local permanent resident, with some possible wandering or migration during winter. This species was observed 33 years on Christmas Counts from 1968 to 2007 in South Dakota, was not reported in Nebraska, but occurred twice in Kansas, 37 years in Oklahoma and 38 years in the Texas panhandle. In South Dakota fall migration appears to be in October and November (Pettingill and Whitney, 1965). It is a permanent resident in western Oklahoma and the Texas panhandle (Woods and Schnell, 1984; Seyffert, 2001).

*Habitats.* The combination of steep, rocky walls and running water are a major part of this species' year-around habitat requirements.

*National Population.* Breeding Bird Surveys between 1966 and 2012 indicate that this species exhibited a survey-wide population decrease (0.04% annually) during that period.

The estimated 1990s continental population north of Mexico was about 330,000 birds (Rich *et al.,* 2004). The species breeds south to southern Mexico.

*Further Reading.* Bent, 1948; Jones and Dieni, 1995 (*The Birds of North America:* No. 197).

## Carolina Wren
*Thryothorus ludovicianus*

Carolina wrens appear to be increasing in numbers and expanding their ranges northward in Nebraska, at least along the Missouri River valley. Because they sing so loudly, at almost any time of the year, their occurrence and abundance can be easily detected. The birds establish permanent pair-bonds and defend their territories throughout the winter, which explains the high incidence of winter singing. These songs are highly diverse and include mimicry of several other species, including bluebirds and meadowlarks. Carolina wrens do not flock in winter, and the presence of suitable roosting cavities may be important near the northern edges of their range.

*Winter Distribution.* Root (1988) mapped the winter range of this wren as extending from northeastern Nebraska through Kansas to western Oklahoma and the Texas panhandle. Our data indicate very few birds in either Nebraska or the Texas panhandle, but a relatively large and apparently increasing population in Oklahoma, as well as a smaller but possibly increasing population in Kansas. Such increasing numbers would be in accordance with an increasing national population, as indicated by Breeding Bird Survey data.

*Seasonality and Migrations.* The Carolina wren is a local permanent resident, with some wandering during winter. This species was observed one year on Christmas Counts from 1968 to 2007 in North Dakota (which is well outside its nearest known breeding range), two years in South Dakota, 31 years in Nebraska, all 40 years in Kansas and Oklahoma, and 13 years in the Texas panhandle. From southern Nebraska to the Texas panhandle, it is a uncommon permanent resident. The average number of birds observed per party-hour for the 2013–2014 Christmas Bird Count were: N. Dakota 0, S. Dakota 0.03, Nebraska 0.07, Kansas 0.31, Oklahoma 0.59, northwest Texas 0. For complete species abundance data, see Appendix 1:88

*Habitats.* During the breeding season and probably also the rest of the year this species occupies river bottom forests, forest edges, cutover forests, and cultivated areas with brush heaps, and suburban parks and gardens. It is more closely associated with bottomland forests than is the Bewick's wren or house wren. The three species all overlap in their ecological distributions.

*National Population.* Breeding Bird Surveys between 1966 and 2012 indicate that this species exhibited a survey-wide population increase (1.14% annually) during that period.

. The estimated 1990s continental population north of Mexico was about 15 million birds (Rich *et al.*, 2004). The species breeds south to Nicaragua.

*Further Reading.* Bent, 1948; Haggerty and Morton, 1995 (*The Birds of North America:* No. 188); Dunn and Tessaglia-Hymes, 1999.

## Bewick's Wren
*Thryomanes bewickii*

Bewick's wrens are relatively rare in the northern parts of their Great Plains range, where they are migratory. Farther south, where they are residential, males typically maintain their territories through the winter. However, when competing for territories against the larger house wrens and Carolina wrens the Bewick's wrens are likely to lose out, and this interspecific competition may be one reason why the species has been in a gradual state of decline for several decades.

*Winter Distribution.* Root (1988) mapped the winter range of this wren as reaching its northern limits in northern Kansas, and attaining its greatest abundance in the Texas panhandle. That general pattern may still apply, although out data for Kansas indicates that the species if quite rare during winter. The species' breeding range has retracted from Nebraska during the past half-century, and its population has been generally declining east of the Rocky Mountains, a trend that is evident in our data from both Oklahoma and Texas.

*Seasonality and Migrations.* The Bewick's wren is a Great Plains seasonal migrant and local breeder north to Nebraska, wintering southwardly. This species was not observed on Christmas Counts from 1968 to 2007 north of Kansas, but was observed 39 years in Kansas, and all 40 years in Oklahoma and the Texas panhandle. Forty-four initial spring Nebraska sightings dating from the mid-1930s to the late 1970s (Johnsgard, 1980) range from March 26 to May 28, with a median of April 24. Nine final autumn sightings range from August 11 to October 3, with a mean of September 20. It is uncommon during winter in southern Kansas (Thompson *et al.*, 2011), but is considered a common permanent resident in Oklahoma (Woods and Schnell, 1984) and the Texas panhandle (Seyffert, 2001). The average number of birds observed per party-hour for the 2013–2014 Christmas Bird Count were: N. Dakota 0, S. Dakota 0, Nebraska 0, Kansas 0.19, Oklahoma 0.09, northwest Texas 0. For complete species abundance data, see Appendix 1:89.

*Habitats.* Habitats used during the breeding season include open woodlands, brushy habitats, farmsteads, and towns. In Colorado the birds are mostly associated with dry canyons and scrubby forests, but farther east they overlap with the house wren in their habitats.

*National Population.* Breeding Bird Surveys between 1966 and 2012 indicate that this species exhibited a survey-wide population decrease (1.03%

annually) during that period. The estimated 1990s continental population north of Mexico was about 4,550,000 birds (Rich *et al.*, 2004). The species breeds south to southern Mexico.

*Further Reading.* Bent, 1948; Kennedy and White, 1997 (*The Birds of North America:* No. 325); Dunn and Tessaglia-Hymes, 1999.

## House Wren
### *Troglodytes aedon*

House wrens are one of the best known of America songbirds and, at least to many, one of the most beloved. However, they exhibit intense territorial behavior and tend to destroy the eggs and young of other cavity-nesting songbirds, at least when nest sites are limited. This attribute makes them very unwelcome guests for people trying to attract and encourage nesting by bluebirds, tree swallows, or other cavity-nesting species. Although house wrens are highly vocal during the breeding season, afterwards the males become quiet and secretive, and they remain quite inconspicuous through the winter.

*Winter Distribution.* Root (1988) mapped south-central Oklahoma as the only wintering area for house wrens in the Great Plains states. Her indicated range didn't include the Texas panhandle, but she noted that southern Texas is an area of secondary national abundance. We found only small numbers to be present in Oklahoma and the Texas panhandle, as well as a very few in Kansas during the final decade of the study period. Judging from our data, there has not been any measurable movement northward in the species' wintering range. There is also no evidence of increased numbers at the northern end of its range, in spite of a probable population increase nationally, as judged by Breeding Bird Survey data.

*Seasonality and Migrations.* The house wren is an abundant Great Plains seasonal migrant and widespread breeder, wintering southwardly. This species was observed nine years on Christmas Counts from 1968 to 2007 in Kansas, 33 years in Oklahoma and 14 years in the Texas panhandle. South Dakota records extend to mid-October (Tallman, Swanson and Palmer, 2002). The range of 131 final autumn Nebraska sightings is from July 24 to October 22, with a median of September 26 (Johnsgard, 1980). It is very rare during winter in Kansas (Thompson *et al.*, 2011)., but is considered a permanent resident in Oklahoma (Woods and Schnell, 1984), and is a breeder and very rare winter visitor in the Texas panhandle (Seyffert, 2001).

*Habitats.* Originally associated with deciduous forests and open woods, this species now is also city-adapted, and frequently nests in birdhouses. However, it is also abundant in riverbottom forests, cottonwood groves, and wooded hillsides or canyons.

*National Population.* Breeding Bird Surveys between 1966 and 2012 indicate that this species exhibited a survey-wide population increase (0.12% annually) during that period.

The estimated 1990s continental population north of Mexico was about 19 million birds (Rich *et al.,* 2004). The species breeds south to Chile and Argentina.

*Further Reading.* Bent, 1948; Johnson, 1998 (*The Birds of North America:* No. 380).

## Winter Wren
### *Troglodytes troglodytes*

My most memorable memory of a winter wren was hearing and seeing a male singing lustily at a spring in a soggy rainforest in northwestern Oregon, at precisely the same location (Fort Clatsop) where Lewis and Clark described finding a singing male almost exactly two centuries previously. It nearly brought shivers to my spine to think that this site had probably supported hundreds of generations of winter wrens occupying it between those two events, every male singing with all the endless enthusiasm that only a winter wren can generate. The observation is also a testimonial to the values preserving rare habitats, thereby preserving the unique species on which they depend.

*Winter Distribution.* Root (1988) mapped this wren's winter range as having an isolated South Dakota (Black Hills region) population, and a more general eastern population reaching west to a line extending from northeastern Nebraska to southwestern Oklahoma. Our data suggest that the entire Great Plains winter population is quite small, with only Kansas and Oklahoma having more than minimal numbers. In spite of an apparent nationally increasing population, our data from the Great Plains states do not show such a trend.

*Seasonality and Migrations.* The winter wren is a Great Plains seasonal migrant and local breeder, wintering southwardly. This species was observed 15 years on Christmas Counts from 1968 to 2007 in South Dakota, 32 years in Nebraska, all 40 years in Kansas and Oklahoma, and 13 years in the Texas panhandle. In South Dakota wintering is very local and very rare (Tallman, Swanson and Palmer, 2002). The range of 38 initial autumn

Nebraska sightings dating from the mid-1930s to the late 1970s (Johnsgard, 1980) is from August 30 to December 26, with a median of October 16. The spring records are from January 21 to May 29, with a median of April 13, suggestive of some wintering. It is rare but regular during winter in Kansas, mostly reported from October to March in eastern counties (Thompson *et al.*, 2011). It is considered an uncommon during winter in Oklahoma (Woods and Schnell, 1984), and is a rare winter visitor in the Texas panhandle (Seyffert, 2001).

*Habitats.* While wintering, this inconspicuous species is usually found among dense ravine thickets along streams, but sometimes also occurs in suburban gardens, parks and other similar *Habitats.*

*National Population.* Breeding Bird Surveys between 1966 and 2012 indicate that this species exhibited a survey-wide population increase (1.11% annually) during that period. The estimated 1990s continental population was about 18 million birds (Rich *et al.*, 2004). There are also Icelandic and Eurasian populations.

*Further Reading.* Bent, 1948; Jejl, Holmes and Kroodsma, 2002 (*The Birds of North America:* No. 623).

### Sedge Wren
#### *Cistothorus platensis*

Sedge wrens are another of the wrens that manage to become highly inconspicuous after the breeding season is over, making their winter movements and behavior little known. It is believed that, after rearing a brood fairly far north in their breeding range, the birds may migrate several hundred miles south in late summer, and begin another nesting cycle there. This seems to be the case in Nebraska, where there is a sudden appearance of territorial males at wetlands around Lincoln in August. Nests with eggs have been found in Nebraska as late as early September, adding support to the possibility of dual nesting seasons.

*Winter Distribution.* The only region shown by Root (1988) as has supporting sedge wrens in the Great Plains during winter is centered in south-central Kansas and north-central Oklahoma. Our data do not bear out this pattern, as there were very few sightings for Kansas and only minimal numbers for Oklahoma, in spite of an apparent nationally increasing population.

*Seasonality and Migrations.* The sedge wren is an inconspicuous Great Plains seasonal migrant and local breeder, wintering southwardly. This species was observed one year on Christmas Counts from 1968 to 2007 in Nebraska, seven years in Kansas, 22 years in Oklahoma, and one year in the

Texas panhandle. South Dakota records extend to early November (Tallman, Swanson and Palmer, 2002). Seventeen final autumn Nebraska sightings are from July 29 to October 22, with a median of September 28 (Johnsgard, 1980). It is a very rare during winter in Kansas (Thompson *et al.*, 2011)., and a rare winter visitor in Oklahoma (Woods and Schnell, 1984), and is a very rare spring and autumn migrant in the Texas panhandle, with records extending to mid-December (Seyffert, 2001).

*Habitats.* In the northern plains these birds breed in wet meadows, typically those dominated by sedges and tall grasses, and less often in the emergent vegetation of marshes, retired croplands and hayfields. Migration and wintering habitats are similar.

*National Population.* Breeding Bird Surveys between 1966 and 2012 indicate that this species exhibited a survey-wide population increase (1.38% annually) during that period. The estimated 1990s continental population was about 6.5 million birds (Rich *et al.,* 2004).

*Further Reading.* Bent, 1948; Herkert, Kroodsma and Gibbs, 2001 (*The Birds of North America:* No. 582); Dechant *et al.,* 2001.

### Marsh Wren
*Cistothorus palustris*

Although the breeding biology of this polygynous wren is well known, its autumn and winter behavior is much more poorly understood. The Great Plains population (which perhaps consist of two sibling species that can best be distinguished by their songs) evidently mostly winters along the Gulf Coast. While there, the birds are non-territorial and move about silently in small groups amid marshy vegetation.

*Winter Distribution.* Root (1988) mapped the winter distribution of the marsh wren as extending east to include southwestern Nebraska, western Kansas, and central Oklahoma, and to encompass the entire Texas panhandle. That pattern still applies, although increasing numbers seen during recent decades in Kansas and Nebraska may indicate a northern shift in winter distribution, There is also evidence of an increasing Texas population, which is in accordance with an apparently increasing national population, judging from national Breeding Bird Survey data.

*Seasonality and Migrations.* The marsh wren is a Great Plains seasonal migrant and local breeder, wintering southwardly. This species was observed 10 years on Christmas Counts from 1968 to 2007 in South Dakota, 18 years in Nebraska, 31 years in Kansas, 38 years in Oklahoma, and 39 years in the

Texas panhandle. In South Dakota this species very rarely winters (Tallman, Swanson and Palmer, 2002). Thirty-two final autumn Nebraska sightings are from August 9 to November 22, with a median of October 2 (Johnsgard, 1980). It is rare during winter in Kansas (Thompson *et al.*, 2011), and is a winter resident in Oklahoma, with continuous monthly records extending from August to May (Woods and Schnell, 1984). It is a fairly common winter visitor in the Texas panhandle, with records extending from August to May (Seyffert, 2001). The average number of birds observed per party-hour for the 2013–2014 Christmas Bird Count were: N. Dakota 0, S. Dakota 0.04, Nebraska 0.01, Kansas 0.09, Oklahoma 0.01, northwest Texas 0.4. For complete species abundance data, see Appendix 1:90.

*Habitats.* These birds are primarily found in freshwater marshes having extensive tall emergent vegetation, such as bulrushes (*Scirpus*) and cattails (*Typha*) throughout the year.

*National Population.* Breeding Bird Surveys between 1966 and 2012 indicate that this species exhibited a survey-wide population increase (2.02% annually) during that period.

Breeding Bird Surveys between 1966 and 2006 indicate that this species exhibited a significant national population increase (2.8% annually) during that period. The estimated 1990s continental population was about 7.7 million birds (Rich *et al.*, 2004).

*Further Reading.* Bent, 1948; Kroodsma and Verner, 1997 (*The Birds of North America:* No. 308).

## FAMILY CINCLIDAE: DIPPERS

### American Dipper
*Cinclus mexicanus*

Likes wrens, dippers are strongly attached to their territories and are likely to remain within them all year long if temperatures allow. However, ice conditions during winter may force the birds to abandon their breeding territories and tolerate limited contacts with others. Under these conditions dominance hierarchies may develop that are not site-specific.

*Winter Distribution.* The only Great Plains population is the resident population in South Dakota's Black Hills, where the species is highly localized in Lawrence County. A few records from Nebraska probably represent vagrant movements from that population, rather than true migrations.

*Seasonality and Migrations.* The American dipper is an extremely local permanent resident (Spearfish Canyon) in South Dakota, with some probable limited winter movements to surrounding ice-free areas. This species has occurred at least once on two South Dakota Christmas Bird Count locations (Spearfish and Rapid City) between 1949 and 1998 (Tallman, Swanson and Palmer, 2002), but was not recorded during our survey period.

*Habitats.* This species is associated with swift, clear mountain streams throughout the year.

*National Population.* Breeding Bird Surveys between 1966 and 2012 indicate that this species exhibited a survey-wide population increase (0.15% annually) during that period.

The estimated 1990s continental population north of Mexico was about 586,000 birds (Rich *et al.,* 2004). The species breeds south to Panama.

*Further Reading.* Bent, 1948; Kingery, 1996 (*The Birds of North America:* No. 229).

# FAMILY REGULIDAE: KINGLETS

## Golden-crowned Kinglet
*Regulus satrapa*

Golden-crowned kinglets are the tiniest of wintering birds on the northern plains, but nevertheless manage to survive through the worst of winters. Part of their wintering survival strategy is to join flocks of ruby-crowned kinglets, chickadees, creepers and downy woodpeckers, which adds to the flocks' collective predator-awareness effectiveness. Kinglets will drink tree sap when it is available, but their regular diet almost entirely consists of insects and their eggs, such as aphids, scale insects and bark beetles.

*Winter Distribution.* Root (1988) mapped this species as wintering throughout the Great Plains states with the exception of some parts of North Dakota, South Dakota and western Nebraska, but at fairly low numbers throughout the region. Our data show a now-broadly distributed species, from North Dakota to Texas, and a probably increasing winter population in the two Dakotas. In contrast, the Texas panhandle and Oklahoma populations may be declining, suggesting a possibly northward wintering shift over the past four decades. Root (1988) suggested that this species avoids regions where the January temperature is colder than 0° F. (-18° C.). Heinrich (2003) determined that the overall weight of an adult female golden-crowned kinglet with a featherless body weight of 5.43 grams, had 0.4 gram of body feathers (about nine percent of its total weight) and 0.095 gram of wing and tail feathers. With less than a half a gram of body feathers the birds can nevertheless maintain a normal body temperature of 44° C., even when the surrounding temperature is as much as 78°C. colder. This ability is largely achieved by the insulation provided by air trapped below the maximally fluffed-out body feathers, allowing these tiny birds to survive sub-zero temperatures.

*Seasonality and Migrations.* The golden-crowned kinglet is a seasonal migrant, wintering southwardly. This species was observed 33 years on Christmas Counts from 1968 to 2007 in North Dakota, 39 years in South Dakota, all 40 years in Nebraska, Kansas and Oklahoma, and 38 years in the Texas panhandle. In South Dakota wintering is uncommon (Tallman, Swanson and Palmer, 2002). Fifty-nine final autumn Nebraska sightings are from November 6 to December 31, with a median of December 26, suggesting that this species should be considered a winter resident (Johnsgard, 1980). It is common (east) to rare (west) during winter in Kansas (Thompson *et al.*, 2011)., with most records extending from October to April (Thompson *et al.*, 2011), and is a winter resident in Oklahoma, with continuous monthly records extending from October to April (Woods and Schnell, 1984). It is an uncommon to common winter visitor in the Texas panhandle, with records extending from October to June (Seyffert, 2001). The average number of birds observed per party-hour for the 2013–2014 Christmas Bird Count were: N. Dakota 0.03, S. Dakota 0.11, Nebraska 0.09, Kansas 0.14, Oklahoma 0.09, northwest Texas 0.06. For complete species abundance data, see Appendix 1:91

*Habitats.* During winter this species occupies a wide variety of woodlands, forests and scrubby habitats, including both coniferous and hardwoods but especially the former.

*National Population.* Breeding Bird Surveys between 1966 and 2012 indicate that this species exhibited a survey-wide population decrease (0.76% annually) during that period.

The estimated 1990s continental population was about 34 million birds (Rich *et al.,* 2004). There is also a Eurasian ("goldcrest") population.

*Further Reading.* Bent, 1949; Godfrey, 1986; Ingold and Galati, 1997 (*The Birds of North America:* No. 301); Heinrich, 2003.

## Ruby-crowned Kinglet
### *Regulus calendula*

The ruby-crowned kinglet is very slightly larger than the golden-crowned, and like it often joins mixed winter foraging flocks that might include chickadees, titmice, nuthatches, warblers and creepers. It is somewhat more diverse in its diet than is the golden-crowned's, with various insects, spiders, pseudoscorpions, berries and small seeds all being consumed in varying amounts.

*Winter Distribution.* The ruby-crowned kinglet is not so cold-tolerant as the golden-crowned, and has a distinctly more southward-centered winter distribution. According to Root (1988) it is mostly located from Kansas southward, with a few isolated pockets in eastern Nebraska and southeastern South Dakota. Our data show that a few birds still remain north to North Dakota until at least late December, and that while populations in Kansas and Oklahoma may be stable, those in the Texas panhandle are apparently increasing. This apparently increasing Texas trend is counter to the national declining trend indicated by Breeding Bird Surveys, and is also counter to the prediction that the population should be shifting northward as winter climates ameliorate. The Texas Christmas data are based on only five sites, so such trend indications may be unreliable. The average number of birds observed per party-hour for the 2013–2014 Christmas Bird Count were: N. Dakota 0, S. Dakota 0, Nebraska 0, Kansas 0.31, Oklahoma 0.18, northwest Texas 0.21. For complete species abundance data, see Appendix 1:92.

*Seasonality and Migrations.* The ruby-crowned kinglet is an uncommon seasonal migrant, wintering southwardly. This species was observed five years on Christmas Counts from 1968 to 2007 in North Dakota, six years in South Dakota, 23 years in Nebraska, and all 40 years in Kansas, Oklahoma and the Texas panhandle. In South Dakota there are no records for January or February (Tallman, Swanson and Palmer, 2002). Sixty-nine final autumn Nebraska records are from August 16 to December 31, with a median of October 28. Less than a fourth of the final autumn records are for December, suggesting that this species winters only rather rarely in Nebraska (Johnsgard, 1980). It is rare during winter in southwestern and south-central Kansas (Thompson *et al.*, 2011), and is a winter resident in Oklahoma, with continuous monthly records extending from September to May (Woods and Schnell, 1984). It is a variably common winter visitor in the Texas panhandle, with records extending from August to May (Seyffert, 2001).

*Habitats.* During migration and winter this species occurs in a wide variety of forested and shrubby habitats, including gardens and parks. It occurs in both deciduous and coniferous vegetation, showing no apparent preference for the latter.

*National Population.* Breeding Bird Surveys between 1966 and 2012 indicate that this species exhibited a survey-wide population increase (0.08% annually) during that period.

Niven *et al.* (2004) estimated a national annual increase rate of 1.1 percent, based on Christmas Counts from 1965–67 to 2002–2003. The estimated 1990s continental population was about 72 million birds (Rich *et al.,* 2004).

*Further Reading.* Bent, 1949; Lepthien and Bock, 1976; Laurenzi, Anderson and Ohmart, 1982; Godfrey, 1986; Dunn and Tessaglia-Hymes, 1999; Ingold and Wallace, 1994 (*The Birds of North America:* No. 119).

## FAMILY TURDIDAE:
## THRUSHES AND SOLITAIRES

### Eastern Bluebird
*Sialia sialis*

Eastern bluebirds are surprisingly cold-tolerant, and their winter range seems to have expanded north somewhat in recent decades. Their population has also greatly improved, certainly in part because of a major nationwide effort to provide nesting boxes for the birds. In Nebraska, for example, as many as 20,000 bluebirds are being raised each year in nesting boxes, as well as large numbers of tree swallows (*Tachycineta bicolor*). In the winter these family groups tend to remain intact, merging with other families to form flocks up to a few dozen birds. In very cold weather, bluebirds will cluster together to form communal roosts in a tree cavity, leaving the roost site during the day only long enough to search for and find enough berries and fruits to survive.

*Winter Distribution.* Root (1988) mapped the winter distribution of this bluebird as extending from western Nebraska southwardly and increasing progressively in density from Nebraska to southeastern Oklahoma. Our data are in general agreement with this pattern, but wintering now occurs north to South Dakota, and even to North Dakota in recent years. Although Root described the species' sharp decline that occurred during the pesticide area ending in the early 1970s, since then bluebirds have made a remarkable recovery. Our data suggest that they are now at least twice as common as they were in the later 1960s.

*Seasonality and Migrations.* The eastern bluebird is a common Great Plains seasonal migrant and breeder, wintering southwardly. This species was observed four years on Christmas Counts from 1968 to 2007 in North Dakota, 18 years in South Dakota, 36 years in Nebraska, all 40 years in Kansas and Oklahoma, and 39 years in the Texas panhandle. In South Dakota wintering is rare (Tallman, Swanson and Palmer, 2002). Seventy-four final autumn Nebraska sightings are from August 14 to December 31, with a median of November 5. and nearly a third are for December. This species occasionally winters in Nebraska (Johnsgard, 1980), and is regular through winter in eastern Kansas (Thompson *et al.*, 2011), and a permanent resident in Oklahoma (Woods and Schnell, 1984), and is a fairly common to common breeder and winter visitor in the Texas panhandle, with year-around records (Seyffert, 2001). The average number of birds observed per party-hour for the 2013–2014 Christmas Bird Count were: N. Dakota

0.02, S. Dakota 0.25, Nebraska 0.92, Kansas 2.06, Oklahoma 3.44, northwest Texas 3.13. For complete species abundance data, see Appendix 1:93.

*Habitats.* Open hardwood forests, forest edges, shelterbelts, city parks and farmsteads are used by wintering birds and migrants.

*National Population.* Breeding Bird Surveys between 1966 and 2012 indicate that this species exhibited a survey-wide population increase (2.01% annually) during that period.

The estimated 1990s continental population north of Mexico was about eight million birds (Rich *et al.,* 2004). The species breeds south to Nicaragua.

*Further Reading.* Bent, 1949; Gowaty and Plissner, 1998 (*The Birds of North America:* No. 381); Dunn and Tessaglia-Hymes, 1999; Clement, 2000.

## Western Bluebird
*Sialia mexicana*

There are relatively few Christmas Count records of this vagrant or rare migrant in the Great Plains states. Root (1988) believed it to occur only within a rather narrow range of winter temperatures and moisture, mostly in those regions averaging warmer than 20° F. in January.

*Winter Distribution.* Root's (1988) map does not show any of the Great Plains states as falling within the winter range of this southwestern species, but it has appeared on counts north to Kansas, and is regular in the Texas panhandle. These areas are all well outside the breeding range of the species, and probably represent autumn and winter vagrants.

*Seasonality and Migrations.* The western bluebird is a rare to extremely rare Great Plains seasonal migrant in the southwestern parts of the region. This species was observed four years on Christmas Counts from 1968 to 2007 in Kansas, five years in Oklahoma, and 11 years in the Texas panhandle. Five Nebraska sightings dating from the mid-1930s to the late 1970s of this rare migrant range from April 11 to October 16 (Johnsgard, 2007a). Most records in Kansas range from October to January (Thompson *et al.,* 2011). There are a few December records for Oklahoma (Woods and Schnell, 1984), and is a very rare winter visitor in the Texas panhandle, with records extending from September to May (Seyffert, 2001).

*Habitats.* Migrants and wintering birds are associated with open plains and foothills, or similar habitats to those used by mountain bluebirds.

*National Population.* Breeding Bird Surveys between 1966 and 2012 indicate that this species exhibited a survey-wide population increase (0.79%

annually) during that period.

The estimated 1990s continental population north of Mexico was about 1.2 million birds (Rich *et al.*, 2004). The species breeds south to central Mexico.

*Further Reading.* Bent, 1949; Guinan, Gowaty and Eltzroth, 2000 (*The Birds of North America:* No. 510); Clement, 2000.


## Mountain Bluebird
### *Sialia currucoides*

One of the joys of autumn travel in the high plains is the possibility of coming across a flock of migrating mountain bluebirds. They move across the landscape like a loose cloud of blue confetti, often skipping from fencepost to fencepost, with individuals stopping occasionally to capture a grasshopper in the grass below. Eventually the group moves out of sight to the south, leaving the observer with a forlorn sense that the land has suddenly become bleak and empty, and that winter is approaching all too soon.

*Winter Distribution.* Root (1988) mapped the winter range of the mountain bluebird as ending along the northern border of Kansas, but rapidly increasing in abundance from there to a peak in the Oklahoma panhandle and adjacent parts of Kansas and Colorado, and declining again in the Texas panhandle. Our data indicate a dense population in the Texas panhandle, but with some birds extending north to Nebraska and recently even to South Dakota. In all regions there have been increasing numbers in recent decades, providing supporting evidence for a nationally increasing population, as indicated by national Breeding Bird Surveys.

*Seasonality and Migrations.* The mountain bluebird is an uncommon Great Plains seasonal migrant and breeder in western regions, wintering southwardly. This species was observed one year on Christmas Counts from 1968 to 2007 in South Dakota, 13 years in Nebraska, 23 years in Kansas, 39 years in Oklahoma, and 37 years in the Texas panhandle. In South Dakota, wintering is very rare, and occurs in southwestern parts of the state (Tallman, Swanson and Palmer, 2002). Thirty-five final autumn sightings in Nebraska are from July 21 to December 31, with a median of October 16 (Johnsgard, 1980). It is uncommon to locally common during winter in Kansas from October to April (Thompson *et al.*, 2011), and is a local permanent resident in Oklahoma (Woods and Schnell, 1984; Reinking, 2004). It is a fairly common to abundant winter visitor in the Texas panhandle, with records extending from September to May (Seyffert, 2001). The average number of birds observed per party-hour for the 2013–2014 Christmas

Bird Count were: N. Dakota 0, S. Dakota 0, Nebraska 0, Kansas 14.02, Oklahoma 0.92, northwest Texas 6.18. For complete species abundance data, see Appendix 1:94.

*Habitats.* While on migration and during winter this species often occurs as scattered flocks in open country, usually perching on roadside fences or telephone wires.

*National Population.* Breeding Bird Surveys between 1966 and 2012 indicate that this species exhibited a survey-wide population decrease (0.74% annually) during that period.

The estimated 1990s continental population was about 5.2 million birds (Rich *et al.,* 2004).

*Further Reading.* Bent, 1949; Power and Lombardo, 1996 (*The Birds of North America:* No. 222); Clement, 2000.

## Townsend's Solitaire
### *Myadestes townsendi*

The name "solitaire" has never seemed appropriate to me for this marvelous western bird. Although territorial during the breeding season, large flocks of solitaires may gather during winter at favored foraging areas, such as in juniper groves. Even then the species' wonderful song may be heard, reminding one of the forest-covered mountain slopes that are its summer home.

*Winter Distribution.* According to Root (1988), this western species extends east during winter to central parts of the Dakotas, the Missouri Valley of Nebraska and Kansas, and most of Oklahoma, with a peak in southwestern Kansas and the panhandles of Oklahoma and Texas. Our data suggest there have been increasing populations in the two Dakotas, Nebraska and Kansas, a population in Oklahoma that initially rose and then declined, and declining population in the Texas panhandle. These rather odd results might be explained if there were a major northward shift in Great Plains wintering between the 1960s and the early 2000s. In any case, western Nebraska would now appear to be this species' center of winter distribution in the Great Plains, rather than Kansas or Oklahoma.

*Seasonality and Migrations.* Townsend's solitaire is an uncommon to common Great Plains seasonal migrant and western breeder, wintering southwardly and eastwardly. This species was observed 35 years on Christmas Counts from 1968 to 2007 in North Dakota, all 40 years in South Dakota, 38 years in Nebraska, all 40 years in Kansas, 37 years in Oklahoma, and 36 years in the Texas panhandle. In South Dakota wintering occurs rarely except in the Black Hills. Fifty initial autumn Nebraska sightings dating from the mid-1930s to the late 1970s (Johnsgard, 1980) are from August 23 to December 5, with a median of September 26. Forty-five final spring Nebraska sightings are from January 10 to May 25, with a median of March 20. The species rarely breeds in northwestern Nebraska, but is a regular winter visitor over much of the state, especially in western juniper-rich canyons. It is uncommon during winter in eastern Kansas from October to March (Thompson *et al.*, 2011), and a winter resident in Oklahoma, with most records extending from September to May (Woods and Schnell, 1984). It is an uncommon to fairly common winter visitor in the Texas panhandle, with records extending from September to May (Seyffert, 2001). The average number of birds observed per party-hour for the 2013–2014 Christmas Bird Count were: N. Dakota 0.07, S. Dakota 0.31, Nebraska 0.57, Kansas 0.13, Oklahoma 0.63, northwest Texas 0.25. For complete species abundance data, see Appendix 1:94.

*Habitats.* During migration and winter the birds are often found on wooded slopes rich in juniper berries.

*National Population.* Breeding Bird Surveys between 1966 and 2012 indicate that this species exhibited a survey-wide population increase (0.72% annually) during that period. The estimated 1990s continental population north of Mexico was about 730,000 birds (Rich *et al.,* 2004). The species breeds south to northern Mexico.

*Further Reading.* Bent, 1949; Bock, 1982; Bowen, 1997 (*The Birds of North America:* No. 269); Clement, 2000.

### Hermit Thrush
*Catharus guttatus*

Like the other forest thrushes, this is a species that is easier to appreciate through hearing, rather than visually. Luckily males not only sing while on their remote northern breeding territories but sometimes also do so while on migration and at their wintering sites. Like robins, the birds consume many berries during fall, after insect populations become unavailable. The hermit thrush is the most cold-tolerant of the forest thrushes. It gradually shifts to berries as winter approaches, especially the berries of evergreens such as junipers. Some wintering hermit thrushes return to the same wintering area year after year. There they establish temporary winter territories, which is the likely purpose of winter singing.

*Winter Distribution.* Root (1988) mapped this winter distribution of this thrush to reach north to the Texas panhandle, most of Oklahoma, and southeastern Kansas. It would appear from our data that the birds are now fairly common north to Oklahoma in late December, with a few still remaining in Nebraska, and a very few in North and South Dakota. Additionally, populations appear to be increasing in both Texas and Oklahoma, which is in accordance with an apparent national population increase, judging from Breeding Bird Surveys.

*Seasonality and Migrations.* The hermit thrush is an uncommon Great Plains seasonal migrant, wintering southwardly. This species was observed seven years on Christmas Counts from 1968 to 2007 in North Dakota, one year in South Dakota, 13 years in Nebraska, 31 years in Kansas, 40 years in Oklahoma, and 30 years in the Texas panhandle. South Dakota records extend to late November (Tallman, Swanson and Palmer, 2002). Twelve final autumn Nebraska sightings are from September 11 to December 14 with a median of October 16 There is at least one January record (Johnsgard, 1980). There are few winter records in Kansas (Thompson *et al.*, 2011). It is regarded as a winter resident in Oklahoma, with continuous monthly records extending from September to June (Woods and Schnell, 1984), and is a fairly common to common winter visitor in the Texas panhandle, with records extending from September to July (Seyffert, 2001).

*Habitats.* Migrants are found in dense to semi-open areas of woodland, shrubbery, and vine-draped tangles, but occasionally moving into more open areas. Fairly heavy deciduous woods are a favored habitat in Nebraska.

*National Population.* Breeding Bird Surveys between 1966 and 2012 indicate that this species exhibited a survey-wide population increase (0.93% annually) during that period. Niven *et al.* (2004) estimated a national annual increase rate of 2.2 percent, based on Christmas Counts from 1965–67 to 2002–2003. The estimated 1990s continental population was about 56 million birds (Rich *et al.*, 2004).

*Further Reading.* Bent, 1949; Morse, 1971; Jones and Donovan, 1996 (*The Birds of North America:* No. 261); Dunn and Tessaglia-Hymes, 1999; Clement, 2000.


### American Robin
*Turdus migratorius*


Robins are almost too common to be fully appreciated; we tend to ignore them as they walk about our lawns in search of foods, or sing endlessly from our backyard trees. If they had more beautiful and typical thrush-like songs, or were as brilliantly colored as bluebirds, they might get more of the attention they deserve. Fruits and berries are the primary winter foods of robins, and a very wide array of fruit types are consumed. Flocking is common among migrating and wintering robins, with the birds scattering to forage during the day, but then gathering at favorite roost sites to spend the night.

*Winter Distribution.* Root (1988) noted that, because of the nomadic nature of this species, her map might not represent its actual regional winter abundance. She showed its northern Plains limits in southern North Dakota, and apparent peaks in eastern Nebraska and southwestern Oklahoma. Our data show a broad distribution across the all Plains states from North Dakota to Texas, with a peak in Kansas, but with large numbers extending from South Dakota to Oklahoma, at least during the final decade of study. Steady increases in numbers occurred in North Dakota and Nebraska, while both Kansas and Oklahoma showed marked fluctuations, reflecting the species' vagabond migratory tendency. Judging from the five-state averages, the Plains population may have trebled since the 1960s, when pesticides were causing great mortality to robins across the country.

*Seasonality and Migrations.* The American robin is an abundant Great Plains seasonal migrant and breeder, wintering southwardly. This species was observed all 40 years on Christmas Counts from 1968 to 2007 from South Dakota to the Texas panhandle. In South Dakota wintering varies from rare to fairly common (Tallman, Swanson and Palmer, 2002). Fifty-four final autumn Nebraska sightings are from September 1 to December 31, with a median of November 19. Over a third of the final records are for December, suggesting that this species commonly winters in Nebraska (Johnsgard, 1980). It is variably common during winter in eastern Kansas (Thompson *et al.*, 2011), and is a permanent resident in Oklahoma (Woods and Schnell, 1984). It is a common to abundant winter visitor and uncommon breeder in the Texas panhandle, with records extending from September to May (Seyffert, 2001). The average number of birds observed per party-hour for the 2013–2014 Christmas Bird Count were: N. Dakota 2.61, S. Dakota 8.48, Nebraska 10.55, Kansas 12,7, Oklahoma 36.44, northwest Texas 0.41. For complete species abundance data, see Appendix 1:96.

*Habitats.* Although this species was originally associated with open woodlands, it is probably now most common in cities, suburbs, parks, gardens, and farmlands. During winter the birds often move to more wooded areas.

*National Population.* Breeding Bird Surveys between 1966 and 2012 indicate that this species exhibited a survey-wide population increase (0.20%

annually) during that period. The estimated 1990s continental population north of Mexico was about 307 million birds (Rich *et al.*, 2004). This total would make it the most common of North American land birds, a surprising statistic in view of the multimillions of red-winged blackbirds reported each year as Christmas count maximums, as compared with the relatively modest numbers of robins, which usually number in the tens of thousands. The species breeds south to Guatemala.

*Further Reading.* Bent, 1949; Sallabanks and James, 1999 (*The Birds of North America:* No. 462); Dunn and Tessaglia-Hymes, 1999; Clement, 2000.

# FAMILY MIMIDAE:
## MOCKINGBIRDS, THRASHERS AND CATBIRDS

### Gray Catbird
*Dumetella carolinensis*

Most catbirds winter south to Central America, so those seen in the Great Plains during winter must be thought of as stragglers. Such late birds gradually shift from insects to fruit and berries as temperatures decline; wild grapes are a favorite food, and raisins or currants are popular foods at feeding stations.

*Winter Distribution.* The only winter location mapped by Root (1988) for the gray catbird was central Kansas and the Missouri River valley of Kansas and adjoining Missouri. During our study trace numbers have shown up regularly on Christmas Counts from North Dakota to the Texas panhandle, but there is no evidence of an increasing Plains population.

*Seasonality and Migrations.* The gray catbird is a common Great Plains seasonal migrant and widespread breeder, wintering southwardly. This species was observed four years on Christmas Counts from 1968 to 2007 in North Dakota, two years in South Dakota, one year in Nebraska, 13 years in Kansas and Oklahoma, and four years in the Texas panhandle. In South Dakota there are December but no January or February records (Tallman, Swanson and Palmer, 2002). The range of 128 final autumn Nebraska sightings is from July 22 to December 11, with a median of September 24 (Johnsgard, 1980). There are a few November records for Kansas (Thompson *et al.*, 2011). and a December record for Oklahoma (Woods and Schnell, 1984). It is a regular migrant and very rare winter visitor in the Texas panhandle, with records for all months but March (Seyffert, 2001).

*Habitats.* Typical breeding and non-breeding habitats include thickets, woodland edges, shrubby marsh borders, orchards, parks and similar brushy *Habitats.*

*National Population.* Breeding Bird Surveys between 1966 and 2012 indicate that this species exhibited a survey-wide population decrease (0.06% annually) during that period.

The estimated 1990s continental population was about ten million birds (Rich *et al.,* 2004).

*Further Reading.* Laskey, 1936; Bent, 1948; Doughty, 1988; Cimprich and Moore, 1995 (*The Birds of North America:* No. 167).

## Northern Mockingbird

*Mimus polyglottos*

Mockingbirds have been breeding in Nebraska with regularity for several decades, and a few breeding records exist for South Dakota. There are some possible breeding records for North Dakota. The birds may also be wintering farther north as a result of these range expansions. In common with thrushes and other mimids, fruit becomes increasingly important to mockingbirds as winter approaches, and territories are either maintained or established around trees or bushes that provide a reliable winter food supply. These food stores are defended against all other berry- or fruit-eating species. Autumn and winter territories thus are centered around food, while spring territories center on a nest site. Only males sing in spring, but during autumn both sexes might sing, and the pair may subdivide the defended area.

*Winter Distribution.* The northern mockingbird may be somewhat better adapted to cold weather than is the gray catbird, as it regularly winters north to Kansas and extreme southeastern Nebraska, with larger numbers centered in southeastern Oklahoma, according to Root's (1988) map. Our data show birds appearing occasionally as far north as the Dakotas, but with the largest numbers in Oklahoma. Oklahoma numbers may be gradually increasing, while those in the Texas panhandle seem to be declining, perhaps indicating a regional eastward or northward shift in populations.

*Seasonality and Migrations.* The northern mockingbird is a common to uncommon Great Plains seasonal migrant and widespread breeder, wintering southwardly. This species was observed three years on Christmas Counts from 1968 to 2007 in North Dakota, four years in South Dakota, ten years in Nebraska, and all 40 years in Kansas, Oklahoma, and the Texas panhandle. In South Dakota this species is extremely rare during winter (Tallman, Swanson and Palmer, 2002). Sixty-one final autumn Nebraska sightings are from July 22 to December 31, with a median of September 11. The range of 132 initial spring Nebraska sightings is from January 1 to June 10, with a median of May 2. The data suggest that this species winters occasionally in Nebraska (Johnsgard, 1980). It is irregular during winter in western Kansas (Thompson *et al.*, 2011), is a permanent resident in Oklahoma (Woods and Schnell, 1984), and is also a common resident in the Texas panhandle (Seyffert, 2001). The average number of birds observed per party-hour for the 2013–2014 Christmas Bird Count were: N. Dakota 0, S. Dakota 0, Nebraska 0, Kansas 0.3, Oklahoma 0.99, northwest Texas 0.06. For complete species abundance data, see Appendix 1:97.

PART 2. WINTER BIRDS OF THE GREAT PLAINS

*Habitats.* A variety of habitats, ranging from open woodlands, forest edges and farmlands to parks and cities are utilized, but treeless plains and heavy forests are avoided.

*National Population.* Breeding Bird Surveys between 1966 and 2012 indicate that this species exhibited a survey-wide population decrease (0.64% annually) during that period. The estimated 1990s continental population north of Mexico was about 37 million birds (Rich *et al.,* 2004). The species breeds south to southern Mexico.

*Further Reading.* Bent, 1948; Derrickson and Breitwisch, 1992 (*The Birds of North America:* No. 7); Dunn and Tessaglia-Hymes, 1999.


### Sage Thrasher
*Oreoscoptes montanus*

During autumn and winter this thrasher becomes more difficult to find, after it has stopped singing. It then moves into brushy habitats where it can find a reliable supply of fruits and berries such as currants (*Ribes*), serviceberries (*Amelanchier*) or junipers. At times it may also enter yards and forage in gardens and shrubbery.

*Winter Distribution.* Root (1988) does not includes any part of the study region as within the sage thrasher's wintering range, although she does show it in western Texas. Our data indicate a small but regularly encountered winter population in the Texas panhandle, which is evidently stable. There are also a few Christmas Count records north to North Dakota, indicating a surprising tolerance for cold by this species.

*Seasonality and Migrations.* The sage thrasher is an uncommon Great Plains seasonal migrant and local breeder in western areas, wintering southwardly. This species was observed 14 years on Christmas Counts from 1968 to 2007 in North Dakota, none in South Dakota or Nebraska, one year in Kansas, 16 years in Oklahoma, and 15 years in the Texas panhandle. No autumn records exist for South Dakota after mid-September (Tallman, Swanson and Palmer, 2002). Seven final autumn sightings in Nebraska from the mid-1930s to the late 1970s are from August 24 to October 12, with a mean of September 16 (Johnsgard, 1980). Most records in Kansas extend only to October (Thompson and Ely, 1992). It is considered a rare winter visitor in Oklahoma, with continuous monthly records extending from September to April (Woods and Schnell, 1984), and is a regular migrant and rare winter visitor in the Texas panhandle, with records for all months except June (Seyffert, 2001).

*Habitats.* Although during the breeding season this species is closely associated with sage-dominated grasslands and similar shrubby arid lands, it has a broader winter distribution, occurring in open prairies and also in ponderosa pine woodlands.

*National Population.* Breeding Bird Surveys between 1966 and 2012 indicate that this species exhibited a survey-wide population decrease (1.09% annually) during that period. The estimated 1990s continental population was about 7.9 million birds (Rich *et al.,* 2004).

*Further Reading.* Bent, 1948; Reynolds, Rich and Stephens, 1999 (*The Birds of North America:* No. 463).

### Brown Thrasher
*Toxostoma rufum*

Winter territories are established after brown thrashers have arrived on their winter range, or resident birds have completed their brood-rearing and post-breeding molts. These territories are usually located in areas of heavy brush, often close to water. Bouts of morning and evening song by males announce these territories, which are marked by a very high degree of vocal variability, much like the songs of mockingbirds.

*Winter Distribution.* Root (1988) mapped the winter distribution of this thrasher as occurring south of a line connecting northeastern and southwestern Nebraska, with larger numbers in southeastern Kansas and eastern Oklahoma. Our data show Oklahoma to have the highest average numbers, but with some sightings north to North Dakota. Overall population trends appear to be downward, in accordance with apparent national trends as shown by Breeding Bird Surveys.

*Seasonality and Migrations.* The brown thrasher is a common Great Plains seasonal migrant and widespread breeder, wintering southwardly. This species was observed 14 years on Christmas Counts from 1968 to 2007 in North Dakota, 15 years in South Dakota, 26 years in Nebraska, all 40 years in Kansas and Oklahoma, and 38 years in the Texas panhandle. In South Dakota wintering is only very rare (Tallman, Swanson and Palmer, 2002). The range of 164 final autumn Nebraska records is from July 22 to December 31, with a median of September 28. Over ten percent of the records are for December, suggesting that this species may overwinter occasionally in Nebraska (Johnsgard, 1980). It is very rare during winter in Kansas (Thompson *et al.,* 2011), and is a permanent resident in Oklahoma (Woods and Schnell, 1984). It is an uncommon to fairly common migrant and winter visitor in the western Texas panhandle, and a resident in the east (Seyffert, 2001). The average number of birds observed per party-hour for the 2013–2014 Christmas Bird Count were: N. Dakota 0.01, S. Dakota 0, Nebraska 0.01, Kansas 0.02, Oklahoma 0.1, northwest Texas 0.06. For complete species abundance data, see Appendix 1:98.

*Habitats.* This species frequents open brushy woods, scattered patches of brush and small trees in open environments, shelterbelts, woodlands and shrubby residential areas.

*National Population.* Breeding Bird Surveys between 1966 and 2012 indicate that this species exhibited a survey-wide population decrease (1.09% annually) during that period. The estimated 1990s continental population was about 7.3 million birds (Rich *et al.,* 2004).

*Further Reading.* Bent, 1948; Fisher, 1981b; Dunn and Tessaglia-Hymes, 1999; Cavitt and Haas, 2000 (*The Birds of North America:* No. 557).

## Curve-billed Thrasher
*Toxostoma curvirostre*

Most curve-billed thrashers live far enough south to be residential throughout the year, and thus can maintain permanent territories, which are often characterized by the presence of cholla cactus (*Opuntia imbricata*). The cactus not only provides a relatively safe nesting site, but is also used for winter roosting Winter roost sites are simple twig platforms, which later might be converted into a more substantial nest for breeding.

*Winter Distribution.* Root (1988) mapped the winter range of this southwestern species as barely reaching the Texas panhandle. Our data show a significant population in the Texas panhandle, and a few regularly occurring in Oklahoma as well. The species has moved north into Kansas as a breeding bird in recent decades, especially in areas where cholla cactus occurs, and has even been seen a few times in Nebraska and even in South Dakota, where it has nested at least once.

*Seasonality and Migrations.* The curve-billed thrasher is a local resident in the southern part of its range, but the northernmost birds are probably somewhat migratory. This species was observed 17 years on Christmas Counts from 1968 to 2007 in Kansas, 32 years in Oklahoma, and 39 years in the Texas panhandle. This species is a permanent resident in the Texas panhandle, ranging from common in the west to rare in the east. First observed in a 1967 Amarillo Christmas Count, the species appeared on 20 of 28 later counts (Seyffert, 2001). It is also a permanent resident in southwestern and western Oklahoma (Woods and Schnell, 1984), and a very rare resident in southwestern Kansas, with most sighting between November and April (Thompson and Ely, 1992; Thompson *et al.*, 2011). There have been several Nebraska sightings since 1962. The average number of birds observed per party-hour for the 2013–2014 Christmas Bird Count were: N. Dakota 0, S. Dakota 0, Nebraska 0, Kansas 0, Oklahoma 0.11, northwest Texas 0.18. For complete species abundance data, see Appendix 1:99.

*Habitats.* Sandsage (*Artemisia filifolia*) and cholla cactus grasslands are breeding habitats, but during winter the birds often occupy barnyards, windbreaks, cemeteries, brushy ravines and backyards

*National Population.* Breeding Bird Surveys between 1966 and 2012 indicate that this species exhibited a survey-wide population decrease (1.35% annu-ally) during that period. The Texas population has been in long-term de-cline (3.4%)(Benson and Arnold, 2001; Lockwood and Freeman, 2004). The estimated 1990s continental population north of Mexico was about 1.2 million birds (Rich *et al.*, 2004). The species breeds south to southern Mexico.

*Further Reading.* Bent, 1948; Fischer, 1980, 1981a; Tweit, 1996 (*The Birds of North America:* No. 235); Dunn and Tessaglia-Hymes, 1999.

## FAMILY STURNIDAE: STARLINGS

### European Starling
*Sturnus vulgaris*

Starlings are the bane of bird-lovers; they confiscate nests of cavity-nesting birds, gather by the countless thousands in cities during winter, and move about in restless gangs in search of backyard bird-feeders, especially those providing suet or meat wastes. A flock to 20 or 30 can strip a chicken skeleton bare in a surprisingly short time. Then, when the food is gone, the flock quickly disappears. Toward evening they usually gather in favored roosting sites, often where the temperature may be somewhat warmer than elsewhere. In spite of their bad behavior, the iridescent plumage of adult starlings is quite spectacular, and provides a slight visual upside to their presence.

*Winter Distribution.* Root's (1988) map of the starling's winter distribution shows some birds occurring as far north as northeastern North Dakota, but no regions of clearly high densities. Our data suggest that Kansas is probably the region of highest density in the Great Plains states, although the species is so closely associated with humans that count numbers are heavily influenced by the number of city-centered versus rural-based counts. No clear population trends are evident in our data, and they do not follow a general downward trend, as is suggested by national Breeding Bird Surveys.

*Seasonality and Migrations.* The European starling is an abundant permanent resident throughout the five Great Plains states, with some possible limited winter migration out of northern areas, judging from banding results in South Dakota. It appeared all years on Christmas Counts from 1968 to 2007 from North Dakota to the Texas panhandle. The average number of birds observed per party-hour for the 2013–2014 Christmas Bird Count were: N. Dakota 6.29, S. Dakota 25.01, Nebraska 138.58, Kansas 94.2, Oklahoma 584.6, northwest Texas 7.14. For complete species abundance data, see Appendix 1:100.

*Habitats.* This species is found virtually everywhere throughout the year, but is especially associated with human habitations such as cities, suburbs and farms.

*National Population.* Breeding Bird Surveys between 1966 and 2012 indicate that this species exhibited a survey-wide population decrease (1.26 annually) during that period. A recent North American population estimate is 120 million birds (Rich *et al.,* 2004). The species breeds south to northern Mexico and the West Indies.

*Further Reading.* Cabe, 1993 (*The Birds of North America:* No. 48); Dunn and Tessaglia-Hymes, 1999.

FAMILY MOTACILLIDAE: PIPITS

## American Pipit
*Anthus rubescens*

Pipits breed on remote mountain tundras and arctic coasts across North America. During autumn they migrate to prairies and lowlands, including coastal and inland beaches, south into Mexico. Those occurring on the Great Plains during winter must represent only a miniscule portion of the overall continental population.

*Winter Distribution.* Root (1988) showed only a few areas of winter occurrence of this species in the Great Plains, namely some scattered regions from South Dakota to Oklahoma and the Texas panhandle, with somewhat larger numbers in Oklahoma. Our data suggest a broadly distributed distribution that is centered in Oklahoma, and rapidly grading off in Kansas and northwestern Texas. The numbers are too small and too variable to allow for any inferences regarding possible population trends.

*Seasonality and Migrations.* The American pipit is an uncommon seasonal migrant, wintering southwardly. This species was observed one year on Christmas Counts from 1968 to 2007 in South Dakota, one year in Nebraska, five years in Kansas, 28 years in Oklahoma, and 17 years in the Texas panhandle. In South Dakota this species has been reported as late as December, but January or February records are lacking (Tallman, Swanson and Palmer, 2002). Sixteen final autumn Nebraska sightings are from September 14 to December 31, with a median of October 26. Wintering is probably quite rare in Nebraska, judging from the limited number of late autumn records (Johnsgard, 1980). It is occasional during winter in Kansas (Thompson *et al.*, 2011), and a rare winter visitor in Oklahoma, with continuous monthly records extending from September to May (Woods and Schnell, 1984). It is an uncommon to rather common winter visitor in the Texas panhandle, with records extending from September to May (Seyffert, 2001). The average number of birds observed per party-hour for the 2013–2014 Christmas Bird Count were: N. Dakota 0, S. Dakota 0, Nebraska 0.06, Kansas 0, Oklahoma 0.75, northwest Texas 0.06. For complete species abundance data, see Appendix 1:101.

*Habitats.* Migrating and wintering birds are found in open plains, fields, and bare shorelines, generally favoring moist to wet environments over dry ones.

*National Population.* Breeding Bird Survey trend data are not available for this alpine breeder. The estimated 1990s continental population was about 19.8 million birds (Rich *et al.,* 2004).

*Further Reading.* Bent, 1950; Godfrey, 1986; Verbeek and Hendricks, 1994 (*The Birds of North America:* No. 95).

## Sprague's Pipit
### *Anthus spragueii*

Sprague's pipits are as hard to find on migration as they are on their breeding grounds, so their winter ecology and behavior is relatively little-known. Fields overgrown with weeds seems to be their typical winter habitat, but probably northern Mexico is their major wintering region.

*Winter Distribution.* Root (1988) doubted the significance of her map of this elusive species, since it was based on small sample sizes, but did show some birds in central Kansas and central Oklahoma. Our data are equally non-informative, but suggest that Oklahoma may be the best Great Plains location for seeing these inconspicuous birds in early winter.

*Seasonality and Migrations.* The Sprague's pipit is an uncommon and inconspicuous seasonal migrant, and a local breeder in western North Dakota, wintering southwardly. This species was observed one year on Christmas Counts from 1968 to 2007 in Nebraska, three years in Kansas, 16 years in Oklahoma, and none in the Texas panhandle. South Dakota records extend to late October (Tallman, Swanson and Palmer, 2002). Eleven final autumn Nebraska sightings are from October 2 to November 8, with a median of October 23 (Johnsgard, 1980). It is a very rare during winter in Kansas, with a few December and January records (Thompson and Ely, 1992; Thompson *et al.*, 2011), and a rare winter visitor in Oklahoma, with continuous monthly records extending from September to April (Woods and Schnell, 1984). It is a rare migrant in the Texas panhandle, with autumn records extending from September to November (Seyffert, 2001).

*Habitats.* Associated with dense, grassy vegetation of plains and prairies, Unlike the American pipit, the Sprague's is not often found in bare areas close to water. It also differs from the American by not usually moving in flocks, and is thus more easily overlooked.

*National Population.* Breeding Bird Surveys between 1966 and 2012 indicate that this species exhibited a survey-wide population decrease (3.47% annually) during that period. It has been designated a species of continental conservation importance, with the Prairie Avifaunal Biome supporting an estimated 18 percent of the continental winter population (Rich *et al.*, 2004). The estimated 1990s continental population was about 870,000 birds (Rich *et al.*, 2004).

*Further Reading.* Bent, 1950; Godfrey, 1986; Dechant *et al.*,1999i; Robbins and Dale, 1999 (*The Birds of North America:* No. 439); Johnsgard, 2001b.

# FAMILY BOMBYCILLIDAE: WAXWINGS

## Bohemian Waxwing
*Bombycilla garrulus*

Like other waxwings, this species depends on fruits and berries to sustain it from autumn to spring. The fruits of raspberries (*Rubus*), chokecherries (*Prunus*), roses (*Rosa*), crabapples (*Pyrus*), hawthorns (*Crataegus*) and junipers *(Juniperus)* all are eaten avidly, and a tree may be stripped of its fruit in a very short time when a flock of waxwings arrives. These flocks may number from the dozens to hundreds of birds. At times a single Bohemian waxwing will appear in a flock of wintering cedar waxwings, especially toward the southern end of the Bohemian's wintering range.

*Winter Distribution.* This species is slightly larger than the cedar waxwing, and has a considerably more northern winter distribution. Root (1988) mapped it as occurring north and west of a line from northeastern North Dakota to the Nebraska panhandle, and noted that occasional irruptions southward may occur. Our data confirm that the same general winter pattern may still persist, but do not indicate any clear population trends. An apparent recent decline in North Dakota Christmas Count numbers may indicate that there has been a recent shift northward into Manitoba. Counts in Manitoba do show a population increase over this same four-decade period, so this apparent trend may provide some evidence pointing in that direction.

*Seasonality and Migrations.* The Bohemian waxwing is a common to rare wintering migrant throughout the Great Plains. This species was observed all 40 years on Christmas Counts from 1968 to 2007 in North Dakota, during 37 years in South Dakota, 11 years in Nebraska, ten years in Kansas, and none in Oklahoma or the Texas panhandle. In South Dakota this species is a common though winter, especially in the Black Hills. It arrives during November and December, and departs between March and early April. Early autumn dates are October 23–28, and late spring dates are April 18–30 (Tallman, Swanson and Palmer, 2002). The range of 11 initial autumn Nebraska sightings dating from the mid-1930s to the late 1970s (Johnsgard, 1980) is from September 25 to December 27, with a median of November 20. Nineteen final spring sightings range from January 2 to May 22, with a median of February 28. It is irregular during winter in Kansas from late November to mid-March (Thompson *et al.*, 2011), and is a rare winter visitor in Oklahoma, with continuous monthly records extending from January to May (Woods and Schnell, 1984). It is a very rare winter visitor in the

Texas panhandle, with records extending from November to May (Seyffert, 2001). The average number of birds observed per party-hour for the 2013–2014 Christmas Bird Count were: N. Dakota 0.1.24, S. Dakota 0.58, Nebraska 0, Kansas 0, Oklahoma 0, northwest Texas 0. For complete species abundance data, see Appendix 1:102.

*Habitats.* Migrants and wintering birds are associated with fruit-bearing trees in woodlands, shelterbelts, and urban parks or gardens, often in association with cedar waxwings.

*National Population.* Breeding Bird Survey trend data are not available for this boreal species. Niven *et al.* (2004) estimated a national annual decline rate of 4.5 percent, based on Christmas Counts from 1965–67 to 2002–2003. The estimated 1990s continental population was about 1.4 million birds (Rich *et al.,* 2004). There is also a Eurasian population (the "waxwing").

*Further Reading.* Bent, 1950; Godfrey, 1986; Dunn and Tessaglia-Hymes, 1999; Witmer, 2003 (*The Birds of North America:* No. 714).

## Cedar Waxwing
### *Bombycilla cedrorum*

One of the most esthetically attractive of all North American birds, the cedar waxwing is also one of the most strongly fruit-dependent. During winter, virtually all of their food comes from fruit sources, especially juniper berries. Junipers have become more common over the Great Plains as fire suppression has allowed many grasslands to grow up to juniper woodlands, providing ever greater food resources for the waxwings. Although waxwings are attracted to tree sap, they generally ignore hummingbird feeders. However, they will eat the flowering parts of various trees in spring, which may help provide them with much-needed protein just prior to the breeding season.

*Winter Distribution.* Root's (1988) map indicates a broadly distributed winter distribution in the Great Plains for this species, with southeastern Oklahoma having the largest numbers, and relatively few in northern North Dakota. Our data similarly show a distinct peak in Oklahoma, but substantial numbers extending north to North Dakota. An increasing population trend is indicated for all states, which is in accordance with data from national Breeding Bird Surveys.

*Seasonality and Migrations.* The cedar waxwing is a common to occasional wintering migrant throughout the Great Plains and a local breeder. This species was observed every year on Christmas Counts from 1968 to 2007 in all states from North Dakota to Oklahoma, and 35 years in the Texas panhandle (five locations).In South Dakota wintering is regular (Tallman, Swanson and Palmer, 2002). The range of 54 initial spring Nebraska sightings dating from the mid-1930s to the late 1970s (Johnsgard, 1980) is from January 2 to May 20, with a median of February 24. Fifty-eight final autumn sightings are all in December, suggesting that this species rather frequently winters in Nebraska. It is common during winter in Kansas, with most records from September to May (Thompson *et al.*, 2011). and a winter resident in Oklahoma, with continuous monthly records extending throughout the year (Woods and Schnell, 1984). It is a rsther common to abundant winter visitor in the Texas panhandle, with records for every month except July (Seyffert, 2001). The average number of birds observed per party-hour for the 2013–2014 Christmas Bird Count were: N. Dakota 2.64, S. Dakota 3.19, Nebraska 2.03, Kansas 8.71, Oklahoma 14.77, northwest Texas 0. For complete species abundance data, see Appendix 1:103.

*Habitats.* Outside the breeding season this species occurs in flocks that congregate in habitats having fruit-bearing trees, such as junipers, hackberries (*Celtis*) and mountain ash (*Sorbus*), and tall shrubs such as pyracantha (*Pyracantha)* and sumac (*Rhus*).

*National Population.* Breeding Bird Surveys between 1966 and 2012 indicate that this species exhibited a survey-wide population increase (0.30% annually) during that period.

The estimated 1990s continental population was about 15 million birds (Rich *et al.*, 2004).

*Further Reading.* Bent, 1950; Witmer, Mountjoy and Elliot, 1997 (*The Birds of North America:* No. 309); Dunn and Tessaglia-Hymes, 1999.

## FAMILY CALCARIIDAE:
## LONGSPURS AND SNOW BUNTINGS

### Lapland Longspur
*Calcarius lapponicus*

Great Plains residents have opportunities to observe Lapland longspurs only while they are in their winter plumage, but what we might miss in visual appeal is often made up for in quantity. Flocks of thousands of longspurs sometimes settle into Great Plains stubble fields and meadows, where they promptly almost disappear among the leaves, and are unlikely to be noticed again until they fly.

*Winter Distribution.* The Lapland longspur is the most northerly breeder of the four longspurs. Root (188) mapped its wintering range as extending from Texas north to southern Saskatchewan and Manitoba, with maximum numbers in South Dakota, Kansas and Oklahoma. Our data indicate that Kansas has had the largest numbers of wintering birds over the entire four-decade period, with considerable year-to-year and decade-to-decade variations in numbers. These variations make it impossible to judge long-term population trends for the species.

*Seasonality and Migrations.* The Lapland longspur is a variably abundant wintering migrant throughout the five Great Plains states. This species was observed all 40 years on Christmas Counts from 1968 to 2007 in North Dakota and South Dakota, 35 years in Nebraska, 40 years in Kansas, 38 years in Oklahoma, and nine years in the Texas panhandle. In South Dakota wintering is regular (Tallman, Swanson and Palmer, 2002). Fifty-six initial autumn Nebraska sightings dating from the mid-1930s to the late 1970s (Johnsgard, 1980) are from September 25 to December 31, with a median of November 12. Forty-four final spring records are from January 3 to May 10, with a median of February 27. It is common to uncommon (east) and common to abundant (west) during winter in Kansas, with most records extending from October to April (Thompson *et al.*, 2011), and is a winter resident in Oklahoma, with continuous monthly records extending from October to April (Woods and Schnell, 1984). It is a common to abundant winter visitor in the Texas panhandle, with records extending from October to March (Seyffert, 2001). The average number of birds observed per party-hour for the 2013–2014 Christmas Bird Count were: N. Dakota 2.25, S. Dakota 3.19, Nebraska 2.03, Kansas 8.71, Oklahoma 14.77, northwest Texas 0. For complete species abundance data, see Appendix 1:104.

*Habitats.* Migrants and wintering birds occur in open, grassy plains, stubble fields, overgrazed pastures, and similar grassy or low-stature herbaceous *Habitats.*

*National Population.* Breeding Bird Survey trend data are not available for this high arctic-breeding breeder. It has been designated a species of continental conservation importance, with the Prairie Avifaunal Biome supporting an estimated 99 percent of the continental winter population (Rich *et al.,* 2004). The estimated 1990s continental population was about 74 million birds (Rich *et al.,* 2004). There is also a Eurasian ("Lapland bunting") population.

*Further Reading.* Bent, 1968; Godfrey, 1986; Byers, Curson and Olsen, 1995; Rising and Beadle, 1996; Hussell and Montgomerie, 2002 (*The Birds of North America:* No. 656).

## Smith's Longspur
### *Calcarius pictus*

Smith's longspurs sometimes appear among flocks of Lapland longspurs as they settle into the Great Plains for the winter, but finding one is like searching for a needle in a haystack. However, farther east they usually occur in flocks of 20 to 30 birds, often feeding in pastures or grassy airports. There they usually remain separate from other wintering birds such as horned larks and other longspurs.

*Winter Distribution.* Like the Lapland longspur, this is a high-latitude breeder, typically nesting in subarctic regions south of the Lapland longspur, and wintering in large, highly mobile flocks. Root (1988) did not attempt to map its winter range, but noted that it was most often seen in Oklahoma and Arkansas. Our data show Oklahoma to have the highest numbers of birds within the Great Plains region, but with some scattering of occurrences north to South Dakota. Our numbers are too small to judge any population trends.

*Seasonality and Migrations.* Smith's longspur is an uncommon to occasional wintering migrant throughout the five Great Plains states. This species was observed two years on Christmas Counts from 1968 to 2007 in South Dakota, three years in Nebraska, 21 years in Kansas, 39 years in Oklahoma, and one year in the Texas panhandle. Other South Dakota records extend only to November 7 (Tallman, Swanson and Palmer, 2002). Ten autumn Nebraska sightings dating from the mid-1930s to the late 1970s (Johnsgard, 1980) range from September 18 to December 17, with a median of

November 5. Six spring records are from February 5 to May 22, with a mean of April 8. It is rare during winter in eastern Kansas (Thompson *et al.*, 2011). and a winter resident in Oklahoma, with continuous monthly records extending from October to April (Woods and Schnell, 1984). There are no confirmed records for the Texas panhandle (Seyffert, 2001). The average number of birds observed per party-hour for the 2013–2014 Christmas Bird Count were: N. Dakota 0, S. Dakota 0, Nebraska 0, Kansas 0.06, Oklahoma 1.25, northwest Texas 0. For complete species abundance data, see Appendix 1:105.

*Habitats.* Migrants and wintering birds are associated with open grassy plains and pastures, preferring those covered by thick, short grass, including airports.

*National Population.* Breeding Bird Survey trend data are not available for this Canadian grassland breeder. Niven *et al.* (2004) estimated a national annual increase rate of 0.8 percent, based on Christmas Counts from 1965–67 to 2002–2003. It has been designated a species of continental conservation importance, with the Prairie Avifaunal Biome supporting an estimated 99 percent of the continental winter population (Rich *et al.*, 2004), the highest estimated incidence of prairie endemicity of any species. Oklahoma would seem to be of special conservation importance for this bird's winter survival. The estimated 1990s continental population was about 75,000 birds (Rich *et al.*, 2004). This species was yellow-listed in the National Audubon Society's 2007 WatchList of rare and declining birds (Butcher *et* al., 2007).

*Further Reading.* Bent, 1968; Godfrey, 1986; Briskie, 1993 (*The Birds of North America:* No. 34); Byers, Curson and Olsen, 1995; Rising and Beadle, 1996.

## Chestnut-collared Longspur
### *Calcarius ornate*

Like other wintering longspurs, this is a generally inconspicuous bird, often remaining almost invisible when seen among dead-grass surroundings, until it takes flight and flashes its partially white tail. Except for their distinctive tail markings wintering longspurs would pass for any of a number of rather nondescript grassland sparrows.

*Winter Distribution.* The mixed-grass species of longspur was mapped by Root (1988) has having a rather confined wintering range extending from Kansas and Oklahoma through western Texas and New Mexico to southeastern Arizona, with a center of abundance near the Texas–New Mexico border. Our data show a rather diffuse wintering area extending from Kansas south, and the usual rather erratic variation in numbers that reflect the wandering and unpredictable nature of longspurs on wintering grounds.

*Seasonality and Migrations.* The chestnut-collared longspur is an uncommon migrant and local breeder from western Nebraska north, wintering southwardly. This species was observed one year on Christmas Counts from 1968 to 2007 in North Dakota, two years in South Dakota, one year in Nebraska, 24 years in Kansas, all 40 years in Oklahoma, and 25 years in the Texas panhandle. South Dakota records extend to November 7 (Tallman, Swanson and Palmer, 2002). Thirty initial spring Nebraska sightings dating from the mid-1930s to the late 1970s (Johnsgard, 1980) in northwestern Nebraska are from March 18 to June 3, with a median of April 12. Sixteen final autumn sightings are from September 22 to October 22, with a median of October 8. Like the McCown's longspur, the migration pattern of this species is extremely difficult to estimate in Nebraska, since in various parts of the state it is a summer resident, a spring and autumn migrant, or a winter visitor. It is rare (east) to uncommon (west) during winter in Kansas, with records extending from September to April (Thompson *et al.*, 2011), and a winter resident in Oklahoma, with continuous monthly records extending from September to April (Woods and Schnell, 1984). It is a common to abundant winter visitor in the Texas panhandle, with records extending from October to April (Seyffert, 2001). The average number of birds observed per party-hour for the 2013–2014 Christmas Bird Count were: N. Dakota 0, S. Dakota 0, Nebraska 0, Kansas 0, Oklahoma 1.85, northwest Texas 0. For complete species abundance data, see Appendix 1:105

*Habitats.* Migrants and wintering birds occur on open plains and grassy fields, including airports. Breeding usually occurs on shortgrass or cut mixed-grass prairies, and less frequently in the low meadow zones around ponds, and disturbed grasslands such as grazed pasturelands.

*National Population.* Breeding Bird Surveys between 1966 and 2012 indicate that this species exhibited a survey-wide population decrease (4.23% annually) during that period.

It has been designated a species of continental conservation importance, with the Prairie Avifaunal Biome supporting an estimated 23 percent of the continental winter population (Rich *et al.*, 2004). The estimated 1990s continental population was about 5.6 million birds (Rich *et al.*, 2004). This species was yellow-listed in the National Audubon Society's 2007 WatchList of rare and declining birds (Butcher *et* al., 2007).

*Further Reading.* Bent, 1968; Byers, Curson and Olsen, 1995; Rising and Beadle, 1996; Hill and Gould, 1997 (*The Birds of North America:* No. 288); Dechant *et al.*, 1999b; Johnsgard, 2001b.

## McCown's Longspur
### *Calcarius mccownii*

The longspurs are classic grassland sparrows, collectively ranging in their breeding habitats from shortgrass prairies (McCown's) to arctic tundra (Lapland). Like horned larks, all the species have lengthened hind claws that perhaps help serve as supports when the birds are walking over soft snow. This species is one of the less conspicuous longspurs even when in breeding plumage, and during winter it becomes nearly invisible.

*Winter Distribution.* Root (1988) provided no map for this species, which is characteristic of the western Great Plains, but our data indicate that only northwestern Texas supports a significant population of this species within our region of analysis. Other sources suggest that western Texas and adjoining northern Mexico are the primary wintering grounds for this species. Our data for the Texas panhandle show varied densities, suggesting that winter distributions might vary considerably from year to year, as in other longspur species.

*Seasonality and Migrations.* McCown's longspur is a common to uncommon seasonal migrant and local breeder in western regions, wintering southwardly. This species was observed one year on Christmas Counts from 1968 to 2007 in Nebraska, 23 years in Kansas, 11 years in Oklahoma, and 38 years in the Texas panhandle, but none in the Dakotas. South Dakota records extend to late November (Tallman, Swanson and Palmer, 2002). Twenty-six initial spring Nebraska sightings dating from the mid-1930s to the late 1970s (Johnsgard, 1980) in northwestern Nebraska are from March 16 to May 21, with a median of April 3. Six final autumn records are from September 5 to November 26, with a mean of October 1. Elsewhere in Nebraska this species is a spring and autumn migrant, and sometimes a winter visitor. It is a very rare (east) to variably common (west) during winter in Kansas, with records extending from September to May (Thompson *et al.*, 2011), and a winter resident in Oklahoma, with most records occurring from September to April (Woods and Schnell, 1984). It is a common to abundant winter visitor in the Texas panhandle, with records extending from October to April (Seyffert, 2001). The average number of birds observed per party-hour for the 2013–2014 Christmas Bird Count were: N. Dakota 0, S. Dakota 0, Nebraska 0, Kansas 0, Oklahoma 0.06, northwest Texas 1.0. For complete species abundance data, see Appendix 1:107.

*Habitats.* Migrants inhabit shortgrass plains, pasturelands, and plowed fields.

*National Population.* Breeding Bird Surveys between 1966 and 2012 indicate that this species exhibited a survey-wide population decrease (4.25% annually) during that period. It has been designated a species of continental conservation importance, with the Prairie Avifaunal Biome supporting an estimated 57 percent of the continental winter population (Rich *et al.*, 2004). The estimated 1990s continental population was about 1.1 million birds (Rich *et al.*, 2004).

*Further Reading.* Bent, 1968; With, 1994 (*The Birds of North America:* No. 96); Byers, Curson and Olsen, 1995; Rising and Beadle, 1996; Dechant *et al.*, 1999f; Johnsgard, 2001b.

## Snow Bunting
*Plectrophenax nivalis*

Snow buntings are true snowbirds; they are rarely seen during winter where no snow is on the ground, they bathe in snow, burrow into snow to stay warm, and can withstand environmental temperatures as low as -50° F., so long as food is available. The few snow buntings that migrate farther south than the Dakotas are often seen among flocks of horned larks or longspurs.

*Winter Distribution.* This is the most northerly-oriented of our wintering Great Plains birds, with a distribution that Root (1988) mapped as centered in North Dakota and adjacent parts of Montana and Saskatchewan. Our data confirm North Dakota as having the highest average numbers in the Great Plains states, and very few birds wintering south of South Dakota. North Dakota numbers show a possible long-term decline, which might mean that they have begun to winter farther north or west. The long-term average Christmas Counts for Manitoba and Saskatchewan have remained stable since a peak in the mid-1980s, while those for Montana show sharp peaks in the mid-1980s and late 1990s, along with a gradually increasing long-term trend. It seems likely that the North Dakota numbers are not indicative of a general downward trend in the species' overall population, but rather only reflect regional fluctuations.

*Seasonality and Migrations.* The snow bunting is a variably common winter visitor throughout the five Great Plains states, mainly in northern regions. This species was observed all 40 years on Christmas Counts from 1968 to 2007 in North Dakota, 37 years in South Dakota, 23 years in Nebraska, 11 years in Kansas, and one year in Oklahoma. In South Dakota wintering is regular (Tallman, Swanson and Palmer, 2002). Eleven initial autumn Nebraska sightings dating from the mid-1930s to the late 1970s (Johnsgard, 1980) are from October 19 to December 24, with a median of November 16. Thirty-one final spring Nebraska sightings dating from the mid-1930s to the late 1970s are from January 1 to March 23, with a median of February 10. It is rare and irregular during winter in Kansas (Thompson *et al.*, 2011), and a rare winter resident in Oklahoma, with most records extending from December to February (Woods and Schnell, 1984). It is an extremely rare winter visitor in the Texas panhandle, with a single December record (Seyffert, 2001). The average number of birds observed per party-hour for the 2013–2014 Christmas Bird Count were: N. Dakota 11.37, S. Dakota 1.41, Nebraska 0.42, Kansas 0.01, Oklahoma 0, northwest Texas 0. For complete species abundance data, see Appendix 1:108.

*Habitats.* Migrants and wintering birds are associated with open plains and snow-covered fields.

*National Population.* Breeding Bird Survey trend data are not available for this high arctic-breeding breeder. The estimated 1990s continental population was about 19 million birds (Rich *et al.,* 2004). There is also a Eurasian population.

*Further Reading.* Bent, 1968; Godfrey, 1986; Lyon and Montgomerie, 1995 (*The Birds of North America:* No. 199); Byers, Curson and Olsen, 1995; Rising and Beadle, 1996.

# FAMILY PARULIDAE: WOOD WARBLERS

## Orange-crowned Warbler
*Vermivora celata*

This rather drab-colored warbler is one of the more cold-tolerant in the entire wood-warbler family, probably because it is able to shift to berry-eating as the insect supply diminishes during autumn. In any case, it is one of the last autumn warbler migrants, and among the earliest to arrive in spring.

*Winter Distribution.* This species of warbler is one of the earliest spring and latest autumn warbler migrants, so it sometimes tarries long enough in autumn to appear on Christmas Counts in the Great Plains. Root (1988) mapped its winter distribution as reaching extreme southeastern Oklahoma, Our data show trace levels of the species from Nebraska southward, but never in numbers large enough to exceed the "present" category.

*Seasonality and Migrations.* The orange-crowned warbler is a common seasonal migrant, wintering southwardly. This species was observed two years on Christmas Counts from 1968 to 2007 in Nebraska, 14 years in Kansas, 28 years in Oklahoma, and ten years in the Texas panhandle. Autumn South Dakota records extend to late October (Tallman, Swanson and Palmer, 2002). Sixty final autumn Nebraska sightings are from September 11 to November 6, with a median of October 15 (Johnsgard, 1980). It is a very rare during winter in Kansas, with records extending to February (Thompson *et al.*, 2011), and a rare winter visitor in Oklahoma, with continuous monthly records extending from August to May (Woods and Schnell, 1984). It is a very rare winter visitor in the Texas panhandle, with records for every month except July (Seyffert, 2001).

*Habitats.* Migrants and wintering birds are associated with deciduous forests, woodlands and brushy thickets. They also forage in stands of tall sunflowers (*Helianthus*), ragweeds (*Ambrosia*) and shrubs, often fairly close to the ground.

*National Population.* Breeding Bird Surveys between 1966 and 2012 indicate that this species exhibited a survey-wide population decrease (1.07% annually) during that period. The estimated 1990s continental population was about 76 million birds (Rich *et al.*, 2004).

*Further Reading.* Bent, 1953; Morse, 1989; Sogge, Gilbert and Van Riper, 1994 (*The Birds of North America:* No. 101); Curson, Quinn and Beadle, 1994; Dunn and Garrett, 1997.

## Yellow-rumped Warbler
*Setophaga coronata*

The old name for the eastern form of yellow-rumped warbler was myrtle warbler, so-named because of the species' fondness for the fruits of wax myrtles (*Myrica spp.*) during winter. These berries are high in energy-rich waxes that the birds are somehow able to digest. Where these sources are unavailable, other fruits such as those of junipers, sumacs (*Rhus*) and poison ivy (*Toxicodendron*) may be consumed. Recent climate-warming has probably allowed insect foods to remain available longer into autumn as well, making more northerly wintering possible for this highly adaptable forager.

*Winter Distribution.* This is the very earliest spring and latest autumn warbler migrant in the Great Plains, Root (1988) mapped the northern limits (of the nominate or *coronata* race) as extending north to southeastern Kansas an west about the Oklahoma–Texas panhandle line. Our data indicate that the birds (both races) are seen with some regularity during Nebraska Christmas Counts, and that they can be expected to occur yearly in Kansas and Oklahoma. Decade-long count averages have increased progressively over the five-state Plains region by about four-fold over the four-decade period, and have increased about ten-fold in northwest Texas.

*Seasonality and Migrations.* The yellow-rumped warbler is common Great Plains seasonal migrant and a local breeder (Nebraska to North Dakota), wintering southwardly. This species was observed one year on Christmas Counts from 1968 to 2007 in North Dakota, six years in South Dakota, 13 years in Nebraska, all 40 years in both Kansas and Oklahoma, and 27 years in the Texas panhandle. In South Dakota this species is very rare during winter (Tallman, Swanson and Palmer, 2002). Seventy-seven final autumn Nebraska sightings are from September 10 to December 18, with a median of October 22 (Johnsgard, 1980). It is occasional during winter in Kansas, especially in the east (Thompson *et al.*, 2011), and is a winter resident in Oklahoma, with continuous monthly records extending from August to June (Woods and Schnell, 1984). It is a rare to uncommon winter visitor and a common to abundant migrant in the Texas panhandle. There the western (Audubon's) race outnumbers the eastern (myrtle) race by about three to one, and is both an earlier autumn migrant and a later spring migrant (Seyffert, 2001). The average number of birds observed per party-hour for the 2013–2014 Christmas Bird Count were: N. Dakota 0, S. Dakota 0.15, Nebraska 0.2, Kansas 0.41, Oklahoma 0.68, northwest Texas 0. For complete species abundance data, see Appendix 1:109.

*Habitats.* The species is widespread in wooded habitats during migration, arriving well before most leaves appear, and wintering as far north as southern Oklahoma, frequently foraging on juniper berries.

*National Population.* Breeding Bird Surveys between 1966 and 2012 indicate that this species exhibited a survey-wide population decrease (0.12% annually) during that period. The estimated 1990s continental population north of Mexico was about 130 million birds (Rich *et al.,* 2004). The species breeds south to Guatemala.

*Further Reading.* Bent, 1953; Morse, 1989; Curson, Quinn and Beadle, 1994; Dunn and Garrett, 1997; Hunt and Flaspohler, 1998 (*The Birds of North America:* No. 376); Dunn and Tessaglia-Hymes, 1999.

## Pine Warbler
### *Setophaga pinus*

Like orange-crowned and yellow-rumped warblers, pine warblers shift away from eating insects during autumn and begin to consume various fruits and berries, but unlike these other two species it has also become adapted to eating pine seeds. This same food source is extensively used by wintering brown-headed nuthatches, and the two species can often be seen foraging together, sometimes also in the company of chickadees, titmice, kinglets, and other warblers.

*Winter Distribution.* Root (1988) mapped the regional winter distribution of this species as being limited to its known breeding habitat in southeastern Oklahoma, where it often associates and forages with brown-headed nuthatches in pine trees. Our data show no clear population trend.

*Seasonality and Migrations.* The pine warbler is a local breeding resident in Oklahoma; farther north it is a local seasonal migrant. Autumn South Dakota records extend to mid-December (Tallman, Swanson and Palmer, 2002). Four autumn Nebraska records are from September 7 to September 22 (Johnsgard, 1980). It is reported rarely during winter in Kansas (Thompson *et al.,* 2011). This species was observed 37 years on Christmas Counts from 1968 to 2007 in Oklahoma, and is a very rare migrant in the Texas panhandle, with autumn records extending from September to December (Seyffert, 2001). The average number of birds observed per party-hour for the 2013–2014 Christmas Bird Count were: N. Dakota 0, S. Dakota 0, Nebraska 0, Kansas 0.03, Oklahoma 0.48, northwest Texas 0. For complete species abundance data, see Appendix 1:110.

*Habitats.* Migrants and wintering birds are often associated with pine forests, but also utilize deciduous forests, orchards and suburban bird feeders, since this warbler is unusual in that it sometimes consumes seeds. Breeding occurs in upland pine and pine-hardwood forests.

*National Population.* Breeding Bird Surveys between 1966 and 2012 indicate that this species exhibited a survey-wide population increase (0.94% annually) during that period. Breeding Bird Surveys between 1966 and 2006 indicate that this species exhibited a significant national population increase (1.1% annually) during that period. The estimated 1990s continental population was about 11 million birds (Rich *et al.,* 2004).

*Further Reading.* Bent, 1953; Morse, 1967b, 1989; Curson, Quinn and Beadle, 1994; Dunn and Garrett, 1997; Rodewald, Withgott and Smith, 1999 (*The Birds of North America:* No. 438); Dunn and Tessaglia-Hymes, 1999.

## Common Yellowthroat
### *Geothlypis trichas*

This is a marsh-loving warbler that is not likely to remain north long after freeze-up in autumn. There is no indication that it shifts to non-insect foods in late fall, and according to Root (1988) the greatest winter concentrations occur where the average January temperatures are above 45 F.

*Winter Distribution.* According to Root (1988), this species winters as far north as southern Oklahoma. Our data show that it is now a vagrant in states all the way north to South Dakota, but has never exceeded the "present" numerical category during the period of study.

*Seasonality and Migrations.* This species is a common Great Plains seasonal migrant and widespread wetland breeder, wintering southwardly. It was not observed on Christmas Counts from 1968 to 2007 in North Dakota, but was seen one year in South Dakota and Nebraska, five years in Kansas, 25 years in Oklahoma, and two years in the Texas panhandle. The range of 114 final autumn Nebraska sightings is from July 20 to October 29, with a median of September 13 (Johnsgard, 1980). Records in Kansas extend to late November (Thompson *et al.,* 2011), but Oklahoma records extend through December and January (Woods and Schnell, 1984). It is an extremely rare winter visitor in the Texas panhandle (Seyffert, 2001).

*Habitats.* While in Nebraska this species is found near moist or aquatic sites, especially among tall grasses, emergent vegetation, and shrubs or trees along shorelines. Occasionally it also inhabits upland shrub thickets, retired croplands, weedy residential areas, and overgrown orchards.

*National Population.* Breeding Bird Surveys between 1966 and 2012 indicate that this species exhibited a survey-wide population decrease (1.01% annually) during that period. Breeding Bird Survey trend data are not available for this species. The estimated 1990s continental population was about 32 million birds (Rich *et al.,* 2004).

*Further Reading.* Bent, 1953; Morse, 1989; Curson, Quinn and Beadle, 1994; Dunn and Garrett, 1997; Guzy and Ritchison, 1999 (*The Birds of North America:* No. 448).

# FAMILY EMBERIZIDAE:
# TOWHEES & SPARROWS

## Spotted Towhee
*Pipilo maculatus*

It seems strange that the two common Great Plains towhees, the spotted and eastern, seem to having different tolerances to cold, or at least surprisingly different migration patterns. For example, in both eastern Nebraska and Iowa the eastern towhee is the breeding form but the spotted towhee is the wintering species, while the eastern towhee usually doesn't winter north of southern Kansas and Missouri. Even in Kansas and Missouri the spotted towhee is more common during winter than the eastern towhee. In spite of these wintering differences, the winter habitats and foods of the two species seem to be essentially identical.

*Winter Distribution.* Because of the merging of the spotted and eastern towhees during the years of her study, Root's (1988) map does not distinguish the two towhee species. She found towhees wintering north to southern Nebraska, and the highest Plains numbers in southwestern Oklahoma. Our data show the Texas panhandle to have the highest average numbers, followed by Oklahoma. These regions as well as Kansas showed a progressive increase in towhee numbers over the four-decade period, with Kansas and Oklahoma collectively averaging a roughly ten-fold increase, and northwest Texas showing a four-fold increase in average numbers.

*Seasonality and Migrations.* The spotted towhee is a common Great Plains seasonal migrant and widespread breeder in western areas, wintering southwardly. This species was observed six years on Christmas Counts between 1983 and 2007 in North Dakota, ten years in South Dakota, 21 years in Nebraska, 23 years in Kansas, and all 24 years in Oklahoma and the Texas panhandle. In South Dakota wintering in the state is very rare (Tallman, Swanson and Palmer, 2002). Wintering by spotted towhees is rare in central and western Nebraska, but fairly common in the southeast. The species is evenly distributed statewide in December except for the panhandle, when there is a general movement toward the southeast (Sharpe, Silcock and Jorgensen, 2001). It is uncommon (east) to rare (west) during winter in Kansas (Thompson *et al.*, 2011), and a winter resident in Oklahoma (Woods and Schnell, 1984). It is a fairly common to common winter visitor in the Texas panhandle, with records for every month but June and August (Seyffert, 2001). The average number of birds observed per party-hour for the 2013–2014 Christmas Bird Count were: N. Dakota 0, S. Dakota 0.01, Nebraska

0.01, Kansas 0.12, Oklahoma 0.17, northwest Texas 0.2. For complete species abundance data, see Appendix 1:111.

*Habitats.* This species' habitats are much like those of the eastern towhee, and include riparian woodland, shrublands, and sometimes city parks or suburbs.

*National Populations.* Breeding Bird Surveys between 1966 and 2012 indicate that this species exhibited a survey-wide population increase (0.07% annually) during that period. The estimated 1990s continental population north of Mexico was about 12.6 million birds (Rich *et al.,* 2004). The species breeds south to Guatemala.

*Further Reading.* Bent, 1968; Davis, 1957; Byers, Curson and Olsen, 1995; Rising and Beadle, 1996; Greenlaw, 1996b (*The Birds of North America:* No. 263); Dunn and Tessaglia-Hymes, 1999.

### Eastern Towhee
*Pipilo erythropthalmus*

Like many other sparrows, towhees forage by hopping and simultaneously scratching the substrate backwards with both feet, exposing seeds and any hidden invertebrates to view. Their relatively long legs make this an effective strategy for foraging in leafy or litter-covered areas, which they typically do alone rather than in flocks.

*Winter Distribution.* Like the spotted towhee, separate data on the eastern towhee was unavailable while both were merged between the 1960s and the mid-1980s, when they were once again recognized as two species. Our data from 1986 onward show only trace numbers of eastern towhees in the Great Plains region during winter, or considerably fewer than are usually believed present in the region during winter.

*Seasonality and Migrations.* The eastern towhee is a common Great Plains seasonal migrant and widespread breeder in eastern areas, wintering southwardly. This species was observed four years on Christmas Counts between 1983 and 2007 in Nebraska, 12 years in Kansas ten years in Oklahoma, and two years in the Texas panhandle. The latest South Dakota record is for September 5 (Tallman, Swanson and Palmer, 2002). Eastern towhees are rare in most of Nebraska during winter, and even where they are most common in the extreme southeast they are outnumbered by spotted towhees (Sharpe, Silcock and Jorgensen, 2001). The species is uncommon during winter in eastern Kansas (Thompson *et al.,* 2011), and is a winter resident in Oklahoma (Woods and Schnell, 1984). It is a very rare winter visitor in the Texas panhandle (Seyffert, 2001).

*Habitats.* Towhees occur in brushy fields, thickets, woodland edges or open-ings, second-growth forests, and city parks or suburbs with trees and tall shrubbery.

*National Population.* Breeding Bird Surveys between 1966 and 2012 indicate that this species exhibited a survey-wide population decrease (1.41% annu-ally) during that period. The estimated 1990s continental population was about 11 million birds (Rich *et al.,* 2004).

*Further Reading.* Barbour, 1941; Bent, 1968; Byers, Curson and Olsen, 1995; Rising and Beadle, 1996; Greenlaw, 1996a *(The Birds of North America:* No. 262); Dunn and Tessaglia-Hymes, 1999.

## Canyon Towhee
### *Pipilo fuscus*

Unlike the two previous species, canyon (previously brown) towhees are year-around residents throughout their range. This means that territories can be maintained permanently, and permanent pair-bonding is typical. The birds also are not such industrious scratchers as are the other towhees, and are more prone to simply pick foods off the surface of the rather barren desert-like hab-itats that they inhabit.

*Winter Distribution.* This southwestern species is topographically limited to canyon and mesa country. Root's (1988) map shows a center of abundance in the Oklahoma panhandle and adjacent southwestern Kansas, grading off in the Texas panhandle. Our data show high but progressively declining numbers for Oklahoma, and much smaller but seemingly increasing num-bers for the Texas panhandle. A declining national population is strongly indicated by Breeding Bird Survey results, so the Texas panhandle trend is probably the result of a recent range expansion into that region (Seyffert, 2001), thereby countering the national downward trend.

*Seasonality and Migrations.* The canyon towhee is a permanent resident throughout it range. The average number of birds observed per party-hour for the 2013–2014 Christmas Bird Count were: N. Dakota 0, S. Dakota 0, Nebraska 0, Kansas 0, Oklahoma 0.22, northwest Texas 0.23. For complete species abundance data, see Appendix 1:112.

*Habitats.* Typical year-around habitats include shrublands of pinyon pines (*Pinus edulis*) and junipers (*Juniperus*), as well as cholla cactus (*Opuntia im-bricata*) grasslands in canyon and mesa habitats of western Oklahoma and the Texas panhandle. Although it breeds in rugged topography, it often can also be seen around ranches, farms and villages, especially during winter.

*National Population.* Breeding Bird Surveys between 1966 and 2012 indicate that this species exhibited a survey-wide population decrease (1.21% annually) during that period. The estimated 1990s continental population north of Mexico was about 1.6 million birds (Rich *et al.,* 2004). The species breeds south to central Mexico.

*Further Reading.* Bent, 1968; Davis, 1957; Byers, Curson and Olsen, 1995; Rising and Beadle, 1996; Johnson and Haight, 1996 (*The Birds of North America:* No. 264); Dunn and Tessaglia-Hymes, 1999.

## Rufous-crowned Sparrow
### *Aimophila ruficeps*

This species is rarely seen far from rocks and brush, where during winter it forages quietly for seeds and generally remains out of sight. The end of its inconspicuous winter behavior is marked by the onset of spring rains, which triggers a resurgence of territoriality and the onset of singing and breeding activities.

*Winter Distribution.* Root (1988) mapped this desert-adapted sparrow as extending south from northern Kansas to the Oklahoma-Texas border, where it is most abundant in southwestern Oklahoma. Our data suggest that only trace-level populations occur in Kansas, and that Texas panhandle population is substantially higher than that of Oklahoma. Both the Texas and Oklahoma populations may be in decline, as is indicated by national Breeding Bird Surveys.

*Seasonality and Migrations.* The rufous-crowned sparrow is a permanent resident throughout its limited range. The average number of birds observed per party-hour for the 2013–2014 Christmas Bird Count were: N. Dakota 0, S. Dakota 0, Nebraska 0, Kansas 0, Oklahoma 0.04, northwest Texas 0.16, For complete species abundance data, see Appendix 1:113.

*Habitats.* In its limited Great Plains range, this species favors rough topography, especially rocky slopes, and shows little or no movement to lower altitudes during winter (Sutton, 1967). This species was observed six years on Christmas Counts from 1968 to 2007 in Kansas, all 40 years in Oklahoma, and 37 years in the Texas panhandle.

*National Population.* Breeding Bird Surveys between 1966 and 2012 indicate that this species exhibited a survey-wide population decrease (1.43% annually) during that period. The estimated 1990s continental population north of Mexico was about 1.2 million birds (Rich *et al.,* 2004). The species breeds south to southern Mexico.

*Further Reading.* Bent, 1968; Byers, Curson and Olsen, 1995; Rising and Beadle, 1996; Collins, 1999 (*The Birds of North America:* No. 472).

## Chipping Sparrow
*Spizella passerina*

Chipping sparrows are usually found close to trees during winter, so they can divide their time between ground foraging and perching or roosting in trees. Like tree sparrows, they selectively eat small seeds, such as those of crab grass (*Digitaria*) and pigeon grass (*Setaria*), and of annual weeds such as amaranth (*Amaranthus*), ragweed (*Ambrosia*) and knotweed *(Polygonum)*. It has been estimated that a single chipping sparrow might consume over two pounds of seeds over the course of a winter.

*Winter Distribution.* Root (1988) mapped this species as having an isolated winter population in Kansas, and another population extending from Oklahoma and the Texas panhandle sooth into Texas. Our data suggest a much broader distribution in the Great Plains, extending from the Dakotas (with only trace levels) south to a peak in Oklahoma, and substantially smaller numbers in the Texas panhandle. There is no clear evidence of a regional population trend.

*Seasonality and Migrations.* The chipping sparrow is a common Great Plains seasonal migrant and widespread breeder, wintering southwardly. This species was observed four years on Christmas Counts from 1968 to 2007 in North Dakota, 13 years in South Dakota, seven years in Nebraska, 15 years in Kansas, 38 years in Oklahoma, and 21 years in the Texas panhandle. In South Dakota wintering is extremely rare (Tallman, Swanson and Palmer, 2002). Ninety-nine final autumn Nebraska records range from July 23 to December 20, with a median of October 2 (Johnsgard, 1980). There are a few winter (December to February) records from Kansas (Thompson and Ely, 1992), but this species is a permanent resident in Oklahoma (Woods and Schnell, 1984). It is a rare and irregular winter visitor in the Texas panhandle, but is an abundant migrant, with records for every month (Seyffert, 2001). The average number of birds observed per party-hour for the 2013–2014 Christmas Bird Count were: N. Dakota 0, S. Dakota 0, Nebraska 0, Kansas 0.34, Oklahoma 0.37, northwest Texas 0. For complete species abundance data, see Appendix 1:114.

*Habitats.* Throughout the year this species is associated with the margins of deciduous forests, parks, gardens, residential areas, farmsteads, orchards and other open areas with nearby or scattered trees and few or no shrubs.

*National Population.* Breeding Bird Surveys between 1966 and 2012 indicate that this species exhibited a survey-wide population decrease (0.54% annually) during that period. The estimated 1990s continental population north of Mexico was about 89 million birds (Rich *et al.,* 2004). The species breeds south to Nicaragua.

*Further Reading.* Bent, 1968; Byers, Curson and Olsen, 1995; Rising and Beadle, 1996; Middleton, 1998 (*The Birds of North America:* No. 334); Dunn and Tessaglia-Hymes, 1999.

## Field Sparrow
*Spizella pusilla*

Field sparrows and chipping sparrows are about the same size; both species weigh less than half an ounce, making them one of the smallest sparrows to winter as far north as the central Great Plains. Like chipping sparrows, field sparrows are open-ground foragers, seeking out very small seeds of grasses and weeds. Winter flock sizes tend to be small, and the birds prefer to remain close to bushes or small trees, into which they can quickly retreat.

*Winter Distribution.* Root's (1998) map shows field sparrows wintering north to southern South Dakota, increasing southwardly, and maximum numbers occurring in southeastern Oklahoma. Our data are consistent with that general pattern. There is a trend toward declining numbers for the five-state area as a whole, and also for northwestern Texas. This trend is in accordance with national Breeding Bird Survey data indicating a declining population.

*Seasonality and Migrations.* The field sparrow is a common Great Plains seasonal migrant and widespread breeder, wintering southwardly. This species was observed one year on Christmas Counts from 1968 to 2007 in North Dakota, 18 years in South Dakota, 15 years in Nebraska, all 40 years in Kansas and Oklahoma, and 37 years in the Texas panhandle. In South Dakota wintering is extremely rare (Tallman, Swanson and Palmer, 2002). Eighty-three final autumn Nebraska sightings are from August 1 to December 26, with a median of October 6 (Johnsgard, 1980). It is rare during winter in Kansas, mainly eastwardly (Thompson *et al.*, 2011), but is considered a permanent resident in Oklahoma (Woods and Schnell, 1984) and the Texas panhandle (Seyffert, 2001). The average number of birds observed per party-hour for the 2013–2014 Christmas Bird Count were: N. Dakota 0, S. Dakota 0, Nebraska 0, Kansas 0.3, Oklahoma 0.65, northwest Texas 0.23. The average number of birds observed per party-hour for the 2013–2014 Christmas Bird Count were: N. Dakota 0, S. Dakota 0, Nebraska 0, Kansas 0.34, Oklahoma 0.37, northwest Texas 0. For complete species abundance data, see Appendix 1:115

*Habitats.* This species occurs in brushy, open woodlands, forest edges, brushy ravines or draws, sagebrush flats, abandoned hayfields, forest clearings, and similar open habitats having scattered shrubs or low trees. It is similar to the chipping sparrow in its habitat needs, but is more often seen in shrubs, and less often in trees.

*National Population.* Breeding Bird Surveys between 1966 and 2012 indicate that this species exhibited a survey-wide population decrease (2.40% annually) during that period. The estimated 1990s continental population was about 8.2 million birds (Rich *et al.,* 2004).

*Further Reading.* Bent, 1968; Fretwell, 1968; Carey, Burhans and Nelson, 1994 (*The Birds of North America:* No. 103); Byers, Curson and Olsen, 1995; Rising and Beadle, 1996; Dunn and Tessaglia-Hymes, 1999.


### American Tree Sparrow
*Spizella arborea*

Tree sparrows are one of the hallmark birds of a Great Plains winter; they often may seen in small flocks huddled in low, leafless trees trying to stay out of the wind, or industriously searching low grasslands mostly covered by snow. They consume vast quantities of agricultural weed seeds (up to nearly a thousand in a single crop, and another 700 in the stomach), especially small seeds such as those of ragweed (*Ambrosia*), lambs-quarters (*Chenopodium*) and annual grasses, and so are a benefit to farmers. They can become quite tame at feeders during heavy snows; I have then watched them through a storm-door window from as little as four feet away, by simply lying prone and remaining perfectly still.

*Winter Distribution.* This is one of the most widely distributed winter birds of the Great Plains region. Root (1988) mapped its distribution as centering in western Nebraska, with a secondary peak in Kansas. Our data show a clear peak of abundance in Kansas, extending north to North Dakota and south to the Texas panhandle at roughly equal reduced rates. The five-state average numbers indicate a progressive population decline over the entire four-decade period, as is also the case in northwestern Texas. This is one of the generally ignored and seemingly common species for which national population data are lacking, and to which conservation attention should probably be directed.

*Seasonality and Migrations.* The American tree sparrow is a common winter visitor throughout the Great Plains. This species was observed all 40 years on Christmas Counts from 1968 to 2007 in all states from North Dakota to Oklahoma, and the Texas panhandle. In South Dakota wintering is common (Tallman, Swanson and Palmer, 2002). The range of 127 initial autumn sightings is from September 3 to December 31, with a median of October 21. Sixty-five final spring sightings range from January 24 to May 27, with a median of April 6. Most records in Kansas are from September to April, and Oklahoma records extend from October to April (Thompson *et al.*, 2011; Woods and Schnell, 1984). It is an uncommon to common winter visitor in the Texas panhandle, with records extending from October to May (Seyffert, 2001). The average number of birds observed per party-hour for the 2013–2014 Christmas Bird Count were: N. Dakota 0.18, S. Dakota 2.79, Nebraska 5.2, Kansas 9.05, Oklahoma 1.0, northwest Texas 0.25. For complete species abundance data, see Appendix 1:116.

*Habitats.* During migration and winter periods this species is found in flocks among thickets, brushy areas, and shrubby or weedy grasslands.

*National Population.* Breeding Bird Survey trend data are not available for this boreal breeder. It has been designated a species of continental conservation importance, with the Prairie Avifaunal Biome supporting an estimated 85 percent of the continental winter population (Rich *et al.*, 2004). The estimated 1990s continental population was about 26 million birds (Rich *et al.*, 2004).

*Further Reading.* Sabine, 1949; Bent, 1968; Godfrey, 1986; Naugler, 1993 (*The Birds of North America:* No. 37); Byers, Curson and Olsen, 1995; Rising and Beadle, 1996; Dunn and Tessaglia-Hymes, 1999.

## Vesper Sparrow
*Pooecetes gramineus*

One of the larger grassland sparrows, with adults weighing about an ounce, this an open-country species that during winter consumes large quantities of weed seeds, as well as waste grain. It feeds by walking or running over fairly open ground, stopping periodically to scratch for seeds or insects. By spring males may begin performing the songs for which the species was poetically named, often from early morning until well into the evening ("vesper") hours.

*Winter Distribution.* Root's (1988) map shows no wintering in the Plains states north of Kansas, although an isolated population is indicated for central

Iowa. Our data show almost no birds remaining in Kansas during the Christmas Count period, and also rather few in Oklahoma and the Texas panhandle. No clear population trend is evident from our data, so we cannot confirm a nationally declining trend, as suggested by Breeding Bird Survey data.

*Seasonality and Migrations.* The vesper sparrow is a common Great Plains seasonal migrant and widespread breeder, wintering southwardly. This species was observed one year on Christmas Counts from 1968 to 2007 in North Dakota, three years in South Dakota and Nebraska, 32 years in Kansas, 39 years in Oklahoma, and 23 years in the Texas panhandle. In South Dakota wintering is extremely rare (Tallman, Swanson and Palmer, 2002). Eighty-three final autumn Nebraska sightings range from August 13 to November 24, with a median of October 9 (Johnsgard, 1980). It is a very rare during winter in Kansas (Thompson *et al.*, 2011), and is a rare winter visitor in Oklahoma, with continuous monthly records extending from September to May (Woods and Schnell, 1984). It is a rare winter visitor in the southern Texas panhandle (Seyffert, 2001). The average number of birds observed per party-hour for the 2013–2014 Christmas Bird Count were: N. Dakota 0, S. Dakota 0, Nebraska 0, Kansas 0.01, Oklahoma 0.01, northwest Texas 0.11. For complete species abundance data, see Appendix 1:117.

*Habitats.* Migrants and wintering birds frequent overgrown fields, prairie edges, and similar habitats where grasslands join or are mixed with shrubs and scattered low trees.

*National Population.* Breeding Bird Surveys between 1966 and 2012 indicate that this species exhibited a survey-wide population decrease (0.89% annually) during that period. The estimated 1990s continental population was about 30 million birds (Rich *et al.*, 2004).

*Further Reading.* Bent, 1968; Byers, Curson and Olsen, 1995; Rising and Beadle, 1996; Johnsgard, 2001b; Jones and Cornely, 2002 (The *Birds of North America:* No. 624).

### Lark Sparrow
*Chondestes grammacus*

Comparable in size to a vesper sparrow, the lark sparrow is similarly fond of open country, and almost bare terrain. Country roads are a favorite foraging location, where grasshoppers or weed seeds are often sought out. Even in summer, the birds are somewhat gregarious, and during winter flocking is regular behavior.

*Winter Distribution.* Root (1988) mapped the northern limit of this grass-land sparrow's winter distribution as following the Texas–Oklahoma bor-der along the Red River valley, and with the highest numbers occurring in southernmost Texas. The very small numbers shown in our data support the likelihood that essentially all of the Great Plains' lark sparrow popula-tion still winters south of the Oklahoma border.

*Seasonality and Migrations.* The lark sparrow is a common Great Plains sea-sonal migrant and widespread breeder, wintering southwardly. This species was observed two years on Christmas Counts from 1968 to 2007 in South Dakota and Nebraska, 10 years in Kansas, 22 years in Oklahoma, and 14 years in the Texas panhandle. In South Dakota, wintering is extremely rare (Tallman, Swanson and Palmer, 2002). Seventy-six final autumn Nebraska sightings range from July 23 to November 13, with a median of Septem-ber 3 (Johnsgard, 1980). It is very rare during winter in Kansas, with a few records extending from late November to mid-February (Thompson and Ely, 1992), but there are monthly records throughout the year in Okla-homa (Woods and Schnell, 1984). It is an extremely rare winter visitor in the Texas panhandle (Seyffert, 2001).

*Habitats.* Throughout the year this species occupies natural grasslands or weedy fields that adjoin or include scattered trees, shrubs, and weeds.

*National Population.* Breeding Bird Surveys between 1966 and 2012 indicate that this species exhibited a survey-wide population decrease (0.88% an-nually) during that period. The estimated 1990s continental population north of Mexico was about 9.9 million birds (Rich *et al.,* 2004). The species breeds south to southern Mexico.

*Further Reading.* Bent, 1968; Byers, Curson and Olsen, 1995; Rising and Bea-dle, 1996; Dechant *et al.,* 1999j; Martin and Parrish. 2000 (*The Birds of North America:* No. 488); Johnsgard, 2001b.

### Lark Bunting
*Calamospiza melanocorys*

Few birds are more evocative of the shortgrass prairies of the high plains than are lark buntings. In the spring the stunningly patterned males' song-flights enliven the sights and sounds of the western grasslands. They probably typi-cally winter in Mexico, but a few sometimes remain fairly late in the fall, and by then the males are dull and female-like in plumage. Even in summer the birds are highly sociable, and during winter some flocks estimated in the thou-sands have been described. During autumn and spring migrations the birds often gather in large numbers at roadsides, probably to search for seeds, insects and grit.

*Winter Distribution.* Root (1988) mapped the lark bunting's winter distribution as barely entering the western edge of the Texas panhandle, and with most birds wintering along the southwestern Texas border. Our data suggest that some birds remain in Kansas until at least late December, and that the Texas panhandle is probably a significant wintering area. The Texas panhandle population may be quite variable, judging from our sample, and no population trend can be detected.

*Seasonality and Migrations.* The lark bunting is a common Great Plains seasonal migrant and widespread breeder in the western plains, wintering southwardly. This species was observed 20 years on Christmas Counts from 1968 to 2007 in Kansas, 11 years in Oklahoma, and 26 years in the Texas panhandle. The latest South Dakota record is for November 9 (Tallman, Swanson and Palmer, 2002). Sixty-five final autumn Nebraska sightings are from July 20 to October 13, with a median of August 30 (Johnsgard, 1980). It is a very rare to regular during winter in southwestern Kansas (Thompson *et al.*, 2011). Year-around records exist in Oklahoma (Woods and Schnell, 1984). It is a rare to common or abundant winter visitor in the Texas panhandle, with records for every month (Seyffert, 2001). The average number of birds observed per party-hour for the 2013–2014 Christmas Bird Count were: N. Dakota 0, S. Dakota 0, Nebraska 0, Kansas 0, Oklahoma 0.01, northwest Texas 1.56. For complete species abundance data, see Appendix 1:118.

*Habitats.* For much of the year this species is usually found in mixed shortgrass prairie and sage-dominated areas, but it also occurs in areas of taller grasses with scattered shrubs and along weedy roadsides, in retired croplands, and in fields of alfalfa or clover. Outside the breeding season it is highly gregarious.

*National Population.* Breeding Bird Surveys between 1966 and 2012 indicate that this species exhibited a survey-wide population decrease (3.55% annually) during that period. It has been designated a species of continental conservation importance, with the Prairie Avifaunal Biome supporting an estimated 31 percent of the continental winter population (Rich *et al.*, 2004). The estimated 1990s continental population was about 27 million birds (Rich *et al.*, 2004). This species was yellow-listed in the National Audubon Society's 2007 WatchList of rare and declining birds (Butcher *et* al., 2007).

*Further Reading.* Bent, 1968; Byers, Curson and Olsen, 1995; Rising and Beadle, 1996; Shane, 1996, 2000 (*The Birds of North America:* No. 542); Dechant *et al.*, 1999e.

## Savannah Sparrow
*Passerculus sandwichensis*

The Savannah sparrow is one of the smaller grassland sparrows, with adults weighing slightly over half an ounce. It is not especially gregarious even during winter, but has been reported to roost in groups within grassy fields. Seeds of weeds and grasses are its autumn and winter foods, which are sometimes obtained by towhee-like scratching.

*Winter Distribution.* Root's (1988) map shows the Savannah sparrow's winter distribution as extending north to southern Kansas, with no clear areas of abundance within the Great Plains states. Our data suggest that Oklahoma usually has a significant population, and the Texas panhandle population may be somewhat variable in size. Certainly the Oklahoma population shows a distinctly increasing trend, which is counter to the national trend indicated by Breeding Bird Surveys. Perhaps this apparent increase in Oklahoma reflects a northward shift in Texas populations.

*Seasonality and Migrations.* The Savannah sparrow is a common Great Plains seasonal migrant and widespread breeder in the northern plains, wintering southwardly. This species was observed one year on Christmas Counts from 1968 to 2007 in South Dakota, two years in Nebraska, 35 years in Kansas, all 40 years in Oklahoma, and 36 years in the Texas panhandle. The latest South Dakota record is for November 18 (Tallman, Swanson and Palmer, 2002). Thirty-nine final autumn Nebraska sightings are from October 2 to November 22, with a median of October 19 (Johnsgard, 1980). It is a very rare during winter in Kansas (Thompson *et al.*, 2011), and a winter resident in Oklahoma, with continuous monthly records extending from September to May (Woods and Schnell, 1984). It is a variably common winter visitor in the Texas panhandle, with records for every month but June (Seyffert, 2001). The average number of birds observed per party-hour for the 2013–2014 Christmas Bird Count were: N. Dakota 0, S. Dakota 0, Nebraska 0, Kansas 0.59, Oklahoma 0.9, northwest Texas 0.38. For complete species abundance data, see Appendix 1:119.

*Habitats.* Migrants and wintering birds are usually found in open grasslands, lightly grazed pastures, and brushy edges.

*National Population.* Breeding Bird Surveys between 1966 and 2012 indicate that this species exhibited a survey-wide population decrease (1.26% annually) during that period. The estimated 1990s continental population north of Mexico was about 79.5 million birds (Rich *et al.,* 2004). The species breeds south to Guatemala.

*Further Reading.* Bent, 1968; Wheelwright and Rising, 1993 (*The Birds of North*

*America:* No. 45); Byers, Curson and Olsen, 1995; Rising and Beadle, 1996; Dechant *et al.,* 1999g; Johnsgard, 2001b._

## Grasshopper Sparrow
### *Ammodramus savannarum*

Another small sparrow, with adults weighing slightly over a half-ounce, these birds are inconspicuous during summer and even more so during autumn and winter. As autumn passes, the birds shift from their primary food of grasshoppers to a mixture of weed, grass and sedge seeds.

*Winter Distribution.* Root's (1988) map of this species' winter distribution shows its northern limits in northern Texas, and its maximum numbers centered along Texas' Gulf coast. That situation may still apply, as there were no regions where counts exceeded the minimal "present' category over the entire four-decade period of study.

*Seasonality and Migrations.* The grasshopper sparrow is a common Great Plains seasonal migrant and widespread breeder, wintering southwardly. This species was observed six years on Christmas Counts from 1968 to 2007 in Oklahoma, and one year in the Texas panhandle. The latest South Dakota record is for October 26 (Tallman, Swanson and Palmer, 2002). Sixty-seven final autumn Nebraska sightings range from July 26 to November 6, with a median of September 9 (Johnsgard, 1980). It is a vagrant during winter in Kansas (Thompson *et al.*, 2011). A few winter records (December and February) exist for Oklahoma (Woods and Schnell, 1984), but it hasn't been reported after October in the Texas panhandle (Seyffert, 2001).

*Habitats.* This species occurs in mixed-grass prairies, pasturelands, shortgrass prairies, sage prairies, and to a limited extent tallgrass prairies. Areas that have grown up to shrubs are avoided, but scattered trees in grassland are sometimes used for song perches.

*National Population.* Breeding Bird Surveys between 1966 and 2012 indicate that this species exhibited a survey-wide population decrease (2.86% annually) during that period. It has been designated a species of continental conservation importance, with the Prairie Avifaunal Biome supporting an estimated six percent of the continental winter population (Rich *et al.,* 2004). The estimated 1990s continental population north of Mexico was about 14 million birds (Rich *et al.,* 2004). The species breeds south to Panama.

*Further Reading.* Bent, 1968; Byers, Curson and Olsen, 1995; Rising and Beadle, 1996; Vickery, 1996 (*The Birds of North America:* No. 239); Dechant *et al.,* 1999d; Johnsgard, 2001b.

## Le Conte's Sparrow
*Ammospiza leconteii*

This species is inconspicuous at all seasons, and its winter behavior is very poorly known. It is rarely found far from water, and prefers to skulk and hide rather than expose itself to easy view. While on migration it inhabits damp fields and marshes having thick grasses and sedges, where it is prone to hide rather than fly if disturbed.

*Winter Distribution.* Root (1988) mapped the winter distribution of this species as extending north to northern Kansas, and exhibiting a maximum density along the Red River valley of southern Oklahoma. Our data show only very few birds present in Oklahoma over the entire four-decade period, and still fewer in Kansas or the Texas panhandle. Root noted that even the area of highest recorded national density (Shreveport, Louisiana) had an average abundance of only 0.29 bird per party-hour, so the overall winter abundance of the species would seem to be quite low.

*Seasonality and Migrations.* The Le Conte's sparrow is an uncommon and inconspicuous seasonal migrant and local breeder in the northern plains, wintering southwardly. This species was observed 27 years on Christmas Counts from 1968 to 2007 in Kansas, 37 years in Oklahoma, and five years in the Texas panhandle. The latest South Dakota record is for October 28 (Tallman, Swanson and Palmer, 2002). Seventeen final autumn Nebraska sightings are from July 26 to November 9, with a median of October 20 (Johnsgard, 1980). It is uncommon during winter in eastern and southern Kansas (Thompson *et al.*, 2011), and a winter resident in Oklahoma, with continuous monthly records extending from September to May (Woods and Schnell, 1984). It is a very rare winter visitor in the Texas panhandle, with records extending from October to June (Seyffert, 2001). This species was yellow-listed in the National Audubon Society's 2007 WatchList of rare and declining birds (Butcher *et* al., 2007).

*Habitats.* Migrants and wintering birds are found in wet meadows and marshy edges with sedges, cattails and deep grasses.

*National Population.* Breeding Bird Surveys between 1966 and 2012 indicate that this species exhibited a survey-wide population decrease (1.76% annually) during that period. Niven *et al.* (2004) estimated a national annual decline rate of 0.9 percent, based on Christmas Counts from 1965–67 to 2002–2003. The estimated 1990s continental population was about 2.9 million birds (Rich *et al.,* 2004).

*Further Reading.* Bent, 1968; Byers, Curson and Olsen, 1995; Rising and Beadle, 1996; Lowther, 1996 (*The Birds of North America:* No. 224).

## Fox Sparrow
*Passerella iliaca*

The fox sparrow is one of the largest of the Great Plains sparrows, weighing slightly more than an ounce. It is never found far from brush, thickets, or other rather heavy cover. It has strong legs, and forages in the towhee-like manner of kicking backwards with both feet simultaneously. Like other sparrows, its winter foods are mostly seeds, but it also consumes berries whenever they are found. The birds are not gregarious while on their wintering grounds, but sometimes may be seen feeding in the company of song sparrows or various *Zonotrichia* sparrows.

*Winter Distribution.* Root's (1988) map shows fox sparrows wintering north to southeastern Nebraska, eastern Kansas, western Oklahoma and the entire Texas panhandle, with a peak abundance in southeastern Oklahoma, along the Red River valley. Our data suggest a peak concentration in Oklahoma, but with rather few extending south into the Texas panhandle or north to Kansas. A scattering of records have occurred north to North Dakota throughout the entire four-decade period. Our data suggest that a fairly stable winter population exists in the Great Plains states.

*Seasonality and Migrations.* The fox sparrow is an uncommon seasonal migrant, wintering southwardly. This species was observed nine years on Christmas Counts from 1968 to 2007 in North Dakota, three years in South Dakota, 14 years in Nebraska, 39 years in Kansas, 40 years in Oklahoma and 25 years in the Texas panhandle. In South Dakota wintering is only very rare (Tallman, Swanson and Palmer, 2002). Twenty-eight final autumn Nebraska sightings are from August 18 to December 31, with a median of November 11 (Johnsgard, 1980). It is uncommon (east) to rare (west) during winter in Kansas (Thompson *et al.*, 2011), and a winter resident in Oklahoma, with continuous monthly records extending from October to April (Woods and Schnell, 1984). It is a rare to uncommon winter visitor in the Texas panhandle, with records extending from October to May (Seyffert, 2001). The average number of birds observed per party-hour for the 2013–2014 Christmas Bird Count were: N. Dakota 0, S. Dakota 0.02, Nebraska 0.05, Kansas 0.10, Oklahoma 0.25, northwest Texas 0.06. For complete species abundance data, see Appendix 1:120.

*Habitats.* Migrants and wintering birds are usually associated with brushy woodlands, streamside thickets and sometimes with residential shrubbery.

*National Population.* Breeding Bird Surveys between 1966 and 2012 indicate that this species exhibited a survey-wide population decrease (0.66%

annually) during that period. The estimated 1990s continental population was about 16 million birds (Rich *et al.,* 2004).

*Further Reading.* Bent, 1968; Byers, Curson and Olsen, 1995; Rising and Beadle, 1996; Dunn and Tessaglia-Hymes, 1999; Weckstein, Kroodsma and Faucett. 2003 (*The Birds of North America:* No. 715).

## Song Sparrow
### *Melospiza melodia*

This is one of the mostly thoroughly studied of all North American sparrows, and one of the most widespread. The birds are reluctant migrants, with males trending to remain on their breeding territories as long as possible, accounting for the Christmas occurrences as far north as North Dakota. Females and young are more migratory, and are more likely to occur as flocks during winter. Males often use feeding stations during winter, and defend them against other song sparrows, as well and even some of the slightly larger species, such as the *Zonotrichia* sparrows.

*Winter Distribution.* This relatively cold-tolerant sparrow was mapped by Root (1988) as extending north to the Red River valley of North Dakota, with locally denser populations in northwestern and southwestern Kansas, and southeastern Oklahoma. Root noted the very large number of Christmas Count sites at which this species was observed; the sixth-highest number, exceeded only (in increasing frequency) by the downy woodpecker, American robin, European starling, American goldfinch and house sparrow. We also observed a very high level of distributional occurrence throughout the Great Plains, with the highest abundance reported in Oklahoma, followed closely by Kansas, and thirdly by the Texas panhandle. Nearly all of the states in our study show an apparently increasing population trend, in contrast to a national declining trend that is indicated by Breeding Bird Survey data.

*Seasonality and Migrations.* The song sparrow is a common Great Plains seasonal migrant and widespread breeder, wintering variably southwardly. This species was observed 24 years on Christmas Counts from 1968 to 2007 in North Dakota, and all 40 years in South Dakota, Nebraska, Kansas, Oklahoma, and the Texas panhandle. In South Dakota wintering is rare (Tallman, Swanson and Palmer, 2002). Forty-four final autumn Nebraska sightings are from October 6 to December 31, with a median of December 20 (Johnsgard, 1980). The data suggest that this species commonly winters in Nebraska, and that its migration tendencies are very poorly defined. It is common (east) to uncommon (west) during winter in Kansas (Thompson *et al.*, 2011), and a winter resident in Oklahoma, with most records occurring from September to May (Woods and Schnell, 1984). It is a variably common winter visitor in the Texas panhandle, with records for every month (Seyffert, 2001). The average number of birds observed per party-hour for the 2013–2014 Christmas Bird Count were: N. Dakota 0, S. Dakota 0.6, Nebraska 0.22, Kansas 0.73, Oklahoma 1.74, northwest Texas 0.55. For complete species abundance data, see Appendix 1:121.

*Habitats.* Migrants and wintering birds occur in weedy areas, thickets and streamside woodland edges. Breeding occurs in similar habitats, including forest margins, shrubby swamps, the brushy edges of ponds, shelterbelts and farmsteads.

*National Population.* Breeding Bird Surveys between 1966 and 2012 indicate that this species exhibited a survey-wide population decrease (0.69% annually) during that period. The estimated 1990s continental population was about 53 million birds (Rich *et al.,* 2004).

*Further Reading.* Bent, 1968; Knapton and Crebs, 1976; Byers, Curson and Olsen, 1995; Rising and Beadle, 1996; Dunn and Tessaglia-Hymes, 1999; Arcese *et al.* 2003 (*The Birds of North America:* No. 704).

## Lincoln's Sparrow
*Melospiza lincolnii*

The tiny Lincoln's sparrow, weighing a bit over half an ounce, is a shy and retiring species that is likely to spend most of the winter period hidden under brushy vegetation. It typically emerges just long enough to scratch the ground and pick up a seed or two, then rapidly return to the relative safety of shrub cover. It is a only a migrant in the Great Plains, and breeds in boreal and montane forests of northern and western North America.

*Winter Distribution.* Root's (1988) map shows this species reaching its northern limits near the Kansas–Nebraska border, but increasing southwardly to a peak in west-central Texas. Only small numbers were recorded in our region, with Oklahoma having slightly higher numbers than the other states. No clear population trend is evident from our data.

*Seasonality and Migrations.* The Lincoln's sparrow is a common seasonal migrant, wintering southwardly. This species was observed 15 years on Christmas Counts from 1968 to 2007 in South Dakota, three years in Nebraska, all 40 years in Kansas and Oklahoma, and 27 years in the Texas panhandle. In South Dakota wintering has not been reported (Tallman, Swanson and Palmer, 2002). Twenty-five final autumn Nebraska sightings are from September 20 to December 29, with a median of October 19 (Johnsgard, 1980). It is rare and local during winter in eastern and southeastern Kansas (Thompson *et al.*, 2011) and a winter resident in Oklahoma, with continuous monthly records extending from September to June (Woods and Schnell, 1984). It is an uncommon winter visitor in the Texas panhandle, with monthly records extending from September to June (Seyffert, 2001). The average number of birds observed per party-hour for the 2013–2014 Christmas Bird Count were: N. Dakota 0, S. Dakota 0.01, Nebraska 0.02, Kansas 0.08, Oklahoma 0.15, northwest Texas 0.06. For complete species abundance data, see Appendix 1:122.

*Habitats.* Migrants and wintering birds are associated with streamside thickets, thick weedy areas, and other rather dense grassy or weedy areas close to water, and occur less frequently in residential shrubbery.

*National Population.* Breeding Bird Surveys between 1966 and 2012 indicate that this species exhibited a survey-wide population decrease (1.16% annually) during that period. Niven *et al.* (2004) estimated a national annual increase rate of 0.2 percent, based on Christmas Counts from 1965–67 to 2002–2003. The estimated 1990s continental population was about 39 million birds (Rich *et al.,* 2004).

*Further Reading.* Bent, 1968; Ammon, 1995 (*The Birds of North America:* No. 191); Byers, Curson and Olsen, 1995; Rising and Beadle, 1996.


### Swamp Sparrow
*Melospiza georgiana*

Swamp sparrows live up to their common name during the breeding season and inhabit fairly deep wetland marshes from Nebraska northward. However, during winter they move to upland fields, brushy pastures and meadows, where they mix inconspicuously with other sparrows.

*Winter Distribution.* Relative to the other Great Plains sparrows, this species' winter distribution is therefore somewhat farther to the south and east than those of many others. Root's (1988) map shows it reaching north to southeastern Nebraska, eastern Kansas and eastern Oklahoma, with an isolated population in western Kansas and adjacent southwestern Nebraska. Our data show virtually no birds in Nebraska, peak numbers in Oklahoma, and rather few in Kansas and northwestern Texas. In both Kansas and Oklahoma the population appears to be increasing, which supports the results of data from national Breeding Bird Surveys.

*Seasonality and Migrations.* The swamp sparrow is an uncommon seasonal migrant and local wetland breeder from Nebraska north, wintering southwardly. This species was observed three years on Christmas Counts from
1968 to 2007 in North Dakota, four years in South Dakota, 22 years in Nebraska, all 40 years in Kansas and Oklahoma, and 25 years in the Texas
panhandle. South Dakota records extend to early January (Tallman, Swanson and Palmer, 2002). Thirteen final autumn Nebraska sightings are
from October 2 to December 29, with a median of October 24 (Johnsgard, 1980). It is uncommon (east) to rare (west) during winter in Kansas (Thompson *et al.*, 2011), and a common winter resident in Oklahoma,
with continuous monthly records extending from October to May (Woods
and Schnell, 1984). It is a rare winter visitor in the Texas panhandle, with
monthly records extending from September to May (Seyffert, 2001). The
average number of birds observed per party-hour for the 2013–2014 Christmas Bird Count were: N. Dakota 0, S. Dakota 0,02, Nebraska 0.01, Kansas 0.08 Oklahoma 0.15, northwest Texas 0.6. For complete species abundance data, see Appendix 1:123.

*Habitats.* Migrants and wintering birds are found in marshes or other wetlands
having such vegetation as cattails (*Typha*), rushes (*Phragmites*), shrubs or
small trees.

*National Population.* Breeding Bird Surveys between 1966 and 2012 indicate
that this species exhibited a survey-wide population increase (1.01% annually) during that period. Niven *et al.* (2004) estimated a national annual
decline rate of 0.4 percent, based on Christmas Counts from 1965–67 to
2002–2003. The estimated 1990s continental population was about nine
million birds (Rich *et al.,* 2004).

*Further Reading.* Bent, 1968; Byers, Curson and Olsen, 1995; Rising and Beadle, 1996; Mowbray, 1997 (*The Birds of North America:* No. 279).

### White-throated Sparrow
*Zonotrichia albicollis*

Of the three *Zonotrichia* sparrows of the Great Plains, this is the most easterly
in its winter distribution. It is also the only one to exhibit two definite plumage
types in both sexes. Adults of one are clearly white-striped above the eyes and
have whiter throats, while the other morph adults are tan in these areas. Males
of the white-striped morph are the more aggressive of the two types, and they
selectively mate with tan-striped females, which probably keep the govern-
ing genes in balance. Both plumage morphs are common on the Great Plains.

*Winter Distribution.* Root (1988) mapped the northwestern winter limits of this species as extending along a line from southeastern Nebraska through eastern Kansas, central Oklahoma and the eastern Texas panhandle, with maximum numbers in southern Oklahoma. This description generally is consistent with our data, although some birds are now consistently seen as far north as the Dakotas, and the numbers in the Texas panhandle are rather low. Additionally, there has been a clear increase (roughly three-fold) in numbers over the four-decade period, especially in Oklahoma and Kansas, consistent with trends indicated by Niven *et al.* (2004), but in contrast to the national population decline suggested by Breeding Bird Survey data.

*Seasonality and Migrations.* The white-throated sparrow is a common wintering migrant in the Great Plains states. This species was observed 32 years on Christmas Counts from 1968 to 2007 in North Dakota, 14 years in South Dakota, 37 years in Nebraska, all 40 years in Kansas and Oklahoma, and 36 years in the Texas panhandle. In South Dakota wintering is rare (Tallman, Swanson and Palmer, 2002). Sixty-five initial autumn Nebraska sightings dating from the mid-1930s to the late 1970s (Johnsgard, 1980) range from September 18 to November 25, with a median of October 3. Fifty-two final spring sightings are from February 2 to June 4, with a median of May 12. It is uncommon (east) to rare (west) during winter in Kansas (Thompson *et al.*, 2011), and a winter resident in Oklahoma, with continuous monthly records extending from September to June (Woods and Schnell, 1984). It is an uncommon to fairly common winter visitor in the Texas panhandle, with monthly records extending from September to May (Seyffert, 2001). The average number of birds observed per party-hour for the 2013–2014 Christmas Bird Count were: N. Dakota 0,01, S. Dakota 0.02, Nebraska 0.01, Kansas 0.23, Oklahoma 2.18, northwest Texas 0.06. For complete species abundance data, see Appendix 1:124.

*Habitats.* Migrants and wintering birds are associated with woodland edges, thickets, weedy fields, and sheltered areas near water, sometimes using feeding stations during winter.

*National Population.* Breeding Bird Surveys between 1966 and 2012 indicate that this species exhibited a survey-wide population decrease (0.51% annually) during that period. Niven *et al.* (2004) estimated a national annual increase rate of 0.1 percent, based on Christmas Counts from 1965–67 to 2002–2003. The estimated 1990s continental population was about 140 million birds (Rich *et al.,* 2004).

*Further Reading.* Bent, 1968; Ficken, Ficken and Hailman, 1978; Watt, Ralph and Atkinson, 1984; Godfrey, 1986; Falls and Kopachena, 1994 (*The Birds of North America:* No. 128); Byers, Curson and Olsen, 1995; Rising and Beadle, 1996; Dunn and Tessaglia-Hymes, 1999.

### Harris' Sparrow
*Zonotrichia querula*

The Harris' sparrow is the most distinctive wintering bird of the Great Plains; the multi-state region covered by this study closely matches its *Winter Distribution.* In addition it is one of the most handsome of the *Zonotrichia* sparrows, and has a memorable plaintive whistle that somehow catches the loneliness of a Great Plains winter. When these birds arrive in autumn they lack their full black-throated breeding plumage, but by the time they leave in spring they are simply stunning. They tend to return to the same wintering areas in successive years, and prefer generally brushy and open wooded *Habitats.*

*Winter Distribution.* Of all the species studied here, this has a winter distribution that most closely corresponds to the limits of the Great Plains. Root (1988) mapped it as extending from northern South Dakota south to the Texas Gulf cost, with population centers in southern Kansas, south-central Oklahoma and adjacent Texas. The eastern edge of its range extends into Iowa, Missouri and Arkansas. Our data indicate a region of high density in Kansas. Most of the Great Plains states show long-term population decline. Unfortunately, no breeding-range data are available to test this apparently declining trend.

*Seasonality and Migrations.* The Harris' sparrow is a common wintering migrant throughout the Great Plains states. This species was observed 32 years on Christmas Counts from 1968 to 2007 in North Dakota, 39 years in South Dakota, all 40 years in Nebraska, Kansas, and Oklahoma, and 30 years in the Texas panhandle. In South Dakota wintering is rare (Tallman, Swanson and Palmer, 2002). The range of 115 initial autumn Nebraska sightings dating from the mid-1930s to the late 1970s (Johnsgard, 1980) is from August 13 to December 31, with a median of October 14. Ninety-five final spring sightings are from February 8 to June 10, with a median of May 12. It is common (east) to uncommon (west) during winter in Kansas with records extending from September to June (Thompson and Ely, 1992), and a winter resident in Oklahoma, with continuous monthly records extending from September to May (Woods and Schnell, 1984). It is an uncommon to fairly common winter visitor in the Texas panhandle, with records extending from

October to May (Seyffert, 2001). The average number of birds observed per party-hour for the 2013–2014 Christmas Bird Count were: N. Dakota 0.02, S. Dakota 0.02, Nebraska 0.07, Kansas 4.07, Oklahoma 3.18, northwest Texas 0. For complete species abundance data, see Appendix 1:125.

*Habitats.* Migrants and wintering birds occur in rural, suburban or urban areas having shrubs, low trees and tall weedy plants, often near streamside woodland edges or thickets.

*National Population.* Breeding Bird Survey trend data for this subarctic breeder are not available. Niven *et al.* (2004) estimated a national annual decline rate of 1.8 percent, based on Christmas Counts from 1965–67 to 2002–2003. It has been designated a species of continental conservation importance, with the Prairie Avifaunal Biome supporting an estimated 97 percent of the continental winter population (Rich *et al.,* 2004), a percentage second only to the Smith's longspur as indicating a true winter prairie endemic. The estimated 1990s continental population was about 3.7 million birds (Rich *et al.,* 2004).

*Further Reading.* Harking, 1937; Bridgewater, 1966; Bent, 1968; Godfrey, 1986; Watt, 1986; Norment and Shackleton, 1993 (*The Birds of North America: No. 64*); Byers, Curson and Olsen, 1995; Rising and Beadle, 1996; Dunn and Tessaglia-Hymes, 1999.

## White-crowned Sparrow
### *Zonotrichia leucophrys*

This is a large, mostly western-oriented sparrow that weighs about an ounce and forms fairly large wintering flocks that tend to return to the same wintering area each year. Within thee flocks there is a definite social hierarchy, the adult and more brightly patterned males the most dominant. The females rank below adult males, and immatures are the most subordinate, a common songbird characteristic. In this species, however, variations in brightness of plumage determine relative dominance status, rather than status being dependent upon the outcome of individual aggressive encounters.

*Winter Distribution.* Root's (1988) map show the winter range of this species extending north to southern Nebraska, and having its greatest Great Plains abundance in the Texas panhandle. That pattern is consistent with our data. Our data also indicates a gradually increasing Great Plains population, contrary to reported national trends. The rate of apparent increase has been highest in Kansas, and least in northwestern Texas, suggesting a northward shift has occurred in wintering birds.

*Seasonality and Migrations.* The white-crowned sparrow is a common wintering migrant throughout the five Great Plains states, especially in western areas. This species was observed 11 years on Christmas Counts from 1968 to 2007 in North Dakota, 12 years in South Dakota, and all 40 years in Nebraska, Kansas, Oklahoma and the Texas panhandle. In South Dakota wintering is very rare (Tallman, Swanson and Palmer, 2002). Ninety-eight initial autumn Nebraska sightings dating from the mid-1930s to the late 1970s (Johnsgard, 1980) range from August 25 to December 29, with a median of October 3. Eighty-two final spring sightings are from February 1 to May 27, with a median of May 15. It is local (east) to common (west) during winter in Kansas, with records extending from September to May (Thompson and Ely, 1992), and a winter resident in Oklahoma, with continuous monthly records extending from September to June (Woods and Schnell, 1984). It is a common to abundant winter visitor in the Texas panhandle, with records for every month but July (Seyffert, 2001). The average number of birds observed per party-hour for the 2013–2014 Christmas Bird Count were: N. Dakota 0.01, S. Dakota 0.13, Nebraska 3.97, Kansas 2.80, Oklahoma 1.70, northwest Texas 3.45. For complete species abundance data, see Appendix 1:126.

*Habitats.* Migrants and wintering birds are associated with thickets, woodland edges, and weedy areas, sometimes moving to farmyards and feeding stations in winter.

*National Population.* Breeding Bird Surveys between 1966 and 2012 indicate that this species exhibited a survey-wide population decrease (0.74% annually) during that period. The estimated 1990s continental population was about 72 million birds (Rich *et al.,* 2004).

*Further Reading.* Bent, 1968; Mewaldt, 1976; Godfrey, 1986; Chilton, *et al.,* 1995 *(The Birds of North America:* No. 183); Byers, Curson and Olsen, 1995; Rising and Beadle, 1996; Dunn and Tessaglia-Hymes, 1999.

### Dark-eyed Junco
*Junco hyemalis*

Juncos may be the most familiar of winter birds for most Great Plains residents; thy are abundant, fairly tame, and are easily attracted to feeding stations that offer small seeds such as millet and cracked corn. Junco flocks are somewhat cohesive; their memberships tend to be stable through the winter, so that definite dominance relationships are established. In descending order, these dominance ranks are: adult male, immature male, adult female, and immature female. Perhaps in part to avoid male dominance and food competition, females tend to migrate farther south than do males. Thus, northern wintering flocks are mostly of males, while more southerly flocks are predominantly females.

*Winter Distribution.* According to Root (1988), this species' winter distribution extends north to southwestern North Dakota, and south to central Texas, with higher numbers in western parts of Kansas and Oklahoma, and the Texas panhandle. We found the highest numbers to occur in the Texas panhandle, with Kansas having the second-highest long-term average, but substantial populations occurring north to North Dakota, especially in recent decades. Throughout the region our numbers show a progressive increase, counter to national population trends as judged by the Breeding Bird Survey. Long-term Christmas count trends for the entire state of Texas are also upward, as are those for Manitoba, so it seems unlikely that this apparent trend is the result of a purely regional influence.

*Seasonality and Migrations.* The dark-eyed junco is an abundant wintering migrant throughout the five Great Plains states, and breeds locally in western South Dakota. This species was observed nearly year in all states between North Dakota and the Texas panhandle on Christmas Counts from 1968 to 2007. In South Dakota wintering is regular (Tallman, Swanson and Palmer, 2002). The range of 105 initial autumn Nebraska sightings dating from the mid-1930s to the late 1970s (Johnsgard, 1980) is from September 1 to December 31, with a median of October 6. Seventy-five final spring sightings are from January 1 to May 20, with a median of April 15. Twenty final spring sightings are from January 1 to May 18, with a median of March 23. It is common during winter in Kansas, mostly from late September to mid-April (Thompson and Ely, 1992), and a winter resident in Oklahoma, with most records occurring from September to May (Woods and Schnell, 1984). It is a common to abundant winter visitor in the Texas panhandle, with records extending from September to May (Seyffert, 2001). The average number of birds observed per party-hour for the 2013–2014 Christmas Bird Count were: N. Dakota 0.98, S. Dakota 3.37, Nebraska 6.13, Kansas 6.14, Oklahoma 6.36, northwest Texas 7.34. For complete species abundance data, see Appendix 1:127.

*Habitats.* Migrants and wintering birds are widely distributed in woodlands, suburbs and residential areas, foraging on the ground and often visiting feeding stations. Breeding in the Black Hills occurs in coniferous forests, aspen (*Populus*) groves and deciduous woodlands in hollows, canyons and gulches. Brushy forested canyons are used for nesting in Nebraska's Pine Ridge region.

*National Population.* Breeding Bird Surveys between 1966 and 2012 indicate that this species exhibited a survey-wide population decrease (1.20% annually) during that period. The estimated 1990s continental population was about 260 million birds (Rich *et al.,* 2004), making it one of the most abundant of all North American land birds.

*Further Reading.* Sabine, 1949, 1955, 1956, 1959; Bent, 1968; Fretwell, 1969; Davis, 1973; Balph and Balph, 1977, 1979; Byers, Curson and Olsen, 1995; Rising and Beadle, 1996; Dunn and Tessaglia-Hymes, 1999; Nolan *et al.,* 2003 *(The Birds of North America:* No. 716).

# FAMILY CARDINALIDAE:
# CARDINALS AND GROSBEAKS

## Northern Cardinal
### *Cardinalis cardinalis*

The most colorful and widely recognized of our winter birds, no other species can brighten a winter day quite so well as a cardinal. Winter amelioration and feeding efforts of residents in northern states have no doubt helped cardinals to expand their range northward and westward in recent decades. They have even been reported from northwestern North Dakota (Minot), a state where they were still unproven to breed less than 50 years ago (Stewart, 1975). During winter as many as two or three pairs or families might share a Nebraska yard if there is a well-stocked feeding station. By early in the calendar year singing gets underway as day-lengths begin to increase, and breeding territories are re-established.

*Winter Distribution.* Root (1988) noted how the northern cardinal has expanded its range during the past century. She mapped its western limits as of the 1960s as extending from eastern North Dakota south through most of South Dakota to encompass nearly all of Nebraska and Kansas, all of Oklahoma, and Texas west to the New Mexico border, with maximum numbers in eastern Oklahoma. Our data also indicate a peak population in Oklahoma, with a substantial number extending into Kansas, and "present' levels reaching North Dakota. No clear population trend is apparent from our data, which for Kansas and Oklahoma suggest peak numbers were reached in the 1990s, followed by a slight decline. National population trends are also ambiguous; perhaps the cardinal has reached its northern and western physiological distributional limits, and its population is possibly now controlled by weather extremes or other periodic stress factors.

*Seasonality and Migrations.* The northern cardinal is a common permanent resident throughout its range. This species was observed 32 years on Christmas Counts from 1968 to 2007 in North Dakota, and all 40 years in all the states from South Dakota to Oklahoma, and the Texas panhandle. The average number of birds observed per party-hour for the 2013–2014 Christmas Bird Count were: N. Dakota 0.09, S. Dakota 0.36, Nebraska 1.59, Kansas 2.66, Oklahoma 2.99, northwest Texas 0.37. For complete species abundance data, see Appendix 1:128.

*Habitats.* Throughout the year this species is associated with forest edges or brushy forest openings, parks and residential areas planted to shrubs and low trees, second-growth woods, and river bottom gallery forests in grasslands.

*National Population.* Breeding Bird Surveys between 1966 and 2012 indicate that this species exhibited a survey-wide population increase (0.29% annually) during that period.

The estimated 1990s continental population north of Mexico was about 82 million birds (Rich *et al.,* 2004). The species breeds south to Belize.

*Further Reading.* Nice, 1927 Bent, 1968; Osborne, 1992; Dunn and Tessaglia-Hymes, 1999; Halkin and Linville, 1999 (*The Birds of North America:* No. 440); Beadle and Rising, 2006.

## Pyrrhuloxia
### *Cardinalis sinuatus*

Pyrrhuloxias are the southwestern desert counterpart of northern cardinals, and have many of the same attributes. Winter flocking occurs in both species, and the two sometimes roost and forage together. During winter these flocks tend to wander, but concentrate near bird-feeding stations.

*Winter Distribution.* The desert-adapted and cardinal-like sparrow is closely associated with mesquite (*Prosopis*) and acacia (*Acacia*), and its distribution is likely to be influenced by the distribution of those thorny plants. Root (1988) mapped it as occurring in the Texas panhandle, but a range expansion into Texas began in the late 1970s.

*Seasonality and Migrations.* The pyrrhuloxia is a rare to uncommon resident in the Texas panhandle, and has been reported every month except July (Seyffert, 2001). It was recorded 11 years during Christmas Counts from 1968 to 2007 in the Texas panhandle. It has been reported once during January from Oklahoma (Woods and Schnell, 1984).

*Habitats.* Associated with mesquite thickets and acacia thorn brush, as well as arid ranchlands

*National Population.* Breeding Bird Surveys between 1966 and 2012 indicate that this species exhibited a survey-wide population decrease (1.17% annually) during that period. The estimated 1990s continental population north of Mexico was about 1.9 million birds (Rich *et al.,* 2004). The species breeds south to central Mexico.

*Further Reading.* Bent, 1968; Dunn and Tessaglia-Hymes, 1999; Twiet and Thompson, 1999 (*The Birds of North America:* No. 391); Beadle and Rising, 2006.

# FAMILY ICTERIDAE:
# BLACKBIRDS, ORIOLES AND MEADOWLARKS

## Red-winged Blackbird
*Agelaius phoeniceus*

Red-winged blackbirds may prove the adage that there can be too much of a good thing; a few males singing from cattails in a wetland can be a beautiful and memorable sight, but several hundred thousand blackbirds passing by overhead tend to remind one of an Alfred Hitchcock horror movie. Some roosts have been estimated to contain more than a million birds! Like many other songbird species, flocks of these blackbirds are often sexually unbalanced. Females and young migrate south earlier in autumn and tend to winter farther south than do adult males. Return flights in spring are initially composed almost entirely of adult males, which seem eager to set up their territories as soon as possible, and often begin singing while still on migration.

*Winter Distribution.* This is perhaps the most abundant land bird of North America, and its winter distribution is virtually ubiquitous. Because of the species' high degree of flocking and winter mobility, Root (1983) considered her distribution map to be unreliable. Her map shows very high numbers in Nebraska, Kansas and the Texas panhandle, while our data indicate extremely high numbers in Kansas, followed by Oklahoma and the Texas panhandle. Vast year-to-year and decade-to-decade and state-to-state differences in numbers seen on Christmas Counts are the result of highly localized distributions of enormous flocks that strongly influence overall count averages. No population trends can be interpreted from such highly variable data.

*Seasonality and Migrations.* The red-winged blackbird is an abundant seasonal migrant and widespread breeder throughout the Great Plains, wintering southwardly. This species was observed all 40 years on Christmas Counts from 1968 to 2007 in all the states from North Dakota to Oklahoma, and the Texas panhandle. In South Dakota wintering is uncommon (Tallman, Swanson and Palmer, 2002). Eighty final autumn Nebraska sightings range from August 8 to December 31, with a median of November 21 (Johnsgard, 1980). There are many January sightings. It is locally abundant (especially at Quivira and Cheyenne Bottoms) during winter in Kansas (Thompson *et al.*, 2011), and is a permanent resident in Oklahoma (Woods and Schnell, 1984) and the Texas panhandle (Seyffert, 2001). The average number of birds observed per party-hour for the 2013–2014 Christmas Bird Count were: N. Dakota 2.02, S. Dakota 71.79, Nebraska 31.15, Kansas 1.451.4, Oklahoma 200.05, northwest Texas 867.0 For complete species abundance data, see Appendix 1:129.

*Habitats.* This adaptable species uses a wide range of habitats, from deep marshes or the emergent zones of lakes and impoundments, through progressively drier habitats such as wet meadows, ditches, brushy patches in prairie, hayfields, and weedy croplands or roadsides. Sometimes vast numbers are found in Great Plains wetlands during winter; 12.5 million were reported at Quivira National Wildlife Refuge, Kansas, during the 2002–2003 Christmas Count, and eight million were seen at Sooner Lake, Oklahoma, during the 2006–2007 count Migrants often are seen in flocks of other blackbird species, feeding in fields or elsewhere, but roosting is typically done in wetland areas rather than in residential locations.

*National Population.* Breeding Bird Surveys between 1966 and 2012 indicate that this species exhibited a survey-wide population decrease (0.95% annually) during that period.

The estimated 1990s continental population north of Mexico was about 193 million birds (Rich *et al.,* 2004). The species breeds south to Costa Rica.

*Further Reading.* Bent, 1958; Orians, 1985; Jamarillo and Burke, 1995; Yasukawa and Searcy, 1995 (*The Birds of North America:* No. 184); Dunn and Tessaglia-Hymes, 1999.

### Eastern Meadowlark
*Sturnella magna*

The western and eastern meadowlarks are something like the towhees in the central Great Plains, the westerns winter farther north than the easterns, so that wintering meadowlarks in Nebraska are much more likely to be westerns than easterns. Both species are late autumn and early spring migrants, and it is not rare to see flocks that have been caught up in early autumn or late spring snowstorms. Then they tend to gather in snow-free areas along snow-free gravel roadsides, trying to find enough waste grain or weed seeds to get them through these emergency periods.

*Winter Distribution.* Root (1988) mapped the northwestern limits of this species as reaching eastern South Dakota, central Nebraska, west-central Kansas, and western Oklahoma, ending at about the Texas panhandle border, and with the highest numbers in Oklahoma. Our data show a similar pattern, with Oklahoma having substantially the highest average numbers.

Additionally, at least Oklahoma and Kansas data show slowly declining numbers over the four-decade period, with the five-state average count declining by about two-thirds. This depressing trend is consistent with national trends, based on Breeding Bird Survey data.

*Seasonality and Migrations.* The eastern meadowlark is a common Great Plains seasonal migrant and widespread breeder, wintering southwardly. This species was observed five years on Christmas Counts from 1968 to 2007 in South Dakota, 12 years in Nebraska, all 40 years in Kansas and Oklahoma, and 12 years in the Texas panhandle. Other South Dakota records extend to mid-October (Tallman, Swanson and Palmer, 2002). Thirty final autumn Nebraska sightings are from August 2 to December 31, with a median of October 10, so wintering in Nebraska must be rare. It is uncommon statewide during winter in Kansas (Thompson *et al.*, 2011), and is considered a permanent resident in Oklahoma (Woods and Schnell, 1984) and the Texas panhandle (Seyffert, 2001). The average number of birds observed per party-hour for the 2013–2014 Christmas Bird Count were: N. Dakota 0, S. Dakota 0, Nebraska 0, Kansas 0, Oklahoma 1.50, northwest Texas 0. For complete species abundance data, see Appendix 1:130.

*Habitats.* This species is usually associated with tall-grass prairies, meadows, and open croplands of small grain, as well as weedy orchards and similar open, grass-dominated *Habitats.*

*National Population.* Breeding Bird Surveys between 1966 and 2012 indicate that this species exhibited a survey-wide population decrease (3.41% annually) during that period. The estimated 1990s continental population north of Mexico was about eight million birds (Rich *et al.*, 2004). The species breeds south to Brazil.

*Further Reading.* Bent, 1958; Orians, 1985; Jamarillo and Burke, 1995; Lanyon, 1995 (*The Birds of North America:* No. 160); Dunn and Tessaglia-Hymes, 1999; Johnsgard, 2001b.

## Western Meadowlark
### *Sturnella neglecta*

One of the favorite birds of rural residents of the northern Great Plains, the western meadowlark's melodic song is one of the most anticipated sounds of late winter, as it signals the final end of a long winter. During winter, western meadowlarks are found in much the same tallgrass habitats in which they breed, namely grass-dominated fields. Waste grain and weed seeds are especially important winter foods for meadowlarks.

*Winter Distribution.* Root (1988) mapped the winter distribution of this iconic grassland species as extending north to central South Dakota, and south to southern Texas, with the nation's highest density occurring in central Oklahoma, and with a second center of abundance in the Texas panhandle. Our data show that the species' highest density now is centered in the Texas panhandle and has a secondary center of abundance in Kansas. These three regions, Kansas, Oklahoma and northwestern Texas, all appear to have somewhat declining population trends, which is consistent with the results of national Breeding Bird Surveys. It is disturbing that this species, the state bird of several states and an iconic species of the tallgrass prairie, is slowly declining.

*Seasonality and Migrations.* The western meadowlark is a common Great Plains seasonal migrant and widespread breeder, wintering southwardly. This species was observed 24 years on Christmas Counts from 1968 to 2007 in North Dakota, 35 years in South Dakota, 38 years in Nebraska, and all 40 years in Kansas, Oklahoma, and the Texas panhandle. In South Dakota wintering is uncommon (Tallman, Swanson and Palmer, 2002). Forty-three final autumn Nebraska sightings are from August 20 to December 31, with a median of October 28. In Nebraska this meadowlark is an earlier spring and later fall migrant than is the eastern meadowlark, and may overwinter (Johnsgard, 1980). In Kansas it is common in the west but less common in the east during winter (Thompson *et al.*, 2011). It is a permanent resident in Oklahoma (Woods and Schnell, 1984) and the Texas panhandle (Seyffert, 2001). The average number of birds observed per party-hour for the 2013–2014 Christmas Bird Count were: N. Dakota 0, S. Dakota 0.015, Nebraska 0,62, Kansas 4.66, Oklahoma 1.1, northwest Texas 1.36. For complete species abundance data, see Appendix 1:131.

*Habitats.* This species is associated with tall-grass and mixed-grass prairies, hayfields, wet meadows, the weedy borders of croplands, retired croplands, and to a limited extent with shortgrass and sage-dominated plains, where it is limited to moister situations.

*National Population.* Breeding Bird Surveys between 1966 and 2012 indicate that this species exhibited a survey-wide population decrease (1.39% annually) during that period. The estimated 1990s continental population north of Mexico was about 29.5 million birds (Rich *et al.*, 2004). The species breeds south to central Mexico.

*Further Reading.* Bent, 1958; Orians, 1985; Lanyon, 1994 (*The Birds of North America:* No. 104); Jamarillo and Burke, 1995; Johnsgard, 2001b; Dinkins *et al.* 2001.

## Yellow-headed Blackbird
*Xanthocephalus xanthocephalus*

Yellow-headed blackbirds are spectacular birds, especially in flight, when their white wing markings show to best advantage. During autumn migration they often congregate in large flocks that may settle to forage in feedlots, pastures, or other open lands. At times a large flock will land in leafless trees, the males' yellow heads causing the trees to suddenly appear covered by blossoms.

*Winter Distribution.* This is a highly gregarious species whose normal winter range lies to the southwest of the Great Plains states (Root, 1988). Its occurrences within the area studied are the result of vagrant birds, which have been observed on Christmas Counts as far north as North Dakota.

*Seasonality and Migrations.* The yellow-headed blackbird is a variably common Great Plains seasonal migrant and local wetland breeder, wintering southwardly. This species was observed 11 years on Christmas Counts from 1968 to 2007 in North Dakota, 25 years in South Dakota, 12 years in Nebraska, 22 years in Kansas, ten years in Oklahoma, and seven years in the Texas panhandle. In South Dakota wintering is extremely rare (Tallman, Swanson and Palmer, 2002). Eighty-two final autumn Nebraska sightings range from July 23 to December 28, with a median of September 18 (Johnsgard, 1980). Records in Kansas extend to November, but Oklahoma records extend through December and January ((Thompson *et al.*, 2011; Woods and Schnell, 1984). It is a rare winter visitor in the Texas panhandle, with records throughout the year (Seyffert, 2001). The average number of birds observed per party-hour for the 2013–2014 Christmas Bird Count were: N. Dakota 0.04, S. Dakota 0.03, Nebraska 0,13, Kansas 0.04, Oklahoma 0.4, northwest Texas 0. For complete species abundance data, see Appendix 1:132.

*Habitats.* During the breeding season this species occurs in deep marshes, the marsh zones of lakes or shallow impounds, and elsewhere where there are extensive stands of cattails, bulrushes or phragmites. It is often found breeding in association with red-winged blackbirds, utilizing the deeper portions of the marsh. Migrants are sometimes seen flying or perching with groups of red-winged blackbirds, but more often remain separate from them.

*National Population.* Breeding Bird Surveys between 1966 and 2012 indicate that this species exhibited a survey-wide population decrease (0.36% annually) during that period. The estimated 1990s continental population was about 23 million birds (Rich *et al.,* 2004).

*Further Reading.* Bent, 1958; Orians, 1985; Jamarillo and Burke, 1995; Twedt and Crawford, 1995 (*The Birds of North America:* No. 192).

## Brewer's Blackbird
*Euphagus cyanocephalus*

A less gregarious blackbird species than those just described, the Brewer's is attracted to agricultural lands and their associated waste grains during autumn and winter. These birds can also be found in cities and villages, where they may be seen foraging on lawns and dodging local traffic, as they seek out weed seeds and any insects that might still be moving about.

*Winter Distribution.* Root (1988) mapped a rather irregular winter distribution for this species, extending north to southwestern Nebraska and having a center of abundance near the Texas–New Mexico border along the Pecos River. Our data show Oklahoma to have the highest average counts over the entire four-decade period, with the Texas panhandle having secondary abundance levels. There are no clear abundance trends, as is typical of gregarious blackbird species, but perhaps suggests a slight long-term decline. Such a trend fits with national Breeding Bird Survey data.

*Seasonality and Migrations.* The Brewer's blackbird is a common to very common seasonal Great Plains widespread breeder from western Nebraska north, wintering southwardly. This species was observed 28 years on Christmas Counts from 1968 to 2007 in North Dakota, 35 years in South Dakota, 32 years in Nebraska, all 40 years in Kansas and Oklahoma, and 38 years in the Texas panhandle. In South Dakota wintering is very rare (Tallman, Swanson and Palmer, 2002). Forty-five final autumn Nebraska sightings are from September 1 to December 31, with a median of November 5. There is a much lower proportion of late December records than for the rusty blackbird, suggesting that Nebraska wintering is rather rare in the Brewer's blackbird (Johnsgard, 1980). It is local during winter in Kansas, with records extending from September to May (Thompson and Ely, 1992), and a winter resident in Oklahoma, with most records extending from August to May (Woods and Schnell, 1984). It is a common to abundant winter visitor in the Texas panhandle, with records throughout the year (Seyffert, 2001). The average number of birds observed per party-hour for the 2013–2014 Christmas Bird Count were: N. Dakota 0.07, S. Dakota 0.07, Nebraska 0,02, Kansas 4.65, Oklahoma 12.96, northwest Texas 13.3 For complete species abundance data, see Appendix 1:133.

*Habitats.* Migrants and wintering birds are usually seen in pastures, barnyards and grain fields, often in the company of other kinds of blackbirds. Generally the birds favor low-stature grasslands, such as mowed roadsides or burned areas near railroads, residential areas, and farmsteads. Areas that have a combination of grassy habitats, scattered shrubs or small trees, and nearby water are especially favored.

*National Population.* Breeding Bird Surveys between 1966 and 2012 indicate that this species exhibited a survey-wide population decrease (2.20% annually) during that period. The estimated 1990s continental population was about 34 million birds (Rich *et al.,* 2004).

*Further Reading.* Bent, 1958; Orians, 1985; Dunn and Tessaglia-Hymes, 1999; Jamarillo and Burke, 1995; Martin, 2002 *(The Birds of North America:* No. 616).

### Rusty Blackbird
*Euphagus carolinus*

Rusty blackbirds breed in northern swamp forests of Canada, but while on migration and wintering they move to stubble fields, roadsides and pasturelands. Like other blackbirds it is highly mobile and gregarious, and so produces problems in determining overall winter ranges and regional population trends.

*Winter Distribution.* Like the yellow-headed and red-winged blackbirds, this is a highly social species, which adds a good deal of local and temporal variations to Christmas Counts. Root (1988) mapped an area of concentration in central Oklahoma and adjacent Kansas as the only wintering grounds in the Great Plains states. Our data show Kansas to have the highest average more recent counts, but a good deal of variability is present, and no clear geographic patterns or trends of relative abundance are evident. Other national data suggest that the species is undergoing a rapid population decline (Niven *et al.* , 2004).

*Seasonality and Migrations.* The rusty blackbird is an uncommon seasonal migrant, wintering southwardly. This species was observed 39 years on Christmas Counts from 1968 to 2007 in North Dakota and South Dakota, 38 years in Nebraska, all 40 years in Kansas and Oklahoma, and six years in the Texas panhandle. In South Dakota wintering is rare (Tallman, Swanson and Palmer, 2002). Twenty-one final autumn Nebraska sightings are from October 4 to December 31, with a median of December 26. The large proportion of final Nebraska sightings in late December suggests that this species winters rather frequently in the state (Johnsgard, 1980). It is uncommon and local during winter in Kansas, with most records extending from September to May (Thompson *et al.*, 2011), and a winter resident in Oklahoma with records extending from September to April (Woods and Schnell, 1984). It is a very rare winter visitor in the Texas panhandle, with records extending from November to February (Seyffert, 2001). The average number of birds observed per party-hour for the 2013–2014 Christmas Bird Count were: N. Dakota 0.09, S. Dakota 0.17, Nebraska 0.5, Kansas 0.96, Oklahoma 3.03, northwest Texas 0.17. For complete species abundance data, see Appendix 1:134.

*Habitats.* Migrants and wintering birds are usually found in deciduous woodlands near streams, rather than in the open marshlands, grasslands and croplands favored by other species of blackbirds in Nebraska.

*National Population.* Breeding Bird Surveys between 1966 and 2012 indicate that this species exhibited a survey-wide population decrease (5.56% annually) during that period. Niven *et al.* (2004) estimated a national annual decline rate of 5.1 percent, based on Christmas Counts from 1965–67 to 2002–2003. It was designated a species of continental conservation importance, with the Prairie Avifaunal Biome supporting an estimated 28 percent of the continental winter population (Rich *et al.,* 2004). The estimated 1990s continental population was about two million birds (Rich *et al.,* 2004). This species was yellow-listed in the National Audubon Society's 2007 WatchList of rare and declining birds (Butcher *et al.,* 2007).

*Further Reading.* Bent, 1958; Orians, 1985; Jamarillo and Burke, 1995; Avery, 1995 (*The Birds of North America:* No. 200).

### Common Grackle
*Quiscalus quiscula*

Like most blackbirds, common grackles become highly gregarious on autumn migration and through winter, when vast roosting flocks may locally develop. Those that pass through the Great Plains don't approach the countless numbers seen farther east, but when a flock of 20 or more individuals descends on a backyard feeder it certainly is enough of a disruption as to displace most other birds. Through autumn and winter, but especially in spring, threat displays often occur between individuals trying to dominate a feeder. One such male display, the ruff-out, involves the male ruffling most of its feathers, spreading its tail and wings, and rising to the maximum on its toes. Seen in full sunlight, the shimmering iridescent purple sheen of its feathers momentarily transforms the bird into an exotic-appearing creature.

*Winter Distribution.* Because of the extremely high local concentration of grackles in the southeastern states, Root's map of this species has no real relevance to the Great Plains, but she mapped a minor concentration in eastern Kansas. Our data show Oklahoma to have the highest long-term

average numbers, but extremely high counts in some years and some local-
ities tend to obscure any geographic or temporal trends. The massive decade-to-decade changes in Oklahoma grackle numbers illustrate this phenomenon very well.

*Seasonality and Migrations.* The common grackle is a common to abundant
Great Plains seasonal migrant and widespread breeder throughout the region, wintering southwardly. This species was observed all 40 years on
Christmas Counts from 1968 to 2007 in every state from North Dakota to
Oklahoma, and 12 years in the Texas panhandle. In South Dakota wintering is rare (Tallman, Swanson and Palmer, 2002). Ninety final autumn Nebraska records are from August 9 to December 30, with a median of October 28. Nearly half of the records are for December, so wintering may
occur fairly frequently in Nebraska (Johnsgard, 1980). It is local during
winter in Kansas (Thompson *et al.*, 2011), and a permanent resident in
Oklahoma (Woods and Schnell, 1984). It is a rare to uncommon winter
visitor in the Texas panhandle, with records extending throughout the year
(Seyffert, 2001). The average number of birds observed per party-hour for
the 2013–2014 Christmas Bird Count were: N. Dakota 0.03, S. Dakota
1.99, Nebraska 0,22, Kansas 29.7, Oklahoma 1,162.4, northwest Texas 2.0.
For complete species abundance data, see Appendix 1:135.

*Habitats.* Migrants and wintering birds are often seen in large flocks in residential and rural areas. For much of the year this species frequents woodland
edges or areas partially planted to trees, such as residential areas, parks,
farmsteads, shelterbelts and the like. Tall shrub thickets near croplands or
marshlands are also used.

*National Population.* Breeding Bird Surveys between 1966 and 2012 indicate
that this species exhibited a survey-wide population decrease (1.72% annually) during that period. The estimated 1990s continental population was
about 97 million birds (Rich *et al.,* 2004).

*Further Reading.* Bent, 1958; Orians, 1985; Jamarillo and Burke, 1995; Peer
and Bollinger, 1997 (*The Birds of North America:* No. 271); Dunn and Tessaglia-Hymes, 1999.

### Great-tailed Grackle
*Quiscalus mexicanus*

Bird-watchers in the Central Plains have met the slow but inexorable advance
northward of the great-tailed grackle with mixed feelings. It is true that the
birds are interesting to watch; their strange postures and displays give them an

other-worldly quality, but they are even more destructive to the eggs and young of other species than common grackles. They are highly adaptable as to their foods, and are attracted to garbage bins and other sources of edible wastes.

*Winter Distribution.* At the time of Root's (1988) analysis, the great-tailed grackle had barely extended its winter range to northern Texas. Four decades later it was more common in Oklahoma than in the Texas panhandle, and was becoming relatively common as far north as Kansas. It was first seen in Oklahoma in 1953, and the first nesting record was in 1958 (Reinking, 2004). Nesting in Kansas was first noted in 1964 (Busby and Zimmerman, 2001), Nebraska nesting had occurred by 1977, and the first known South Dakota nesting was in 1999 (Tallman, Swanson and Palmer, 2002). Its northward breeding range expansion is apparently still underway, but as it moves north it is probably going to gradually increase its seasonal migratory movements. The rate of population increase in Kansas has been especially notable, having increased during Christmas Counts by an average factor of 100-fold within four decades.

*Seasonality and Migrations.* The great-tailed grackle is an increasingly common Great Plains seasonal migrant and widespread breeder north to South Dakota, wintering southwardly. This species was observed two years on Christmas Counts from 1968 to 2007 in South Dakota (starting 1982), nine years in Nebraska (starting in 1984), 31 years in Kansas (starting in 1974), 34 years in Oklahoma (starting in 1973), and 12 years in the Texas panhandle. Other South Dakota records extend to late November (Tallman, Swanson and Palmer, 2002). Most breeding birds depart Nebraska by the end of August; the latest autumn record is November 27 (Johnsgard, 2007a). It is local during winter in Kansas (Thompson *et al.*, 2011), and a permanent resident in Oklahoma (Woods and Schnell, 1984). It is a fairly common to common winter visitor in the Texas panhandle, with records throughout the year (Seyffert, 2001). The average number of birds observed per party-hour for the 2013–2014 Christmas Bird Count were: N. Dakota 0., S. Dakota 0, Nebraska 15.0, Kansas 1.08, Oklahoma 60.06, northwest Texas 1.71. For complete species abundance data, see Appendix 1:136.

*Habitats.* This grackle occurs in a wide variety of habitats, but these usually include both open ground and nearby water, so it is especially common in irrigated croplands. During winter it is attracted to parks and suburbs.

*National Population.* Breeding Bird Surveys between 1966 and 2012 indicate that this species exhibited a survey-wide population increase (2.46% annually) during that period. The estimated 1990s continental population north of Mexico was about 7,750,000 birds (Rich *et al.*, 2004). The species breeds south to Peru.

*Further Reading.* Bent, 1958; Orians, 1985; Jamarillo and Burke, 1995; Dunn and Tessaglia-Hymes, 1999; Johnson and Peer, 2001 (*The Birds of North America:* No. 576).

## Brown-headed Cowbird
*Molothrus ater*

The original "buffalo bird" has in the past century or more transformed itself into a cowbird. In doing so, it has greatly expanded its overall North American range, breeding north in Canada nearly to the Arctic Circle, and south to central Mexico. Presumably cowbirds once followed the bison southward and northward seasonally, and the nearly constant movements needed to keep up with the bison herds have been suggested as a reason for the evolution of the cowbird's brood-parasitism. However, although the northern populations of cowbirds are strongly migratory, while those in the southern Great Plains are residential. In southern regions resident cowbirds mix with migrants, and also with various blackbirds, sometimes forming immense roosting flocks.

*Winter Distribution.* Like several other icterids, the brown-headed cowbird is highly gregarious in winter, and its main concentration occurs well to the east of the Great Plains, even though historically it breeding range was largely confined to the bison-rich grasslands of the Plains. Root (1988) mapped a zone of abundance extending from southern Nebraska to Oklahoma and the Texas panhandle. Over our four decades of study, Kansas has supported by far the largest concentrations of cowbirds at the time of the Christmas Counts, although there are no clear overall population trends evident from these data.

*Seasonality and Migrations.* The brown-headed cowbird is a common to abundant Great Plains seasonal migrant and breeder throughout the region, wintering southwardly. This species was observed 12 years on Christmas Counts from 1968 to 2007 in North Dakota, 28 years in South Dakota, all 40 years in Nebraska, Kansas, and Oklahoma, and 36 years in the Texas panhandle. In South Dakota wintering is rare (Tallman, Swanson and Palmer, 2002). Eighty-five final autumn Nebraska sightings are from August 1 to December 31, with a median of October 7. Nearly 20 percent of the final records are for December, suggesting that some wintering may occur in Nebraska (Johnsgard, 1980). It is local during winter in Kansas (Thompson *et al.*, 2011), and is a permanent resident in Oklahoma (Woods and Schnell, 1984) and the Texas panhandle (Seyffert, 2001). The average number of birds observed per party-hour for the 2013–2014 Christmas Bird

Count were: N. Dakota 0, S. Dakota 0.10, Nebraska 0.10, Kansas 12.27, Oklahoma 14,02, northwest Texas 5.43. For complete species abundance data, see Appendix 1:137.

*Habitats.* Migrants and wintering birds are often found in open grassy fields among livestock. Breeding habitats include woodland edges, brushy thickets and other situations where low and scattered trees are interspersed with grasslands.

*National Population.* Breeding Bird Surveys between 1966 and 2012 indicate that this species exhibited a survey-wide population decrease (0.71% annually) during that period The estimated 1990s continental population north of Mexico was about 53 million birds (Rich *et al.,* 2004). Assuming that half of these are females, and that each female may lay 40-50 parasitic eggs in a season, the reproductive damage that billions of eggs of this species can do annually to native songbirds is staggering. The species breeds south to southern Mexico.

*Further Reading.* Bent, 1958; Orians, 1985; Lowther, 1993 (*The Birds of North America:* No. 47); Jamarillo and Burke, 1995; Dunn and Tessaglia-Hymes, 1999.

# FAMILY FRINGILLIDAE:
# BOREAL FINCHES

## Gray-crowned Rosy-Finch
*Leucosticte tephrocotis*

Rosy-finches are alpine-breeding birds that are forced to lower altitudes during the fall. Typically they then move into the open plains and spread out for varying distances over grasslands and pastures. The birds move about in rather small flocks and forage on small weed seeds, which they somehow manage to locate even under thin blankets of snow. It has been estimated that the home range of a wintering flock may be as great as 500 square miles, In spite of such mobility, the birds typically return night after night to the same roosting site, which might be in a cave, a mine shaft, or an abandoned building.

*Winter Distribution.* Rosy-finches that occasionally visit South Dakota and Nebraska during winter probably originate from the mountains of eastern Wyoming, and at least two that were banded in South Dakota were later recovered in northeastern Wyoming (Tallman, Swanson and Palmer, 2002). Because of the erratic occurrence of these invasions, no population trend conclusions can be made.

*Seasonality and Migrations.* The gray-crowned rosy-finch is a rare winter visitor in the northern and central plains, the records being mostly limited to South Dakota and Nebraska. This species was reported 11 years during Christmas Bird Counts from 1968 to 2007 in North Dakota, 15 years in South Dakota, two years in Nebraska, and none in Kansas, Oklahoma, or the Texas panhandle. In South Dakota, wintering is common in the Black Hills region (Tallman, Swanson and Palmer, 2002). Six autumn Nebraska sighting range from October 1 to November 6, with a median of October 25. Thirteen spring sightings are from January 1 to March 11, with a median of February 12 (Johnsgard, 1980). The average number of birds observed per party-hour for the 2013–2014 Christmas Bird Count were: N. Dakota 0, S. Dakota 0, Nebraska 0.11, Kansas 0, Oklahoma 0, northwest Texas 0. There are no definite Kansas or Oklahoma records. For complete species abundance data, see Appendix 1:138.

*Habitats.* During winter this species is usually found on open plains, fields and weedy areas.

*National Population.* Breeding Bird Survey trend data are not available for this alpine breeder. The estimated 1990s continental population was about 200,000 birds (Rich *et al.,* 2004).

*Further Reading.* King and Wales, 1964; Bent, 1968; Swenson, Jensen and To-epfer; 1988; Clemens, 1989; Dunn and Tessaglia-Hymes, 1999; MacDou-gall-Shackleton, Johnson and Hahn, 2000 (*The Birds of North America:* No. 559); Beadle and Rising, 2006.

## Pine Grosbeak
*Pinicola enucleator*

Pine grosbeaks are very widely distributed Northern Hemisphere seed-eaters that breed in mountain conifer forests. They mostly eat the seeds of conifer-ous and deciduous trees, as well as grass and weed seeds, but can survive on a great variety of alternative foods, including the fruits and buds of various trees. They have well-developed gular pouches for temporarily holding foods, and males courtship-feed seeds to their mates in the manner of many finches. Like other true (fringilline) finches, their young are also raised on regurgitated seeds rather than insects. Winter flocks are usually small, and probably con-sist of related birds.

*Winter Distribution.* Like the rosy-finches, pine grosbeaks are rare and unpre-dictable visitors to the plains from western mountains and possibly also Canada. Root (1988) mapped western Nebraska as a part of a winter pop-ulation associated with Colorado and southeastern Wyoming. Those re-ported from North Dakota more probably originated in Canada.

*Seasonality and Migrations.* The pine grosbeak is an unpredictable winter vis-itor in the Great Plains, appearing during times of poor seed crops in the mountains. This species was observed 36 years on Christmas Counts from 1968 to 2007 in North Dakota, 18 years in South Dakota, four years in Ne-braska, and five years in Kansas. In South Dakota wintering is rare (Tall-man, Swanson and Palmer, 2002). Fourteen initial autumn Nebraska sight-ings dating from the mid-1930s to the late 1970s (Johnsgard, 1980) range from October 21 to December 31, with a median of November 24. Thir-teen final spring sightings are from January 15 to May 22, with a median of March 10. There is no clustering of autumn or spring records. It is a very rare during winter in Kansas (Thompson *et al.*, 2011), and is a rare winter visitor in Oklahoma, with scattered records extending from Decem-ber to May (Woods and Schnell, 1984). It is also a very rare autumn and winter visitor in the Texas panhandle, with records extending from Octo-ber to March (Seyffert, 2001). The average number of birds observed per party-hour for the 2013–2014 Christmas Bird Count were: N. Dakota 0, S. Dakota 0, Nebraska 0.11, Kansas 0, Oklahoma 0, northwest Texas 0. For complete species abundance data, see Appendix 1:139.

*Habitats.* During the winter this species is normally associated with seed-bearing trees, including both coniferous and deciduous species.

*National Population.* Breeding Bird Surveys between 1966 and 2012 indicate that this species exhibited a survey-wide population decrease (1.34% annually) during that period. The estimated 1990s continental population was about 2.2 million birds (Rich *et al.*, 2004). It also breeds in northern Eurasia.

*Further Reading.* Cade, 1952; Bent, 1968; Godfrey, 1986; Adkisson, 1999 (*The Birds of North America:* No. 456); Dunn and Tessaglia-Hymes, 1999; Beadle and Rising, 2006.

## Purple Finch
*Haemorhous purpureus*

Invasions by purple finches across the Great Plains are unexpected but pleasant events. The winter of 2007–2008 was such an occasion, and at least in Nebraska it was accompanied by a similar appearance of red-breasted nuthatches. Purple finches are now much less common during winter in Nebraska than they were 50 years ago. This change may be partly the result of global warming, but the species is thought to be seriously declining in the northeastern states, and perhaps the lower recent numbers in the Midwest are a reflection of a general population reduction.

*Winter Distribution.* Root (1988) mapped the species' eastern distribution as centered near the Mississippi valley, extending west to the eastern edges of Montana, Wyoming, and Colorado, and to the eastern Texas panhandle. Peak numbers were indicated in Oklahoma, Arkansas and Missouri. Our data indicate that in the first decade of the four-decade study period Oklahoma had the highest average numbers, but by the last decade these had declined to about one-tenth of the early counts. Numbers in Kansas have also declined, while those in the Dakotas have been fairly stable or have increased. It would appear that purple finches are now wintering considerably farther north in the Great Plains than they did during the 1960s.

*Seasonality and Migrations.* Purple finches are irregular winter visitors to the Great Plains, especially the northern plains. This species was observed all 40 years on Christmas Counts from 1968 to 2007 in North Dakota, 39 years in both South Dakota and Nebraska, all 40 years in Kansas and Oklahoma, and seven years in the Texas panhandle. In South Dakota wintering is regular (Tallman, Swanson and Palmer, 2002). Thirty-seven initial autumn Nebraska sightings dating from the mid-1930s to the late 1970s

(Johnsgard, 1980) range from August 14 to December 26, with a median of October 27. Forty-nine final spring sightings are from January 2 to June 5, with a median of April 23. It is irregular during winter in Kansas, with records extending from October to May (Thompson and Ely, 1992), and a winter resident in Oklahoma, with continuous monthly records extending from October to May (Woods and Schnell, 1984). It is a very rare to rare winter visitor in the Texas panhandle, with records extending from September to May (Seyffert, 2001). The average number of birds observed per party-hour for the 2013–2014 Christmas Bird Count were: N. Dakota 0.5, S. Dakota 0.53, Nebraska 0.07, Kansas 0.06, Oklahoma 0.1, northwest Texas 0. For complete species abundance data, see Appendix 1:140.

*Habitats.* Wintering birds are associated with woodland streams, and sometimes also appear at bird feeders.

*National Population.* Breeding Bird Survey trend data are not available for this boreal forest breeder. The estimated 1990s continental population was about three million birds (Rich *et al.,* 2004).

*Further Reading.* Bent, 1968; Godfrey, 1986; Wootton, 1996 (*The Birds of North America:* No. 208); Dunn and Tessaglia-Hymes, 1999; Beadle and Rising, 2006.

## Cassin's Finch
### *Haemorhous cassinii*

Like several other conifer-breeding finches, this montane species descends to lower altitudes after the breeding season. It then spreads out in the foothills, sometimes forming flocks with crossbills and evening grosbeaks. These flocks are usually small and loosely organized, but at times may grow to include as many as 5,000 individuals. Like other typical finches it prefers the seeds of conifers, but will also eat the buds of conifers and other trees, especially during winter.

*Winter Distribution.* This finch is a local resident of the Black Hills and an infrequent winter visitor to the rest of the Great Plains, perhaps from sources in the western mountain forests. Root (1988) mapped it as reaching western parts of South Dakota, Nebraska, Kansas and Oklahoma from population centers in the Rocky Mountains. Our data show it to be a very rare visitor throughout the central and southern Great Plains states, most consistently appearing in Oklahoma.

*Seasonality and Migrations.* Cassin's finch is an permanent resident of the Black Hills and winter visitor in the southern Great Plains, from western

Nebraska south. This species has appeared only occasionally during 40 years of the five Great Plains states' Christmas Counts between 1968 and 2007, including five years in Nebraska, one year in Kansas, 16 years in Oklahoma and three years in the Texas panhandle. The only autumn Nebraska sightings between the mid-1930s and the late 1970s were for late October. Thirteen winter and spring records range from January 1 to May 14, with a median of April 12. Records from Kansas range from October to April (Thompson and Ely, 1992), and Oklahoma records extend from November to May (Woods and Schnell, 1984). It is an extremely rare winter visitor in the Texas panhandle (Seyffert, 2001).

*Habitats.* Normally associated with open coniferous forests during winter, this species usually forages on the ground for seeds.

*National Population.* Breeding Bird Surveys between 1966 and 2012 indicate that this species exhibited a survey-wide population decrease (2.70% annually) during that period. The estimated 1990s continental population north of Mexico was about 1.9 million birds (Rich *et al.,* 2004). The species breeds south to northern Mexico.

*Further Reading.* Bent, 1968; Hahn, 1996 (*The Birds of North America:* No. 240); Dunn and Tessaglia-Hymes, 1999; Beadle and Rising, 2006.


## Evening Grosbeak
### *Coccothraustes vespertina*

This is another of the irruptive species that appears on the Great Plains only irregularly, when the birds move out of their coniferous forest breeding areas and spread out over the adjoining plains. In one Pennsylvania study involving 499 winter band recoveries, only 49 birds had returned to the banding site in subsequent winters, while the remainder were scattered among 17 states and four Canadian provinces (Harrison and Harrison, 1990). These grosbeaks are large and gregarious birds, of about the same size as pine grosbeaks, and like them are likely to dominate smaller finches when competing over a source of seeds. Their huge beaks allow them to crush very large seeds, such as the pits of wild cherries (*Prunus serotina*).

*Winter Distribution.* Root (1988) mapped this irruptive species as mostly confined to the Dakotas during Great Plains winters, but with a few reaching western Nebraska. Our data show a much broader distribution that is centered in the northern states but sometimes reaches Texas during irruptive years.

*Seasonality and Migrations.* The evening grosbeak is a rare breeding resident (Black Hills) and irregular winter visitor in the northern Great Plains, south to Nebraska. This species was observed 32 years on Christmas Counts from 1968 to 2007 in North Dakota, 38 years in South Dakota, 15 years in Nebraska, 19 years in Kansas, 13 years in Oklahoma, and five years in the Texas panhandle. In South Dakota this species regularly winters (Tallman, Swanson and Palmer, 2002). Thirty-four initial autumn Nebraska sightings dating from the mid-1930s to the late 1970s (Johnsgard, 1980) are from September 3–December 31, with a median of November 9. Fifty-two final spring sightings are from January 5 to May 28, with a median of April 25. It is irregular during winter in Kansas, with records extending from October 19 to May 20 (Thompson and Ely, 1992), and is a rare winter visitor in Oklahoma, with most records occurring from October to May (Woods and Schnell, 1984). It is a rare winter visitor in the Texas panhandle, with records extending from August to May (Seyffert, 2001). The average number of birds observed per party-hour for the 2013–2014 Christmas Bird Count were: N. Dakota 0, S. Dakota 0, Nebraska 0, Kansas 0, Oklahoma 0, northwest Texas 0. For complete species abundance data, see Appendix 1:141.

*Habitats.* During winter this species is usually associated with streamside woodlands having seed-bearing deciduous trees, and it sometimes also appears at bird-feeding stations.

*National Population.* Breeding Bird Surveys between 1966 and 2006 indicate that this species exhibited a significant national population decline (2.0% annually) during that period. The estimated 1990s continental population north of Mexico was about 5.7 million birds (Rich *et al.,* 2004). The species breeds south to central Mexico.

*Further Reading.* Bent, 1968; Godfrey, 1986; Dunn and Tessaglia-Hymes, 1999; Gillihan and Byers, 2001 *(The Birds of North America:* No. 599); Beadle and Rising, 2006.

## House Finch

*Haemorhous mexicanus*

In about a half a century the house finch has managed to quietly expand and occupy much of the eastern and central United States and southern Canada, without causing noticeable ecological disruption to any other species. It is true that the house sparrow has noticeably declined during that same period but this is not a clear cause-and-effect relationship, as the house sparrow has also been declining in Europe, for still-uncertain reasons.

*Winter Distribution.* The range map shown by Root (1988) shows the then-recently introduced eastern population of house finches, which were released by bird dealers in New York City in the early 1940s, as having by the 1960s expanded west to southeastern Pennsylvania, Virginia and North Carolina. Since the 1960s the population has continued to expand west, meeting the resident western finch population in the central Great Plains a few decades later. By the start of the 21$^{st}$ century there was a virtually continuous distribution of house finches from the Atlantic to the Pacific coast. North Dakota was the last of the Great Plains states to be reached during this remarkable expansion period, but the species' population has since increased rapidly. The Christmas Count data summarized here illustrate the progression of its westward range expansion and population increase, which is still underway. As the rapid westward expansion of the eastern house finch population proceeded, the native populations in western Kansas and Oklahoma were slowly advancing eastwardly. About 70 years were needed for the house finch to advance from Oklahoma's western panhandle, where it first appeared in 1919, to southeastern Oklahoma, where it first reported nesting in 1992 (Reinking, 2004).

*Seasonality and Migrations.* The house finch is a common to abundant permanent resident, except in the northern plains, where it is probably somewhat migratory. This species was observed 19 years on Christmas Counts from 1968 to 2007 in North Dakota (beginning in 1988), 26 years in South Dakota (beginning in 1971), 39 years in Nebraska and Kansas, 37 years in Oklahoma, and all 40 years in the Texas panhandle. The average number of birds observed per party-hour for the 2013–2014 Christmas Bird Count were: N. Dakota 1.51, S. Dakota 1.4, Nebraska 3.94, Kansas 1.56, Oklahoma 1.28, northwest Texas 0.93. For complete species abundance data, see Appendix 1:141.

*Habitats.* The house finch is associated throughout the year with open woods, riverbottom thickets, scrubby vegetation, ranchlands, suburbs and towns. Its westward expansion across Nebraska seemingly involved following riparian forests and town-hopping.

*National Population.* Breeding Bird Surveys between 1966 and 2012 indicate that this species exhibited a survey-wide population increase (0.07% annually) during that period.

However, since the early 1990s the population has been reduced as a result of bacterial infection by a strain of *Mycoplasma gallisepticum.* The estimated 1990s continental population north of Mexico was about 16.6 million birds (Rich *et al.,* 2004). The species breeds south to central Mexico.

*Further Reading.* Bent, 1968; Mundinger and Hope, 1982; Brown and Brown, 1988; Hill, 1993 (*The Birds of North America:* No. 46), Dunn and Tessaglia-Hymes, 1999; Beadle and Rising, 2006.

304    PART 2. WINTER BIRDS OF THE GREAT PLAINS

## Red Crossbill
*Loxia curvirostra*

The red crossbill is a species that should be an icon of avian evolution. It reflects the influences of how a species' specialized foods (conifer seeds) can be effectively extracted from a well-protected source (conifer cones), and how the effects of competition from other similarly adapted species, can shape a bird's morphology. Individual crossbills vary considerably in the exact shapes of their beaks and the degree to which their tips overlap, affecting their seed-extraction abilities, according to work by Beckman (1988) and many others since (Adkisson, 1996). Furthermore, within North America as a whole there are great differences in bill structure, depending upon variations in cone characteristics of different conifer seed sources. These crossbill populations not only vary in bill structures, but also in their flight calls and alarm notes, suggesting that there may be several biologically distinct but morphologically almost inseparable "sibling" species that are currently considered a single taxonomic species. The crossbills of Eurasia are part of this same taxonomically complex assemblage, and vary in corresponding fashions, with evidence for as many as three species present in Scotland.

*Winter Distribution.* Root (1988) mapped the winter range of this crossbill as extending from Oklahoma northward into Canada, including a major population center in the Colorado Rocky Mountains. Our data show the birds rather broadly distributed through the Dakotas and Nebraska in winter, but with few extending any farther south. It is likely that at least some the birds wintering in the eastern Dakotas and Nebraska may originate from western breeding locations within these states.

*Seasonality and Migrations.* The red crossbill is a local permanent resident (Black Hills, western Nebraska and North Dakota) and a periodic winter visitor in the Great Plains. This species was observed 31 years on Christmas Counts from 1968 to 2007 in North Dakota, 36 years in South Dakota, 27 years in Nebraska, 17 years in Kansas, 12 years in Oklahoma, and one year in the Texas panhandle. In South Dakota this species regularly winters (Tallman, Swanson and Palmer, 2002). Thirty-one initial autumn Nebraska sightings dating from the mid-1930s to the late 1970s (Johnsgard, 1980) range from July 26 to December 29, with a median of November 12. Forty-four final spring sightings are from January 1 to June 2, with a median of April 1. It is irregular during winter in Kansas, usually occurring from mid-November to late March (Thompson and Ely, 1992), and a winter resident in Oklahoma, with continuous monthly records extending from September to May (Woods and Schnell, 1984). It is a rare visitor in the

Texas panhandle, with records throughout the year (Seyffert, 2001). The average number of birds observed per party-hour for the 2013–2014 Christmas Bird Count were: N. Dakota 0, S. Dakota 0.07, Nebraska 0.19, Kansas 0, Oklahoma 0, northwest Texas 0.04. For complete species abundance data, see Appendix 1:142.

*Habitats.* Migrants and wintering birds are also largely confined to conifer plantings or forests, but sometimes flocks also may be found foraging in stands of sunflowers or ragweeds.

*National Population.* Breeding Bird Surveys between 1966 and 2012 indicate that this species exhibited a survey-wide population decrease (0.36% annually) during that period. The estimated 1990s continental population north of Mexico was about 5.7 million birds (Rich *et al.,* 2004). The species breeds south to Nicaragua. There is also a Eurasian population (the "crossbill") of uncertain systematic complexity and distinction from the taxonomically complex North American crossbills.

*Further Reading.* Beckman, 1988; Bent, 1968; Bock, 1976; Godfrey, 1986; Beckman, 1988. Adkisson, 1996 (*The Birds of North America:* No. 256); Beadle and Rising, 2006.

## White-winged Crossbill
### *Loxia leucoptera*

This species ranges widely over the coniferous forests of the Northern Hemisphere, breeding in much the same regions as the red crossbill. As in that species, breeding may occur at any time of year, but the birds are less likely to forage in deciduous trees and shrubs, preferring to remain in open coniferous or mixed forests.

*Winter Distribution.* This crossbill is a very rare winter visitor in the Great Plains, which Root (1988) mapped as occurring only in eastern South Dakota and adjacent northeastern Nebraska. Our data suggest that it might appear anywhere in the Great Plains states, but only in very low numbers.

*Seasonality and Migrations.* The white-winged crossbill is a rare to occasional winter visitor in the northern Great Plains, rarely occurring south to Oklahoma. This species was observed 27 years on Christmas Counts from 1968 to 2007 in North Dakota, 12 years in South Dakota, four years in Nebraska, two years in Kansas, one year in Oklahoma, and none in the Texas panhandle. In South Dakota this species regularly winters (Tallman, Swanson and Palmer, 2002). Three autumn Nebraska sightings dating from the mid-1930s to the late 1970s are from October 16 to November 24. Twenty-two spring sightings range from January 1 to June 14, with a median of March 6. A few Kansas records are from September through March, and scattered Oklahoma records extend from August to January (Thompson and Ely, 1992, Woods and Schnell, 1984). It is an extremely rare winter visitor in the Texas panhandle (Seyffert, 2001).

*Habitats.* This crossbill is associated with coniferous forests or plantations throughout the year, especially pines.

*National Population.* Breeding Bird Surveys between 1966 and 2012 indicate that this species exhibited a survey-wide population increase (3.42% annually) during that period. The estimated 1990s continental population was about 21 million birds (Rich *et al.,* 2004). There is also a Eurasian population (the "two-barred crossbill").

*Further Reading.* Bent, 1968; Bock, 1976; Godfrey, 1986; Benkman, 1992 (*The Birds of North America:* No. 27); Beadle and Rising, 2006.

## Common Redpoll
*Carduelis flammeus*

Redpolls are tough little arctic-breeding finches; I have seen them nesting in low willows in arctic Alaska, where they are probably the smallest of all breeding birds. Redpolls have tiny beaks, and their favorite foods are birch (*Betula*) and alder (*Alnus*) seeds. Redpolls are unusual in that the esophagus has a small lateral pouch, comparable to the gular pouch of some other seed-eaters such as jays, which allows for temporary storage of seeds. Redpolls have very little energy to spare, and during winter may roost in cavities under the snow in order to conserve body heat.

*Winter Distribution.* Unlike many of the boreal finches, this is an arctic-breeding species that usually winters north of the U.S.-Canada border, and is likely to visit only the northern Plains States except during occasional irruptive years. Root (1988) mapped it as extending south to Nebraska, but showed its highest Plains densities in North Dakota. Our data support that general distribution pattern, but with a sharp decline in numbers south of North Dakota. As an irruptive and infrequent species it is impossible to be sure of general population trends, but the Great Plains trend since the 1980s might appear to be downward. An examination of Canadian redpoll Christmas Counts between British Columbia and Ontario since the early 1960s shows an apparently stable long-term winter population, but with sharp peaks averaging about three years apart. Thus, the apparently declining recent numbers in North Dakota may simply reflect a northerly wintering shift into southern Canada.

*Seasonality and Migrations.* The common redpoll is a regular and common winter visitor in the northern Great Plains. This species was observed all 40 years on Christmas Counts from 1968 to 2007 in North Dakota, 38 years in South Dakota, 24 years in Nebraska, and 16 years in Kansas. It regularly winters in South Dakota (Tallman, Swanson and Palmer, 2002). Twenty initial autumn Nebraska sightings dating from the mid-1930s to the late 1970s (Johnsgard, 1980) range from August 8 to December 30, with a median of November 26. Thirty final spring sightings are from January 10 to May 30, with a median of March 17. It is irregular during winter in Kansas, with most records extending from November to March (Thompson and Ely, 1992), and a rare winter visitor in Oklahoma, with continuous monthly records extending from November to March (Woods and Schnell, 1984). It is an extremely rare winter visitor in the Texas panhandle (Seyffert, 2001). The average number of birds observed per party-hour for the 2013–2014 Christmas Bird Count were: N. Dakota 0.39, S. Dakota 0.13, Nebraska 0.05, Kansas 0, Oklahoma 0, northwest Texas 0. For complete species abundance data, see Appendix 1:144.

*Habitats.* During winter this species is associated with conifers, deciduous thickets, and weedy fields. It also sometimes visits bird feeders.

*National Population.* Breeding Bird Survey trend data are not available for this arctic breeder. The estimated 1990s continental population was about 28 million birds (Rich *et al.,* 2004). There is also a Eurasian population (the "redpoll").

*Further Reading.* Cade, 1953; Bent, 1968; Godfrey, 1986; Dunn and Tessaglia-Hymes, 1999; Knox and Lowther, 2000a (*The Birds of North America:* No. 543); Beadle and Rising, 2006.

### Hoary Redpoll
*Carduelis hornemanni*

This arctic-adapted relative of the common redpoll is even rarer in the Great Plains than is the common redpoll. It is most likely to be seen among winter flocks of common redpolls in the more northern states. Adults weigh under half an ounce, and are less than half the weight of some other common tundra passerines, such as the Lapland longspur and snow bunting.

*Winter Distribution.* This species is an even more arctic-oriented species than the common redpoll, an correspondingly winters farther north. Root

(1988) mapped its Great Plains winter range is mostly limited to the Dakotas, and our data suggest that it is very rare south of North Dakota. The small numbers reported o Christmas Counts show no clear numerical trend.

*Seasonality and Migrations.* The hoary redpoll is an irregular winter visitor in the northern Great Plains. This species was observed 23 years on Christmas Counts from 1968 to 2007 in North Dakota, and seven years in South Dakota. In South Dakota this species regularly winters (Tallman, Swanson and Palmer, 2002). Five Nebraska sightings dating from the mid-1930s to the late 1970s (Johnsgard, 1980) range from January to May, with the largest number of sightings (3) in February. There are apparently no Kansas, Oklahoma or Texas panhandle records.

*Habitats.* This species is usually found in the same Great Plains habitats as common redpolls, and in company with them.

*National Population.* Breeding Bird Survey trend data are not available for this arctic breeder. The estimated 1990s continental population was about 26 million birds (Rich *et al.,* 2004). There is also a Eurasian population (the "arctic redpoll").

*Further Reading.* Bent, 1968; Godfrey, 1986; Dunn and Tessaglia-Hymes, 1999; Knox and Lowther, 2000b *(The Birds of North America:* No. 544); Beadle and Rising, 2006.


### Pine Siskin
*Carduelis pinus*


Pine and alder seeds are the primary foods of pine siskins, and so long at these foods are abundant on the breeding grounds the birds will winter in that vicinity. However, in some years the seed crops fail, and the birds then scatter widely to spend the winter in regions often hundreds of miles away. In Nebraska the species breeds regularly in the Pine Ridge and Wildcat Hills regions, and irregularly nests elsewhere following periodic winter invasions.

*Winter Distribution.* Root's (1988) map shows this species as widely distributed across the Great Plains, with the fewest in North Dakota and the most in western Kansas and the Oklahoma and Texas panhandles. Our data show declining numbers in the Texas panhandle and Oklahoma, and fluctuating numbers from Kansas north to North Dakota. Like many irruptive finches, long-term population trends are probably only significant when considered from a continental viewpoint.

*Seasonality and Migrations.* The pine siskin is mostly a common but somewhat irregular winter visitor in the Great Plains, sometimes remaining to breed locally. This species was observed all 40 years on Christmas Counts from 1968 to 2007 in all states from North Dakota to Oklahoma and the Texas panhandle. This species regularly winters in South Dakota (Tallman, Swanson and Palmer, 2002). Sixty initial autumn Nebraska sightings dating from the mid-1930s to the late 1970s (Johnsgard, 1980) range from July 25 to December 31, with a median of October 16. Thirty-five final spring sightings range from January 19 to June 9, with a median of May 12. It is common but irregular during winter in Kansas, usually between mid-October and mid-May (Thompson and Ely, 1992), and a winter resident in Oklahoma, with continuous monthly records extending from September to May (Woods and Schnell, 1984). It is a fairly common to abundant winter visitor in the Texas panhandle, with records for every month (Seyffert, 2001). The average number of birds observed per party-hour for the 2013–2014 Christmas Bird Count were: N. Dakota 0.31, S. Dakota 0.08, Nebraska 0.06, Kansas 0.08, Oklahoma 0.02, northwest Texas 0. For complete species abundance data, see Appendix 1:145.

*Habitats.* Wintering birds occur in both wooded and treeless areas, often feeding in small flocks on weed seeds.

*National Population.* Breeding Bird Surveys between 1966 and 2012 indicate that this species exhibited a survey-wide population decrease (2.26% annually) during that period.

The estimated 1990s continental population north of Mexico was about 13 million birds (Rich *et al.,* 2004). The species breeds south to southern Mexico. There is also a Eurasian population (the "siskin").

*Further Reading.* Bent, 1968; Godfrey, 1986; Dawson, 1997 (*The Birds of North America:* No. 280); Dunn and Tessaglia-Hymes, 1999; Beadle and Rising, 2006.

## Lesser Goldfinch
### *Carduelis psaltria*

This finch is a smaller and more arid-adapted relative of the American goldfinch, and like that species is very fond of the small seeds produced by thistles, dandelions and other composite weeds. After their late breeding season they gather in fairly large flocks, which at times may reach several hundred individuals. Adult males have two plumage variants, black-backed and green-backed, the black-backed form being more common in eastern parts of the species' range.

*Winter Distribution.* This is a southwestern species that Root (1988) mapped as reaching only the western parts of the Oklahoma and Texas panhandles. Our data show it as still extremely rare there, and it was never seen in numbers above the minimal "present" category.

*Seasonality and Migrations.* The lesser goldfinch is a rare summer resident in extreme western Oklahoma, recorded from April 27 to December 25 (Sutton, 1967), with scattered winter sightings elsewhere in the southern and western Plains. It is a very rare winter visitor in the Texas panhandle, with records for every month but January (Seyffert, 2001).

*Habitats.* This marginal plains species favors riparian forests and arid woodlands, mostly of Gambel's oak (*Quercus gambelii*), and ponderosa pines. On migration and during winter it also occurs in agricultural areas, arid grasslands, and sometimes bird feeders.

*National Population.* Breeding Bird Surveys between 1966 and 2012 indicate that this species exhibited a survey-wide population increase (1.13% annually) during that period. The estimated 1990s continental population north of Mexico was about 1,550,000 birds (Rich *et al.,* 2004). The species breeds south to Peru.

*Further Reading.* Bent, 1968; Dunn and Tessaglia-Hymes, 1999; Willoughby, 1999 (*The Birds of North America:* No. 392); Beadle and Rising, 2006.

## American Goldfinch
### *Carduelis tristis*

Male goldfinches spend nearly all winter in their dull non-breeding plumage, which in largely related to the fact that they breed very late in the summer, when seeds of thistles dandelions and other weedy composites become available. These tiny seeds are well adapted to the birds' small beaks. Like other true finches, seeds also are the foods provided for the nestlings, rather than the insects that are the basic chick-raising foods for most other American songbirds.

*Winter Distribution.* This abundant and widespread finch has a nearly continent-wide distribution with several centers of abundance. In the Great Plains states it is shown by Root (1988) as having peaks in southwestern Nebraska and adjacent Kansas, and especially in Oklahoma, along the Red River valley, with northern extensions to southeastern Kansas and west to the Texas panhandle. Our data support that general pattern, with high numbers in Nebraska and Oklahoma, declining to fairly small numbers in

North Dakota. Numbers have clearly increased over time in the two Dakotas and Nebraska, while Oklahoma numbers have declined and Kansas's
numbers seem to have increased in the 1980s and 1990s, and then declined.
This pattern would suggest a major movement northward by goldfinches
since the 1960s, with peak numbers shifting from Oklahoma to Nebraska.

*Seasonality and Migrations.* The American goldfinch is a common permanent
resident over much of the central and southern Great Plains, but is variably migratory in the northern plains. This species was observed 39 years
on Christmas Counts from 1968 to 2007 in North Dakota, all 40 years in
the states from South Dakota to Oklahoma, and 25 years in the Texas panhandle. The average number of birds observed per party-hour for the 2013–
2014 Christmas Bird Count were: N. Dakota 3.6, S. Dakota 3.4, Nebraska
5.27, Kansas 2.18, Oklahoma 3.11, northwest Texas 2.70. For complete
species abundance data, see Appendix 1:146.

*Habitats.* Flocks of this species may often be found foraging in fields of tall
weeds such as ragweeds (*Ambrosia*) and sunflowers (*Helianthus*) during autumn and winter. It is often attracted to bird feeders if thistle seeds are
available.

*National Population.* Breeding Bird Surveys between 1966 and 2012 indicate
that this species exhibited a survey-wide population decrease (0.01% annually) during that period. Breeding Bird Surveys between 1966 and 2006 indicate that this species exhibited a statistically nonsignificant national population decline (0.1% annually) during that period. The estimated 1990s
continental population was about 24 million birds (Rich *et al.,* 2004).

*Further Reading.* Bent, 1968; Middleton, 1977, 1993 (*The Birds of North America:* No. 80); Dunn and Tessaglia-Hymes, 1999; Beadle and Rising, 2006.

# FAMILY PASSERIDAE:
# OLD WORLD SPARROWS

## House Sparrow
*Passer domesticus*

House sparrows are all too familiar to most people, and it is a very rare Christmas Count that does not include many individuals of this species. Yet, house sparrows provide some benefits in giving house-bound residents of inner cities something to watch through their windows. The birds are remarkably adjusted to city life around humans, and somehow manage to find roosting spots that are warm enough to get them through the coldest of Great Plains winter nights. However, they are canny birds, and only rarely does a sharp-shinned or Cooper's hawk catch one unawares, in spite of their relative abundance.

*Winter Distribution.* Root's (1988) map of this ubiquitous and sedentary species shows the densest Great Plains concentration in Kansas, with a secondary peak in the vicinity of the North Dakota–South Dakota border. Our data suggest that the Dakotas now (2014) have the highest winter density of Great Plains house sparrows, with Nebraska, and Kansas progressively fewer, and Oklahoma even fewer. The five-state average counts show a marked population decline over the four-decade period, at a much more rapid rate than is indicated by national Breeding Bird Survey trend data. The native Eurasian population has also been in sharp decline during recent decades.

*Seasonality and Migrations.* The house sparrow is a permanent resident throughout it range. The average number of birds observed per party-hour for the 2013–2014 Christmas Bird Count were: N. Dakota 13.38, S. Dakota 14.27, Nebraska 7.8, Kansas 3.24 Oklahoma 1.92, northwest Texas 3.07. For complete species abundance data, see Appendix 1:147.

*Habitats.* This species is always associated with humans, most often occurring in cities, suburbs, and around farm buildings.

*National Population.* Breeding Bird Surveys between 1966 and 2012 indicate that this species exhibited a survey-wide population decrease (3.7% annually) during that period. A recent North American population estimate is 82 million birds (Rich *et al.,* 2004). The species breeds disjunctively south to South America.

*Further Reading.* Summers-Smith, 1963; Lowther and Cink, 1992 (*The Birds of North America:* No. 12); Dunn and Tessaglia-Hymes, 1999.

# References

Adkisson, C. D. 1996. Red crossbill In *The Birds of North America*, No. 256 (A. Poole and F. Gill, eds.). Philadelphia, PA: The Birds of North America, Inc. 24 pp.

_____. 1999. Pine grosbeak. In *The Birds of North America*, No. 456 (A. Poole and F. Gill, eds.). Philadelphia, PA: The Birds of North America, Inc. 20 pp.

Alderfer, J. (ed.) 2006. *Complete Birds of North America.* Washington, D.C.: National Geographic Society,

Allison, P. S., A W. Leary and M. J. Bechard. 1993. Observations on wintering ferruginous hawks (*Buteo regalis*) feeding on prairie dogs (*Cynomys ludovicianus*) in the Texas panhandle. *Texas J. Sci.* 47:235–237.

Ammon, E. M. 1995. Lincoln's sparrow. In *The Birds of North America*, No. 191 (A. Poole and F. Gill, eds.). Philadelphia, PA: The Birds of North America, Inc. 20 pp.

Anderson, A. A., and A. Anderson. 1973. *The Cactus Wren.* Tucson, AZ: Arizona University Press.

Anderson, S. H., and J. R. Squires. 1997. *The Prairie Falcon.* Austin, TX: University of Texas Press.

Andrews, R., and R, Richter. 1992. *Colorado Birds: A Reference to Their Distribution and Habitat.* Denver, CO: Denver Museum of Natural History.

Arbib, R. S. 1982. "Ideal model" Christmas Bird Counts: a start in 1982–83. *Am. Birds* 36:146–148.

_____. 1983. The ideal model Christmas Bird Count: 1982–3, the experimental year. *Am. Birds* 37:366–368.

Arcese, P. M., K. Sogge, A. B. Marr, and M. A. Patten. 2003. Song sparrow. In *The Birds of North America*, No. 704 (A. Poole and F. Gill, eds.). Philadelphia, PA: The Birds of North America, Inc. 40 pp.

Askins, R. A. 1993. Population trends in grassland, shrubland and forest birds in eastern North America. *Current Ornithology 11*:1–34.

Austin, J. E., and M. R. Miller. 1995. Northern pintail. In *The Birds of North America*, No. 163 (A. Poole and F. Gill, eds.). Philadelphia, PA: The Birds of North America, Inc. 32 pp.

_____. C. M. Custer and A. D. Afton. 1998, Lesser scaup. In *The Birds of North America*, No. 338 (A. Poole and F. Gill, eds.). Philadelphia, PA: The Birds of North America, Inc. 32 pp.

Avery, M. L. 1995. Rusty blackbird. In *The Birds of North America*, No. 200 (A. Poole and F. Gill, eds.). Philadelphia, PA: The Birds of North America, Inc. 16 pp.

Baird. P. H. 1994. Black-legged kittiwake. In *The Birds of North America*, No. 92 (A. Poole and F. Gill, eds.). Philadelphia, PA: The Birds of North America, Inc. 28 pp.

Balda, R. P. 2002. Pinyon jay. In *The Birds of North America*, No. 604 (A. Poole and F. Gill, eds.). Philadelphia, PA: The Birds of North America, Inc. 32 pp.

Baldassare, G. 2014. *Ducks, Geese and Swans of North America.* Revised ed. 2 vol. 1027 pp. Baltimore,MD: Johns Hopkins Univ. Press.

Balph, M. H., and D. F. Balph. 1977. Winter social behavior of dark-eyed juncos: communication, social organization and ecological implications. *Anim. Behav.* 23:859–884.

_____. 1979. Flock stability in relation to social dominance and agonistic behavior in wintering dark-eyed juncos. *Auk* 9:714–722.

Barbour, R. W. 1941. Winter habits of the red-eyed towhee in eastern Kentucky. *Amer. Midl. Nat.* 26:583–595.

Baughman, M. (ed.). 2003, *Reference Atlas to the Birds of North America*. Washington, D.C.: National Geographic Society.

Baumgartner, F. M., and A. M. Baumgartner. 1992. *Oklahoma Bird Life*. Norman, OK: University of Oklahoma Press.

Beadle, D., and J. D. Rising. 2006. *Tanagers, Cardinals and Finches of the United States and Canada: The Photographic Guide*. Princeton, NJ: Princeton University Press.

Beason, R. C. 1995. Horned lark. In *The Birds of North America*, No. 195 (A. Poole and F. Gill, eds.). Philadelphia, PA: The Birds of North America, Inc. 24 pp.

Bechard, M. J., and J. K. Schmutz. 1995. Ferruginous hawk In *The Birds of North America*, No. 172 (A. Poole and F. Gill, eds.). Philadelphia, PA: The Birds of North America, Inc. 20 pp.

_____, and T. R. Swem. 2002. Rough-legged hawk. In *The Birds of North America*, No. 641 (A. Poole and F. Gill, eds.). Philadelphia, PA: The Birds of North America, Inc. 32 pp.

Beckman, C. W. 1988. Seed-handling ability, bill structure and the cost of specialization for crossbills. *Auk* 105:715–719.

Bednarz, J. C., and R. J. Raitt. 2002. Chihuahuan raven. In *The Birds of North America*, No. 606 (A. Poole and F. Gill, eds.). Philadelphia, PA: The Birds of North America, Inc. 20 pp.

Bellrose, F. C. 1980. *The Ducks, Geese and Swans of North America*. Revised ed. Harrisburg, PA: Stackpole.

Benkman, C. W. 1992. White-winged crossbill. In *The Birds of North America*, No. 27 (A. Poole and F. Gill, eds.). Philadelphia, PA: The Birds of North America, Inc. 20 pp.

Benson, K. L. P., and K. A. Arnold. (2001). *The Texas Breeding Bird Atlas*. College Station and Corpus Christi, TX: Texas A&M University System. http://tbba.cbi.tamucc.edu

Bent, A. C. 1921, Life histories of North American gulls and terns. Washington, D.C.: *U.S. Natl. Mus. Bull.* 113, 345 pp.

Bent, A. C. 1926. Life histories of North American marsh birds. Washington, D.C.: *U.S. Natl. Mus. Bull.* 135, 385 pp.

_____. 1927. Life histories of North American shore birds, part 1. Washington, D.C.: *U.S. Natl. Mus. Bull.* 142, 420 pp.

_____. 1929. Life histories of North American shore birds, part 2. Washington, D.C.: *U.S. Natl. Mus. Bull.* 146, 412 pp.

_____. 1932. Life histories of North American gallinaceous birds. Washington, D.C.: *U.S. Natl. Mus. Bull.* 162, 490 pp.

_____. 1937. Life histories of North American birds of prey, part 1. Washington, D.C.: *U.S. Natl. Mus. Bull.* 167, 409 pp.

_____. 1938. Life histories of North American birds of prey, part 2. Washington, D.C.: *U.S. Natl. Mus. Bull.* 170, 482 pp.

_____. 1939. Life histories of North American woodpeckers. Washington, D.C.: *U.S. Natl. Mus. Bull.* 174, 334 pp.

_____. 1940. Life histories of North American cuckoos, goatsuckers, hummingbirds and their allies. Washington, D.C.: *U.S. Natl. Mus. Bull.* 176, 506 pp.

_____. 1942. Life histories of North American flycatchers, larks, swallows and their allies. Washington, D.C.: *U.S. Natl. Mus. Bull.* 179, 555 pp.

_____. 1946. Life histories of North American jays, crows and titmice. Washington, D.C.: *U.S. Natl. Mus. Bull.* 191, 495 pp.

_____. 1948. Life histories of North American nuthatches, wrens, thrashers and their allies. Washington, D.C.: *U.S. Natl. Mus. Bull.* 195, 435 pp.

_____. 1949. Life histories of North American thrushes, kinglets and their allies. Washington, D.C.: *U.S. Natl. Mus. Bull.* 196, 437 pp.

_____. 1950. Life histories of North American wagtails, shrikes, vireos, and their allies. Washington, D.C.: *U.S. Natl. Mus. Bull.* 197, 411 pp.

_____. 1953. Life histories of North American wood warblers. Washington, D.C.: *U.S. Natl. Mus. Bull.* 203, 734 pp.

_____. 1958. Life histories of North American blackbirds, orioles, tanagers, and allies. Washington, D.C.: *U.S. Natl. Mus. Bull.* 211, 549 pp.

_____. 1968. Life histories of North American cardinals, grosbeaks, buntings, towhees, finches, sparrows, and allies. Washington, D.C.: *U.S. Natl. Mus. Bull.* 237, 1889 pp.

Bergerud, A. T., and M. W. Gratson (eds.). 1988. *Adaptive Strategies and Population Ecology of Northern Grouse.* Minneapolis, MN: University of Minnesota Press.

Bildstein, K. L. and K. Meyer. 2000. Sharp-shinned hawk. In *The Birds of North America,* No. 482 (A. Poole and F. Gill, eds.). Philadelphia, PA: The Birds of North America, Inc. 28 pp.

Birkhead, T. R. 1991. *The Magpies: The Ecology and Behavior of Black-billed and Yellow-billed Magpies.* London, UK: Academic Press.

Bock C. E. 1976. Synchronous eruptions of boreal seed-eating birds. *Amer. Natur.* 110:559–571.

_____ 1982. Factors influencing winter distribution and abundance of Townsend's solitaire. *Wilson Bull.* 94:297–302.

_____, and L.W. Lepthien. 1972. Winter eruptions of red-breasted nuthatches in North America. *Am. Birds* 26:558–561.

_____., and L.W. Lepthien. 1976. Changing winter distribution and abundance of the blue jay, 1962–1971. *Am. Midl. Nat.* 96:232–236.

Bordage, D., and J-P. L. Savard. 1995. Black scoter. In *The Birds of North America,* No. 177 (A. Poole and F. Gill, eds.). Philadelphia, PA: The Birds of North America, Inc. 20 pp.

Bowen, R. V. 1997. Townsend's solitaire. In *The Birds of North America,* No. 269 (A. Poole and F. Gill, eds.). Philadelphia, PA: The Birds of North America, Inc. 28 pp.

Bowman, R. 2002. Common ground-dove. In *The Birds of North America,* No. 645 (A. Poole and F. Gill, eds.). Philadelphia, PA: The Birds of North America, Inc. 24 pp.

Brennan, L. A. 1999. Northern bobwhite. In *The Birds of North America,* No. 397 (A. Poole and F. Gill, eds.). Philadelphia, PA: The Birds of North America, Inc. 28 pp.

_____.(ed.) 2007. *Texas Quails: Ecology and Management.* College Station, TX: Texas A&M University Press.

Bridgewater, D. D. 1966. Winter movement and habitat use by Harris' sparrow, *Zonotrichia querula* (Nuttall). *Proc. Oklahoma Acad. Sci.* 47:53–59.

Brisbin, I. L., Jr., H. D. Pratt and T. B. Mobray. 2003. Hawaiian coot and American coot In *The Birds of North America,* No. 697 (A. Poole and F. Gill, eds.). Philadelphia, PA: The Birds of North America, Inc. 44 pp.

Briskie, J. V. 1993. Smith's longspur. In *The Birds of North America,* No. 34 (A. Poole and F. Gill, eds.). Philadelphia, PA: The Birds of North America, Inc. 16 pp.

Brogie, M. A., and W. R. Silcock. 2004. Eurasian collared-dove (*Streptopelia decaocto*) expansion in Nebraska: 1997–2003. *Nebraska Bird Review* 72:18–22.

Brown, M. B., and C. R. Brown. 1988. Access to winter food resources by bright- versus dull-colored house finches. *Condor* 90:729–731.

Brown, P. W., and L. H. Fredrickson. 1997. White-winged scoter. In *The Birds of North America,* No. 274. (A. Poole and F. Gill, eds.). Philadelphia, PA: The Birds of North America, Inc. 28 pp.

Brua, R. B. 2003. Ruddy duck. In *The Birds of North America,* No. 696 (A. Poole and F. Gill, eds.). The Birds of North America, Inc. Philadelphia, PA. 32 pp.

Buckley, N. J. 1999, Black vulture. In *The Birds of North America,* No. 411 (A. Poole and F. Gill, eds.). The Birds of North America, Inc. Philadelphia, PA. 24 pp.

Buehler, D. A. 2000. Bald eagle. In *The Birds of North America,* No. 606 (A. Poole and F. Gill, eds.). Philadelphia, PA: The Birds of North America, Inc. 40 pp.

Bull, E. L., and J. A. Jackson. 1995. Pileated woodpecker. In *The Birds of North America*, No. 148(A. Poole and F. Gill, eds.). Philadelphia, PA: The Birds of North America, Inc. 24 pp.

Burger, J., and M. Gochfeld. 2002. Bonaparte's gull. In *The Birds of North America*, No. 634 (A. Poole and F. Gill, eds.). Philadelphia, PA: The Birds of North America, Inc. 24 pp.

Burton, J. F. 1995. *Birds and Climate Change.* London, UK: Christopher Helm.

Busby, W. H., and J. Zimmerman. 2001. *Kansas Breeding Bird Atlas.* Lawrence, KS: University Press of Kansas.

Butcher, G. S.,D. K. Niven, A. O. Punjabi, D. N. Pushley and K. V. Rosenberg. 2007. The 2007 WatchList for United States birds. *American Birds* 61:18-25.

Butler, C. J. 2003. The disproportionate effect of global warming on the arrival dates of short-distance migratory birds in North America. *Ibis* 145:489–496.

Butler R. W. 1992. Great blue heron. In *The Birds of North America*, No. 25 (A. Poole and F. Gill, eds.). Philadelphia, PA: The Birds of North America, Inc. 20 pp.

Byers, C., J. Curson and U. Olsen. 1995. *Sparrows and Buntings: A Guide to the Sparrows and Buntings of North America and the World.* Boston, MA: Houghton Mifflin.

Cabe, P. R. 1993. European starling. In *The Birds of North America*, No. 48(A. Poole and F. Gill, eds.). Philadelphia, PA: The Birds of North America, Inc. 24 pp.

Cable, T. T., S. Seltman and K. J. Cook. 1997. *Birds of Cimarron National Grassland.* Gen. Tech. Rep. RM-GTR-281. Fort Collins, CO. : Rocky Mtn. Range & Experiment Station. 108 pp.

Cade, T. J. 1952. The influence of food abundance on the over-wintering of pine grosbeaks at College, Alaska. *Condor* 54:363.

———. 1953. Sub-nival feeding of the redpoll in interior Alaska: A possible adaptation to the northern winter. *Condor* 55:43–44.

———. 1955, Experiments on the winter territoriality of the American kestrel (*Falco sparverius*). *Wilson Bull.* 67:5–17.

———. 1982. *The Falcons of the World.* Ithaca, NY: Cornell University Press.

Cade, T. J., and E. C. Atkinson. 2002. Northern shrike. In *The Birds of North America*, No. 671 (A. Poole and F. Gill, eds.). Philadelphia, PA: The Birds of North America, Inc. 32 pp.

Campo, J. J., B. C. Thompson, J. C. Barron, R. C. Telfair, H. F. Durocher and S. Gutreuter. 1993. Diet of double-crested cormorants wintering in Texas. *J. Field Ornith.* 64:135–141.

Cannings, R. J. 1993. Northern saw-whet owl. In *The Birds of North America*, No. 42 (A. Poole and F. Gill, eds.). Philadelphia, PA: The Birds of North America, Inc. 20 pp.

———, and T. Angell. 2001. Western screech-owl. In *The Birds of North America*, No. 597. (A. Poole and F. Gill, eds.). Philadelphia, PA: The Birds of North America, Inc. 20 pp.

Carey, M., D. E. Burhans and D. A. Nelson. 1994. Field sparrow. In *The Birds of North America*, No. 103 (A. Poole and F. Gill, eds.). Philadelphia, PA: The Birds of North America, Inc. 20 pp.

Carroll, J. P. 1993. Gray partridge In *The Birds of North America*, No. 59 (A. Poole and F. Gill, eds.). Philadelphia, PA: The Birds of North America, Inc. 20 pp.

Cavitt, J. F., and C. A. Haas. 2000. Brown thrasher. In *The Birds of North America*, No. 557 (A. Poole and F. Gill, eds.). Philadelphia, PA: The Birds of North America, Inc. 28 pp.

Chaplin, S. B. 1974. Daily energetics of the black-capped chickadee, *Parus atricapillus,* in winter. *J. Comp. Physiol.* 89B:321–330.

———. 1982. The energetic significance of huddling behavior in common bushtits (*Psaltriparus minimus*). *Auk* 99:424–430.

Clemens, D. X. 1989. Nocturnal hypothermia in rosy finches. *Condor* 91:739–741.

Chilton, G, M. C. Baker, C. D. Barrentine and M. A. Cunningham. 1995. White-crowned sparrow. In *The Birds of North America*, No. 183 (A. Poole and F. Gill, eds.). Philadelphia, PA: The Birds of North America, Inc. 28 pp.

Cicero, C. 2000. Oak titmouse and juniper titmouse. In *The Birds of North America*, No. 485 (A. Poole and F. Gill, eds.). Philadelphia, PA: The Birds of North America, Inc. 28 pp.

Cimprich, D. A., and F. R. Moore. 1995. Gray catbird. In *The Birds of North America*, No. 167 (A. Poole and F. Gill, eds.). Philadelphia, PA: The Birds of North America, Inc. 20 pp.

Clement, P. 2000. *Thrushes.* Princeton, NJ: Princeton University Press.

Clum. N. J., and T. J. Cade. 1994. Gyrfalcon. In *The Birds of North America*, No. 114 (A. Poole and F. Gill, eds.). Philadelphia, PA: The Birds of North America, Inc. 28 pp.

Collins, P. W. 1999. Rufous-crowned sparrow. In *The Birds of North America*, No. 472 (A. Poole and F. Gill, eds.). Philadelphia, PA: The Birds of North America, Inc. 28 pp.

Colvin, B. A., and S. R. Spaulding. 1983. Winter foraging behavior of short-eared owls (*Asio flammeus*) in Ohio. *Am. Midl. Nat.* 110:124–8.

Condor, R. W., 1970. The winter territories of tufted titmice. *Wilson Bull.* 82:177–183.

Connelly, J. W., M. W. Gratson, and K. P. Reese. 1998. Sharp-tailed grouse. In *The Birds of North America*, No. 354 (A. Poole and F. Gill, eds.). Philadelphia, PA: The Birds of North America, Inc. 20 pp.

Conway, C. J. 1995. Virginia rail. In *The Birds of North America*, No. 173 (A. Poole and F. Gill, eds.). Philadelphia, PA: The Birds of North America, Inc. 20 pp.

Cooch, F. G. 1960. Ecological aspects of the blue-snow goose complex. *Auk* 78:72–79.

Cooper, J. M. 1994. Least sandpiper. In *The Birds of North America*, No. 115 (A. Poole and F. Gill, eds.). Philadelphia, PA: The Birds of North America, Inc. 28 pp.

Cramp, S. (ed.). 1983. *Handbook of the Birds of Europe the Middle East and North Africa. The Birds of the Western Palearctic.* Vol. 3 (Waders to Gulls). Oxford, UK: Oxford University Press.

Crocoll, S. T. 1994. Red-shouldered hawk. In *The Birds of North America*, No. 107 (A. Poole and F. Gill, eds.). Philadelphia, PA: The Birds of North America, Inc. 20 pp.

Cullen, S. A., J. R. Jehl Jr. and G. L. Nuechterlein. 1999. Eared grebe. In *The Birds of North America*, No. 433 (A. Poole and F. Gill, eds.). Philadelphia, PA: The Birds of North America, Inc. 28 pp.

Cunningham, D., and T. Krueger (eds.). 1996. Weather and climate of Nebraska. *Nebraskaland Magazine* 74(1): 1–138.

Curry, R. L., A. T. Peterson and T. A. Langen. 2003. Western scrub-jay. In *The Birds of North America*, No. 712 (A. Poole and F. Gill, eds.). Philadelphia, PA: The Birds of North America, Inc. 36 pp.

Curson, J., D. Quinn and D. Beadle. 1994. *Warblers of the Americas: An Identification Guide.* Boston, MA: Houghton Mifflin Co.

Davis, J. 1957. Comparative foraging behavior of spotted and brown towhees. *Auk* 74:129–166.

———— 1973. Habitat preferences and competition of wintering juncos and golden-crowned sparrows. *Ecology* 54:174–180.

Dawson, W. R. 1997. Pine siskin. In *The Birds of North America*, No. 280 (A. Poole and F. Gill, eds.). Philadelphia, PA: The Birds of North America, Inc. 24 pp.

Dechant, D. J., M. L. Sondreal, D. H. Johnson, L. D. Igl, C. M. Goldade, M. P. Nenneman and B. R. Euliss. 1999a. Effects of Management Practices on Grassland Birds: Burrowing owl. Jamestown, ND: Northern Prairie Wildlife Research Center. (This and the following 11 references by Dechant *et al.*, Dinkins *et al.* (2001) and an assembled collection by Johnson *et al.* (2004) are all on line: http://www.npwrc.usgov/resource/literatr/grassbird

———— *et al.* 1999b. Effects of Management Practices on Grassland Birds: Chestnut-collared longspur. Jamestown, ND: Northern Prairie Wildlife Research Center.

———— *et al* 1999c. Effects of Management Practices on Grassland Birds: Ferruginous hawk. Jamestown, ND: Northern Prairie Wildlife Research Center.

_____ et al. 1999d. Effects of Management Practices on Grassland Birds: Grasshopper sparrow. Jamestown, ND: Northern Prairie Wildlife Research Center.

_____ et al. 1999e (revised 2000). Effects of Management Practices on Grassland Birds: Lark Bunting. Jamestown, ND: Northern Prairie Wildlife Research Center.

_____ et al. 1999f. Effects of Management Practices on Grassland Birds: McCown's longspur. Jamestown, ND: Northern Prairie Wildlife Research Center.

_____ et al. 1999g. Effects of Management Practices on Grassland Birds: Savannah sparrow. Jamestown, ND: Northern Prairie Wildlife Research Center.

_____ et al. 1999h. Effects of Management Practices on Grassland Birds: Sprague's Pipit. Jamestown, ND: Northern Prairie Wildlife Research Center.

_____ et al. 1999i. Effects of Management Practices on Grassland Birds: Short-eared owl. Jamestown, ND: Northern Prairie Wildlife Research Center.

_____ et al. 1999j. Effects of Management Practices on Grassland Birds: Lark sparrow. Jamestown, ND: Northern Prairie Wildlife Research Center.

_____ et al. 1999k. Effects of Management Practices on Grassland Birds: Sedge wren. Jamestown, ND: Northern Prairie Wildlife Research Center.

del Hoyo, J., A. Elliott, and J. Sargatal. 1994. *Handbook of Birds of the World*. Vol. 2. Barcelona, Spain: Lynx Edicions

_____. 1999. *Handbook of Birds of the World*. Vol. 5. Barcelona, Spain: Lynx Edicions.

Derrickson, K. C., and R. Breitwisch. 1992. Northern mockingbird. In *The Birds of North America*, No. 7 (A. Poole and F. Gill, eds.). Philadelphia, PA: The Birds of North America, Inc. 26 pp.

Desrochers, A., S. J. Hannon and K. E. Nordin. 1988. Winter survival and territory acquisition in a northern population of black-capped chickadees. *Auk* 105:727–736.

Dinkins, M. E., A. L. Zimmerman, J. A. Dechant, B. D. Parkins, D. H. Johnson, L. D. Igl, C. M. Goldade and B. R. Euliss. 2001. *Effects of Management Practices on Grassland Birds:* Western Meadowlark. Jamestown, ND: Northern Prairie Wildlife Research Center.

Dinsmore, J. J., and S. J. Dinsmore, 1993. Range expansion of the great tailed grackle in the 1900s. *Jour. Iowa Acad. Sci.* 100:54–59.

Dickson, K. M. (ed.). 2000. *Toward Conservation of the Diversity of Canada Geese in North America*. Occasional Paper 103, Ottawa, Canada: Canadian Wildlife Service, Environment Canada, 164 pp.

Dixon, K. L 1949. Behavior of the plain titmouse. *Condor* 51: 110–136.

Dixon, R. D., and V. A. Saab. 2000. Black-backed woodpecker. In *The Birds of North America*, No. 509 (A. Poole and F. Gill, eds.). Philadelphia, PA: The Birds of North America, Inc. 20 pp.

Doughty, R. W. 1988. *The Mockingbird*. Austin:, TX: University of Texas Press.

Drilling, N., R. Titman and F. McKinney. 2002. Mallard. In *The Birds of North America*, No. 658 (A. Poole and F. Gill, eds.). The Birds of North America, Inc. Philadelphia, PA. 44 pp.

Dubowy, P. J. 1996. Northern shoveler. In *The Birds of North America*, No. 217 (A. Poole and F. Gill, eds.). The Birds of North America, Inc. Philadelphia, PA.24 pp.

Dugger, B. D., K. M. Dugger and L. H. Fredrickson. 1994. Hooded merganser. In *The Birds of North America*, No. 98 (A. Poole and F. Gill, eds.). Philadelphia, PA: The Birds of North America, Inc. 24 pp.

Duncan, J. R., and P. A. Duncan. 1998. Northern hawk owl. In *The Birds of North America*, No. 356 (A. Poole and F. Gill, eds.). Philadelphia, PA: The Birds of North America, Inc. 28 pp.

Dunn, E. H., and D. L. Tessaglia-Hymes. 1999. *Birds at Your Feeder: A Guide to Feeding Habits, Behavior, Distribution and Abundance*. New York, NY: W. W. Norton.

Dunn, J., and K. Garrett. 1997. *A Field Guide to Warblers of North America.* Boston, MA: Houghton Mifflin.

Eadie, J. M, M. L. Mallory and H. G. Lumsden. 1995. Common goldeneye. In *The Birds of North America*, No. 170 (A. Poole and F. Gill, eds.). Philadelphia, PA: The Birds of North America, Inc. 32 pp.

———, J-P. L. Savard and M. L. Mallory. 2000. Barrow's goldeneye. In *The Birds of North America*, No. 548 (A. Poole and F. Gill, eds.). Philadelphia, PA: The Birds of North America, Inc. 32 pp.

Eaton, S. E. 1992. Wild Turkey. In *The Birds of North America*, No. 22 (A. Poole and F. Gill, eds.). Philadelphia, PA: The Birds of North America, Inc. 28 pp.

Elly, C. R., and A. X. Dzubin. 1994. Greater white-fronted goose. In *The Birds of North America*, No. 131 (A. Poole and F. Gill, eds.). Philadelphia, PA: The Birds of North America, Inc. 32 pp.

Elphick, C. S., and T. L. Tibbitts. 1998. Greater yellowlegs. In *The Birds of North America*, No. 355 (A. Poole and F. Gill, eds.). Philadelphia, PA: The Birds of North America, Inc. 24 pp.

Ervin, S. 1977. Flock size, composition, and behavior in a population of bushtits *(Psaltriparus minimus) Bird-Banding* 48:97–109.

Evans, R. E., and F. L. Knopf. 1993. American white pelican. In *The Birds of North America*, No. 57 (A. Poole and F. Gill, eds.). Philadelphia, PA: The Birds of North America, Inc. 24 pp.

Falls, J. B., and J. G. Kopachena. 1994. White-throated sparrow. In *The Birds of North America*, No. 128 (A. Poole and F. Gill, eds.). Philadelphia, PA: The Birds of North America, Inc. 32 pp.

Ferguson-Lees, J., and D. A. Christie. 2001. *Raptors of the World.* Boston, MA: Houghton-Mifflin.

Fischer, D. H. 1980. Breeding biology of curve-billed thrashers and long-billed thrashers in southern Texas. *Condor* 82:392–397.

———. 1981a. Wintering ecology of thrashers in southern Texas. *Condor* 83:340–346.

———. 1981b. Winter time budgets of brown thrashers. *J. Field Ornithol.* 52:304–308.

Ficken, R., M. S. Ficken and J. P. Hailman. 1978, Differential aggression in genetically different morphs of the white-throated sparrow. *Z. Tierpsychol.* 46:43–57.

Franzreb, K. E. 1985. Foraging ecology of brown creepers in a mixed-coniferous forest. *Field Ornithol.* 56: 9–16.

Fretwell, S. 1968. Habitat distribution and survival in the field sparrow (*Spizella pusilla). Bird-Banding* 39: 293–306.

———. 1969. Dominance behavior and winter habitat distribution in juncos (*Junco hyemalis). Bird-Banding* 40: 1–25.

Gauthier, G. 1993. Bufflehead. In *The Birds of North America*, No. 67 (A. Poole and F. Gill, eds.). Philadelphia, PA: The Birds of North America, Inc. 24 pp.

Gehlbach, F. E. 1994. *The Eastern Screech-Owl: Life History, Ecology and Behavior in the Suburbs and Countryside.* College Station, TX: Texas A&M Press.

———. 1995. Eastern Screech-owl. In *The Birds of North America*, No. 165 (A. Poole and F. Gill, eds.). Philadelphia, PA: The Birds of North America, Inc. 24 pp.

Geluso, K. N. 1970. Food and survival problems of Oklahoma roadrunners in winter. *Bull. Okla. Ornithol. Soc.* 2:5–6.

Ghalambo, C. K., and T. E. Martin. 1999. Red-breasted nuthatch. In *The Birds of North America*, No. 459 (A. Poole and F. Gill, eds.). Philadelphia, PA: The Birds of North America, Inc. 28 pp.

Giesen, K. M. 1998. Lesser prairie-chicken. In *The Birds of North America*, No. 364 (A. Poole and F. Gill, eds.). Philadelphia, PA: The Birds of North America, Inc. 20 pp.

Gilchrist, H. G. 2001. Glaucous gull. In *The Birds of North America*, No. 573 (A. Poole and F. Gill, eds.). Philadelphia, PA: The Birds of North America, Inc. 32 pp.

Gillespie, M. 1930. Behavior and local distribution of tufted titmice in winter and spring. *Bird-Banding* 1: 113–145

Gilligan, J., M. Smith, D. Rogers and A. Conteras. 1994. *Birds of Oregon: Status and Distribution*. McMinville, OR: Cinclus Publications.

Gillihan, S. W., and B. Byers. 2001. Evening grosbeak. In *The Birds of North America*, No. 599 (A. Poole and F. Gill, eds.). Philadelphia, PA: The Birds of North America, Inc. 24 pp.

Godfrey, W. E. 1986. *The Birds of Canada*. Revised ed. Ottawa, Canada: National Museum of Natural Sciences.

Good. T. P. 1998. Great black-backed gull. In *The Birds of North America*, No. 330 (A. Poole and F. Gill, eds.). Philadelphia, PA: The Birds of North America, Inc. 32 pp.

Goodwin, D. 1972. *Crows of the World*. Ithaca NY: Cornell University Press.

Gowaty, P. A., and J. H. Plissner. 1998. Eastern bluebird. In *The Birds of North America*, No. 381 (A. Poole and F. Gill, eds.). Philadelphia, PA: The Birds of North America, Inc. 32 pp.

Greene, J. C., and R. B. Janssen. 1975. *Minnesota Birds: Where, When, and How Many*. Minneapolis, MN: University of Minnesota Press.

Greenlaw, J. S. 1996a. Eastern towhee. In *The Birds of North America*, No. 262 (A. Poole and F. Gill, eds.). Philadelphia, PA: The Birds of North America, Inc. 32 pp.

————. 1996b. Spotted towhee. In *The Birds of North America*, No. 263 (A. Poole and F. Gill, eds.). Philadelphia, PA: The Birds of North America, Inc. 32 pp.

Griffen, C. R. 1981. Interactive behavior among bald eagles wintering in north-central Missouri. *Wilson Bull*. 93: 259–264.

Grubb, T. C, Jr. 1977. Weather-dependent foraging behavior of some birds wintering in a deciduous woodland: horizontal adjustments. *Condor* 79:271–274.

————. 1982. On sex-specific foraging behavior in the white-breasted nuthatch. *Field Ornithol* 53:305–314.

Grubb, T. C. Jr., and V.V. Pravosudov. 1994. Tufted titmouse. In *The Birds of North America*, No. 86 (A. Poole and F. Gill, eds.). Philadelphia, PA: The Birds of North America, Inc. 16 pp.

Guidice, J. H., and J. T. Ratti. 2001. Ring-necked pheasant. In *The Birds of North America*, No. 572 (A. Poole and F. Gill, eds.). Philadelphia, PA: The Birds of North America, Inc. 32 pp.

Guinan, J. A., P. A. Gowaty and E. K. Eltzroth. 2000. Western bluebird. In *The Birds of North America*, No. 290 (A. Poole and F. Gill, eds.). Philadelphia, PA: The Birds of North America, Inc. 32 pp.

Guntert, M., D. B. Hay, and R. P. Balda. 1988. Communal roosting in the pygmy nuthatch: A winter survival strategy. *Proc. Internat. Ornithol. Congress* 19:1964–1972.

Guzy, M. J. and G. Ritchison. 1999. Common yellowthroat. In *The Birds of North America*, No. 448 (A. Poole and F. Gill, eds.). Philadelphia, PA: The Birds of North America, Inc. 24 pp.

Haggerty, T. M., and E. S. Morton. 1995. Carolina wren. In *The Birds of North America*, No. 188 (A. Poole and F. Gill, eds.). Philadelphia, PA: The Birds of North America, Inc. 20 pp.

Hahn, T. P. 1996. Cassin's finch. In *The Birds of North America*, No. 240 (A. Poole and F. Gill, eds.). Philadelphia, PA: The Birds of North America, Inc. 20 pp.

Halkin. S. L., and S. U. Linville. 1999. Northern cardinal. In *The Birds of North America*, No. 440 (A. Poole and F. Gill, eds.). Philadelphia, PA: The Birds of North America, Inc. 32 pp.

Hamas, M. J. 1994. Belted kingfisher. In *The Birds of North America*, No. 84 (A. Poole and F. Gill, eds.). Philadelphia, PA: The Birds of North America, Inc. 16 pp.

Hammerstrom, F. 1986. *Harrier: Hawk of the Marshes*. Washington, D.C.: Smithsonian Inst. Press.

Harkins, C. E. 1937. Harris's sparrow in its winter range. *Wilson Bull* 49:286–92.

Harraq, S., and Quin. 1995. *Chickadees, Tits, Nuthatches and Treecreepers*. Princeton, NJ: Princeton University Press.

Harrison, K., and G. Harrison. 1990. *The Birds of Winter*. New York, NY: Random House.

Hatch, J. J., and D.V. Weseloh. 1999. Double-crested cormorant. In *The Birds of North America*, No. 441 (A. Poole and F. Gill, eds.). Philadelphia, PA: The Birds of North America, Inc. 36 pp.

Haug, E. A., B. A. Millsap and M. S. Martell. 1993. Burrowing owl. In *The Birds of North America*, No.61 (A. Poole and F. Gill, eds.). Philadelphia, PA: The Birds of North America, Inc. 20 pp.

Heinrich, B. 2003. *Winter World: The Ingenuity of Animal Survival*. New York, NY: HarperCollins Publishers.

Hejl, S. J., K. R. Newlon, M. E. McFadzen, J. S. Young and C .K. Ghalambor. 2002. Brown creeper. In *The Birds of North America*, No. 669 (A. Poole and F. Gill, eds.). Philadelphia, PA: The Birds of North America, Inc. 32 pp.

Hepp, G. R. and F. C. Bellrose. 1995. Wood duck. In *The Birds of North America*, No. 169 (A. Poole and F. Gill, eds.). Philadelphia, PA: The Birds of North America, Inc. 24 pp.

Herkert, J. R., D. E. Kroodsma and J. P. Gibbs. 2001. Sedge wren. In *The Birds of North America*, No. 290 (A. Poole and F. Gill, eds.). Philadelphia, PA: The Birds of North America, Inc. 20 pp.

Hill, D., and P. Robertson, 1988. *The Pheasant: Ecology, Management and Conservation*. Oxford, UK: BSP Professional Books.

Hill, D. P., and L. K. Gould. 1997. Chestnut-collared longspur. In *The Birds of North America*, No. 288 (A. Poole and F. Gill, eds.). Philadelphia, PA: The Birds of North America, Inc. 20 pp.

Hill, G. H. 1993. House finch. In *The Birds of North America*, No. 46 (A. Poole and F. Gill, eds.). Philadelphia, PA: The Birds of North America, Inc. 24 pp.

Hitch, A. T., and P. L. Leeberg. 2007. Breeding distribution of North American bird species moving north as a result of climate change. *Conservation Biology* 21:54–59.

Hitchcock, C. L., and D. R. Sherry. 1990. Long-term memory for cache sites in the black-capped chickadee. *Anim. Behav.* 40:701–712.

Hogg, R. 2013. *America's Climate Century*. Cedar Rapids, IA: Published by the author. 121 pp.

Hohman, W. H., and R. T. Eberhardt. 1998. Ring-necked duck. In *The Birds of North America*, No. 329 (A. Poole and F. Gill, eds.). Philadelphia, PA: The Birds of North America, Inc. 32 pp.

Holt, D.S., and S. M. Leasure. 1993. Short-eared owl. In *The Birds of North America*, No. 62 (A. Poole and F. Gill, eds.). Philadelphia, PA: The Birds of North America, Inc. 24 pp.

Houston, C. S., D. G. Smith and C. Rohner. 1998. Great horned owl. In *The Birds of North America*, No. 290 (A. Poole and F. Gill, eds.). Philadelphia, PA: The Birds of North America, Inc. 28 pp.

Howell, S. N., and J. Dunn. 2007. *A Reference Guide to Gulls of the Americas*. Boston, MA: Houghton Mifflin Co.

Hughes. J. S. 1996. Greater roadrunner. In *The Birds of North America*, No. 290 (A. Poole and F. Gill, eds.). Philadelphia, PA: The Birds of North America, Inc. 24 pp.

Hunt, P. D., and D. J. Flaspohler. 1998. Yellow-rumped warbler. In *The Birds of North America*, No. 376 (A. Poole and F. Gill, eds.). Philadelphia, PA: The Birds of North America, Inc. 28 pp.

Husak M. S., and T. C. Maxwell. 1998. Golden-fronted woodpecker. In *The Birds of North America*, No. 373 (A. Poole and F. Gill, eds.). Philadelphia, PA: The Birds of North America, Inc. 16 pp.

Hussell, D. J. T., and R. Montgomerie. 2002. Lapland longspur. In *The Birds of North America*, No. 656 (A. Poole and F. Gill, eds.). Philadelphia, PA: The Birds of North America, Inc. 32 pp.

Ingold, J. L., and G. E. Wallace. 1994. Ruby-crowned kinglet. In *The Birds of North America*, No. 119 (A. Poole and F. Gill, eds.). Philadelphia, PA: The Birds of North America, Inc. 24 pp.

_____, and R. Galati. 1997. Golden-crowned kinglet. In *The Birds of North America*, No. 301 (A. Poole and F. Gill, eds.). Philadelphia, PA: The Birds of North America, Inc. 28 pp.

Jackson, J. A., H. R. Ouellet, and B. J. S. Jackson. 2003. Hairy woodpecker. In *The Birds of North America*, No. 702 (A. Poole and F. Gill, eds.). Philadelphia, PA: The Birds of North America, Inc. 32 pp.

Jackson, B. J. S. and J. A. Jackson. 2000. Killdeer. In *The Birds of North America*, No. 517 (A. Poole and F. Gill, eds.). Philadelphia, PA: The Birds of North America, Inc. 28 pp.

Jackson, J. A. 1994. Red-cockaded woodpecker. In *The Birds of North America*, No. 85 (A. Poole and F. Gill, eds.). Philadelphia, PA: The Birds of North America, Inc. 20 pp.

_____, and W. E. Davis, Jr. 1998. Range expansion of the red-bellied woodpecker. *Bird Observer* 26:4–12.

_____. and H. R. Ouellet. 2002. Downy woodpecker. In *The Birds of North America*, No. 613(A. Poole and F. Gill, eds.). Philadelphia, PA: The Birds of North America, Inc. 32 pp.

_____, and B. S. Jackson. 1987. Red-bellied a generalist expanding its range. *Nature Soc. News* 22(1):11.

Jacobs, B. 2001. *Birds in Missouri.* Jefferson City, MO: Missouri Dept, of Conservation.

James, P. C., and T. J. Ethier. 1989. Trends in the winter distribution and abundance of burrowing owls in North America. *Am. Birds* 43:1224–1225.

Jamarillo, A., and P. Burke. 1995. *New World Blackbirds:* The Icterids. Princeton, NJ: Princeton University Press.

Jejl, S. J., J. A. Holmes and D. E. Kroodsma. 2002. Winter wren. In *The Birds of North America*, No. 623 (A. Poole and F. Gill, eds.). Philadelphia, PA: The Birds of North America, Inc. 32 pp.

Johnsgard, P. A. 1973. *Grouse and Quails of North America.* Lincoln, NE: University of Nebraska Press.

_____. 1975. *Waterfowl of North America.* Lincoln, NE: University of Nebraska Press.

_____. 1978. *Ducks, Geese and Swans of the World.* Lincoln, NE: University of Nebraska Press.

_____. 1979. *Birds of the Great Plains: Breeding Species and their Distribution.* University of Nebraska Press, Lincoln. NE. and University of Nebraska Digital Commons. http://digitalcommons.unl.edu/bioscibirdsgreatplains/1/

_____. 1980. Migration schedules of birds in Nebraska. *Nebraska Bird Review* 48:26–36; 46–57.

_____. 1981. *The Plovers, Sandpipers and Snipes of the World.* Lincoln, NE: University of Nebraska Press.

_____. 1987. *Diving Birds of North America.* Lincoln. NE: University of Nebraska Press.

_____. 1990. *Hawks, Eagles and Falcons of North America.* Washington, D.C.: Smithsonian Institution Press.

_____. 1991. *Crane Music.* Washington, D.C.: Smithsonian Institution Press.

_____. 1993. *Cormorants, Darters and Pelicans of the World.* Washington D.C.: Smithsonian Inst. Press.

_____. 1998. A half-century of winter bird surveys at Lincoln and Scottsbluff, Nebraska. *Nebraska Bird Review* 66:74–84.

_____. 2001. *Prairie Birds: Fragile Splendor on the Great Plains.* Lawrence, KS: University Press of Kansas.

_____. 2002a. *Grassland Grouse and their Conservation.* Washington, D.C.: Smithsonian Institution Press.

_____. 2002b. *North American Owls: Biology and Natural History.* Washington, D.C.: Smithsonian Institution Press.

_____. 2002c. Comments on the falconiform and strigiform fauna of Nebraska. *Nebraska Bird Review* 69:80-84. http://digitalcommons.unl.edu/biosciornithology/17

_____.2003. Nebraska's sandhill crane populations: Past, present and future. *Nebraska Bird Review* 71:175–178.

_____. 2006. Recent changes in winter bird numbers at Lincoln, Nebraska. *Nebraska Bird Review* 74:16–22.

_____. 2008. *The Birds of Nebraska.* 2013. Zea E-Books & Univ. of Nebraska Digital Commons. 140 pp. http://digitalcommons.unl.edu/zeabook/17/. Print ed. $19.95 from Lulu.com.

_____. 2010. Snow geese of the Great Plains. *Prairie Fire,* February, 2010, pp. 12-15. http://www.prairiefirenewspaper.com/2010/3/d\snow-geese-of-the-great-plains

_____. 2011a. *Ancient Voices over the American Wetlands: The Sandhill and Whooping Cranes.* Lincoln, NE: University of Nebr. Press, 155 pp.

_____. 2011b. The feathers of winter. *Prairie Fire,* December, 2011, pp. 17-20. http://www.prairiefirenewspaper.com/2011/12/the-feathers-of-winter

_____. 2012a. *Wetland Birds of the Central Plains: South Dakota, Nebraska and Kansas.* 2012. Zea E-Books & Univ. of Nebraska Digital Commons Lincoln, NE: http://digitalcommons.unl.edu/zeabook/8. 276 pp. Print edition $21.95 from Lulu.com.

_____. 2012b. *Wings over the Great Plains: Bird Migrations in the Central Flyway.* 2012. Zea E-Books & Univ. of Nebraska Digital Commons. 245 pp. http://digitalcommons.unl.edu/zeabook/13/. Print edition $19.95 from Lulu.com.

_____. 2012c. The owls of Nebraska. *Prairie Fire,* February, 2012, pp. 15, 20. http://www.prairiefirenewspaper.com/2012/02/the-owls-of-nebraska

_____. 2013a. Changing Great Plains climate and bird migrations. *Prairie Fire,* December, 2013, pp. 1,3,5. http://www.prairiefirenewspaper.com/2013/12/changing-great-plains-climate-and-bird-migrations

_____ 2013b. A plethora of pelicans. *Prairie Fire,* March 2013, pp. 9-11. http://www.prairiefirenewspaper.com/2013/03/a-plethora-of-pelicans

_____. 2014. What are blue Ross's geese? *Nebraska Bird Review,* 82:81–85.

_____. 2015 (in press). *A Chorus of Cranes.* Boulder, CO: University Press of Colorado.

_____, and M. Carbonell. 1996. *Ruddy Ducks and Other Stifftails: Their Behavior and Biology.* Norman, OK: University of Oklahoma Press. University Press of Colorado.

_____, and R. DiSilvestro. 1974. Seventy-five years of changes in mallard-black duck ratios in eastern North America. *American Birds* 30:904–908.

_____, and T. Shane. 2009. *Four Decades of Christmas Bird Counts in the Great Plains: Ornithological Evidence of a Changing Climate.* 334 pp. URL: http://digitalcommons.unl.edu/biosciornithology/46/

Johnson, D. H., L. D. Igl, and J. A. Dechant Shaffer (Series Coordinators). 2004. Effects of management practices on grassland birds. Jamestown, ND: Northern Prairie Wildlife Research Center Online. http://www.npwrc.usgs.gov/resource/literatr/grasbird/index.htm (Version 12AUG2004)

Johnson, K. 1995. Green-winged teal. In *The Birds of North America*, No. 193 (A. Poole and F. Gill, eds.). Philadelphia, PA: The Birds of North America, Inc. 20 pp.

_____, and B. D. Peer. 2001. Great-tailed grackle. In *The Birds of North America*, No. 576 (A. Poole and F. Gill, eds.). Philadelphia, PA: The Birds of North America, Inc. 28 pp.

Johnson, L. S. 1998. House wren. In *The Birds of North America*, No. 380 (A. Poole and F. Gill, eds.). Philadelphia, PA: The Birds of North America, Inc. 32 pp.

Johnson, R. R., and L. T. Haight. 1996. Canyon towhee. In *The Birds of North America*, No. 264 (A. Poole and F. Gill, eds.). Philadelphia, PA: The Birds of North America, Inc. 20 pp.

Johnston, R. F. 1992. Rock dove. In *The Birds of North America*, No. 13 (A. Poole and F. Gill, eds.). Philadelphia, PA: The Birds of North America, Inc. 16 pp.

Jones, P.W., and T. M. Donovan. 1996. Hermit thrush. In *The Birds of North America*, No. 261 (A. Poole and F. Gill, eds.). Philadelphia, PA: The Birds of North America, Inc. 28 pp.

Jones, S. L., and J. E. Cornely. 2002. Vesper sparrow. In *The Birds of North America*, No. 624 (A. Poole and F. Gill, eds.). Philadelphia, PA: The Birds of North America, Inc. 28 pp.

_____, and J. S. Dieni. 1995. Canyon wren. In *The Birds of North America*, No. 197 (A. Poole and F. Gill, eds.). Philadelphia, PA: The Birds of North America, Inc. 12 pp.

Karl, T. R., J. M. Melillo, and T. C. Peterson (eds.). 2009. *Global Climate Change Impacts in the United States*. (U.S. Global Change Research Program. New York, NY: Cambridge University Press.

Kear, J. 2005. *Ducks, Geese and Swans*. 2 Vol. Oxford, UK: Oxford University Press.

Kennedy, E. D., and D. W. White. 1997. Bewick's wren. In *The Birds of North America*, No. 315 (A. Poole and F. Gill, eds.). Philadelphia, PA: The Birds of North America, Inc. 28 pp.

Keppie, D. M., and R. M. Whiting, Jr. 1994. American woodcock. In *The Birds of North America*, No. 100 (A. Poole and F. Gill, eds.). Philadelphia, PA: The Birds of North America, Inc. 28 pp.

Kerlinger, P., M. R. Lein and B. J. Sevick. 1985. Distribution and population fluctuations of wintering snowy owls (*Nyctea scandiaca*) in North America. *Can. J. Zoology* 63:1829–1834.

Kessel, B., D. A. Rocque and J. S. Barclay. 2002. Greater scaup. In *The Birds of North America*, No. 650 (A. Poole and F. Gill, eds.). Philadelphia, PA: The Birds of North America, Inc. 23 pp.

Kilham, L. 1959. Early reproductive behavior of flickers. *Wilson Bull.* 71:323–336.

_____. 1963. Food storing of red-bellied woodpeckers. *Wilson Bull.* 75:227–234.

_____. 1971. Roosting habits of white-breasted nuthatches. *Condor* 73:113–114.

_____.1976. Winter foraging and associated behavior of pileated woodpeckers in Georgia and Florida. *Auk* 93:15–24.

Kilpatrick, H. J., T. P. Husband and C. A. Pringle. 1988. Winter roost site characteristics of eastern wild turkeys. *J. Wildl. Mgmt.* 52:461–463.

King, J. R., and E. E. Wales Jr. 1964. Observations on migration, ecology, and population flux of wintering rosy finches. *Condor* 66:24–31

Kingery, H. 1996. American dipper. In *The Birds of North America*, No. 229 (A. Poole and F. Gill, eds.). Philadelphia, PA: The Birds of North America, Inc. 28 pp.

_____, and C. E. Ghalambor. 2001. Pygmy nuthatch. In *The Birds of North America*, No. 567 (A. Poole and F. Gill, eds.). Philadelphia, PA: The Birds of North America, Inc. 32 pp.

Kirk, D. A., and M. J. Mossman. 1998. Turkey vulture. In *The Birds of North America*, No. 339 (A. Poole and F. Gill, eds.). Philadelphia, PA: The Birds of North America, Inc. 32 pp.

Knapton, E. W., and J. R. Crebs. 1976. Dominance hierarchies in winter song sparrows. *Condor* 78:567–569.

Knox, A. G., and P. E. Lowther. 2000a. Common redpoll. In *The Birds of North America*, No. 543 (A. Poole and F. Gill, eds.). Philadelphia, PA: The Birds of North America, Inc. 24 pp.

_____, and _____. 2000b. Hoary redpoll. In *The Birds of North America*, No. 544 (A. Poole and F. Gill, eds.). Philadelphia, PA: The Birds of North America, Inc. 16 pp.

Kochert, M. N., K. Steenhof, C. L. McIntyre and E. H. Craig. 2003. Golden eagle. In *The Birds of North America*, No. 684 (A. Poole and F. Gill, eds.). Philadelphia, PA: The Birds of North America, Inc. 44 pp.

Kroodsma, D. E., and J. Verner. 1997. Marsh wren. In *The Birds of North America*, No. 308 (A. Poole and F. Gill, eds.). Philadelphia, PA: The Birds of North America, Inc. 32 pp.

Küchler, A. W. 1964. *Potential natural vegetation of the coterminous United States*. American Geographical Society Special Publication No. 36. Washington, D.C.: American Geographical Society.

Kunkel, K. E., *et al.* 2013. Regional climate trend and scenarios for the U.S. National Climate assessment for the U.S Great Plains. NOAA Technical Report NESDIS 143-4.92. http://www.epa.gov/climatechange

Laskey, A. R. 1936. Fall and winter behavior of mockingbirds. *Wilson Bull.* 43:241–255.

La Sorte, F. A., and F. R. Thompson III. 2007. Poleward shifts in winter ranges of North American birds. *Ecology* 88:1803–1812.

Lanyon, W. E. 1994. Western meadowlark. In *The Birds of North America*, No. 104 (A. Poole and F. Gill, eds.). Philadelphia, PA: The Birds of North America, Inc. 20 pp.

_____. 1995. Eastern meadowlark. In *The Birds of North America*, No. 160 (A. Poole and F. Gill, eds.). Philadelphia, PA: The Birds of North America, Inc. 24 pp.

Laurenzi, A. W., B. W. Anderson and R. D. Ohmart. 1982. Wintering biology of ruby-crowned kinglets in the lower Colorado River valley. *Condor* 84:385–398.

Lefrane, N. 1997. *Shrikes: A Guide to the Species of the World*. New Haven, CT: Yale University Press.

Lepthien, L. W, and C. E. Bock. 1976. Winter abundance patterns of North American kinglets. *Wilson Bull* 88:483–485.

Leonard, D. L., Jr. 2001. Three-toed woodpecker. In *The Birds of North America*, No. 588 (A. Poole and F. Gill, eds.). Philadelphia, PA: The Birds of North America, Inc. 24 pp.

LeSchack, C. R., S. K. McKnight and G. R. Hepp. 1997. Gadwall. In *The Birds of North America*, No. 283 (A. Poole and F. Gill, eds.). Philadelphia, PA: The Birds of North America, Inc. 28 pp.

Ligon, J. D. 1978. Reproductive interdependence of pinon jays and pinon pines. *Ecol Monogr.* 48:111–126.

Limpert, R. J., and S. L. Earnst. 1994. Tundra swan In *The Birds of North America*, No. 89 (A. Poole and F. Gill, eds.). The Birds of North America, Inc. Philadelphia, PA. 20 pp.

Lockwood, M. W., and B. Freeman. 2004. *The TOS Handbook of Texas Birds*. Texas A&M Press, College Station, TX.

Longcore, J. R., D. G. Mcauley, G. R. Hepp and J. M. Rhymer. 2000. American black duck. In *The Birds of North America*, No. 481 (A. Poole and F. Gill, eds.). Philadelphia, PA: The Birds of North America, Inc. 36 pp.

Lowther, P. E. 1993. Brown-headed cowbird. In *The Birds of North America*, No. 47 (A. Poole and F. Gill, eds.). Philadelphia, PA: The Birds of North America, Inc. 24 pp.

_____. 1996. Le Conte's sparrow. In *The Birds of North America*, No. 224 (A. Poole and F. Gill, eds.). Philadelphia, PA: The Birds of North America, Inc. 16 pp.

_____. 2001. Ladder-backed woodpecker. In *The Birds of North America*, No. 565 (A. Poole and F. Gill, eds.). Philadelphia, PA: The Birds of North America, Inc. 12 pp.

_____, and C. L. Cink. 1992. House sparrow. In *The Birds of North America*, No. 12 (A. Poole and F. Gill, eds.). Philadelphia, PA: The Birds of North America, Inc. 20 pp.

_____, D. E. Kroodsma, and G. H. Farley. 2000. Rock wren. In *The Birds of North America*, No. 486 (A. Poole and F. Gill, eds.). Philadelphia, PA: The Birds of North America, Inc. 20 pp.

Lynch, W. 2007. *Owls of the United States and Canada: A Complete Guide to their Biology and Behavior*. Baltimore, MD: Johns Hopkins Press.

Lyon, B., and R. Montgomerie. 1995. Snow bunting and McKay's bunting. In *The Birds of North America*, No. 198–199 (A. Poole and F. Gill, eds.). Philadelphia, PA: The Birds of North America, Inc. 28 pp.

MacDougall-Shackleton, S., R. Johnson and T. Hahn. 2000. Gray-crowned rosy-finch. In *The Birds of North America*, No. 559 (A. Poole and F. Gill, eds.). Philadelphia, PA: The Birds of North America, Inc. 16 pp.

MacWhirter, R. B., and K. L. Bildstein. 1996. Northern harrier. In *The Birds of North America*, No. 210 (A. Poole and F. Gill, eds.). Philadelphia, PA: The Birds of North America, Inc. 32 pp.

Madge, S., and H. Burn. 2001. *Crows and Jays*. Princeton, NJ: Princeton University Press.

Mallory, M., and K. Metz. 1999. Common merganser. In *The Birds of North America*, No. 442 (A. Poole and F. Gill, eds.). Philadelphia, PA: The Birds of North America, Inc. 28 pp.

Marks, J. S., D. L. Evans and D. W. Holt. 1994. Long-eared owl. In *The Birds of North America*, No. 133 (A. Poole and F. Gill, eds.). Philadelphia, PA: The Birds of North America, Inc. 24 pp.

Marti, C. D. 1992. Barn owl. In *The Birds of North America*, No. 1 (A. Poole and F. Gill, eds.). Philadelphia, PA: The Birds of North America, Inc. 16 pp.

Martin, A. C, H. S. Zim and A. L. Nelson. 1951. *American Wildlife and Plants*. New York, NY: Dover Publications.

Martin, J. W., and J. R. Parrish. 2000. Lark sparrow. In *The Birds of North America*, No. 488 (A. Poole and F. Gill, eds.). Philadelphia, PA: The Birds of North America, Inc. 20 pp.

Martin, S. G. 2002. Brewer's Blackbird. In *The Birds of North America*, No. 616 (A. Poole and F. Gill, eds.). Philadelphia, PA: The Birds of North America, Inc. 32 pp.

Mazur, K. M., and P. C. James. 2000. Barred owl. In *The Birds of North America*, No. 508 (A. Poole and F. Gill, eds.). Philadelphia, PA: The Birds of North America, Inc. 20 pp.

McCallum, D. A., R. Grundel and D. L. Dahlsten. 1999. Mountain chickadee. In *The Birds of North America*, No. 453 (A. Poole and F. Gill, eds.). Philadelphia, PA: The Birds of North America, Inc. 28 pp.

McGowan, K. J. 2001. Fish crow. In *The Birds of North America*, No. 589 (A. Poole and F. Gill, eds.). Philadelphia, PA: The Birds of North America, Inc. 28 pp.

McIntyre, J. W, 1988. *The Common Loon: Spirit of Northern Lakes*. Minneapolis, MN: University of Minnesota Press.

McIntyre, J. W., and J. F. Barr. 1997. Common loon. In *The Birds of North America*, No. 313 (A. Poole and F. Gill, eds.). Philadelphia, PA: The Birds of North America, Inc. 32 pp.

Mewaldt, L. R. 1976. Winter philopatry in white-crowned sparrows (*Zonotrichia leucophrys*). *N. Amer. Bird Bander* 1:14–20.

Middleton, A. L. A. 1977. Increase in overwintering by the American goldfinch, *Carduelis tristis*, in Ontario. *Can. Field-Natur.* 91:165–172.

_____, 1993. American goldfinch. In *The Birds of North America*, No. 80 (A. Poole and F. Gill, eds.). Philadelphia, PA: The Birds of North America, Inc. 24 pp.

_____. 1998. Chipping sparrow. In *The Birds of North America*, No. 334 (A. Poole and F. Gill, eds.). Philadelphia, PA: The Birds of North America, Inc. 32 pp.

Mikkola, H. 212. *Owls of the World*. London, UK: Firefly Books. 512 pp.

Mills, A. M. 2005. Changes in the timing of spring and fall migration of North American migrant passerines during a period of global warming. *Ibis* 147:259–269.

Mirarchi, R. D. and T. S. Baskett. 1994. Mourning dove. In *The Birds of North America*, No. 117 (A. Poole and F. Gill, eds.). Philadelphia, PA: The Birds of North America, Inc. 32 pp.

Mitchell, C. D. 1994. Trumpeter swan. In *The Birds of North America*, No. 105 (A. Poole and F. Gill, eds.). The Birds of North America, Inc. Philadelphia, PA. 24 pp.

Moore, W. S. 1995. Northern flicker. In *The Birds of North America*, No. 166 (A. Poole and F. Gill, eds.). Philadelphia, PA: The Birds of North America, Inc. 28 pp.

Morrison, R. I., Y. Aubrey, R. W. Butler, G. W. Beyersbergan, C. L. Gatto-Trevor, P. W. Hicklin, V. H. Johnston, and R. K. Ross. 2001a. Declines in North American shorebirds. *Wader Study Group Bull.* 94:34–38.

_____, R. E. Gill, Jr., B. A. Harrington, S. Skagen, G. W. Page, G. L. Gatto-Trevor, and S. M. Haig. 2001b. Estimates of shorebird populations in North America. Ottawa: Canadian Wildlife Service, Environment Canada, Occasional Paper 204, 64 pp.

Morse, D. H. 1967a. Foraging relationships of brown-headed nuthatches and pine warblers. *Ecology* 48:94–103.

_____. 1967b. The use of tools by brown-headed nuthatches. *Wilson Bull.* 80:220–224

_____. 1971. Effects of the arrival of new species upon habitat utilization by two forest thrushes in Maine. *Wilson Bull.* 83:57–65.

_____. 1972. Habitat utilization of the red-cockaded woodpecker during the winter. *Auk* 89:429–435.

_____. 1989. *American Warblers*. Cambridge, MA: Harvard University Press.

Mostrom, A. M., R. L. Curry and B. Lohr. 2002. Carolina chickadee. In *The Birds of North America*, No. 636 (A. Poole and F. Gill, eds.). Philadelphia, PA: The Birds of North America, Inc. 28 pp.

Mowbray, T. B. 1997. Swamp sparrow. In *The Birds of North America*, No. 279 (A. Poole and F. Gill, eds.). Philadelphia, PA: The Birds of North America, Inc. 24 pp.

_____. 1999. American wigeon. In *The Birds of North America*, No. 401 (A. Poole and F. Gill, eds.). The Birds of North America, Inc. Philadelphia, PA. 32 pp.

_____. 2002. Canvasback. In *The Birds of North America*, No. 659 (A. Poole and F. Gill, eds.). The Birds of North America, Inc. Philadelphia, PA. 40 pp.

_____, C.R. Ely, J. S. Sedinger, and R. E. Trost. 2003. Canada goose. In *The Birds of North America*, No. 682 (A. Poole and F. Gill, eds.). Philadelphia, PA: The Birds of North America, Inc. 44 pp.

_____, F. Cooke, and B. Ganter. 2000. Snow goose. In *The Birds of North America*, No. 514 (A. Poole and F. Gill, eds.). The Birds of North America, Inc. Philadelphia, PA. 40 pp.

Mueller, A. J. 1992. Inca dove. In *The Birds of North America*, No. 28 (A. Poole and F. Gill, eds.). Philadelphia, PA: The Birds of North America, Inc. 12 pp.

_____, H. 1999. Common snipe. In *The Birds of North America*, No. 427 (A. Poole and F. Gill, eds.). Philadelphia, PA: The Birds of North America, Inc. 20 pp.

Muller, M. J., and R.W. Storer. 1999. Pied-billed grebe. In *The Birds of North America*, No. 410 (A. Poole and F. Gill, eds.). Philadelphia, PA: The Birds of North America, Inc. 32 pp.

Mundinger, P. C., and S. Hope. 1982. Expansion of the winter range of the house finch: 1947–1979. *American Birds* 36:347–353.

Murphy-Klassen, H. M., T. J. Underwood, S. G. Sealy and A. A. Czernyi. 2005. Long-term trends in spring arrival dates of migrant birds at Delta Marsh, Manitoba, in relation to climate change. *Auk* 122:1130–1148.

National Audubon Society. 2008. Historical summaries of Audubon Christmas Count data website. http://audubon2.org/cbchist/

National Audubon Society. 2014a. Birds and Climate Change Report. *Audubon Magazine* 116(5): 1–102. http://climate.audubon.org/article/audubon-report-glance

National Audubon Society. 2014b. About the Christmas Count (http://birds.audubon.org/about-christmas-bird-count)

National Audubon Society. 2014c. Data and Research (http://birds.audubon.org/data-research).

Martin, A. C, H. S. Zim, and A. L. Nelson. 1951. *American Wildlife and Plants.* New York, NY: Dover Publications.

Naugler, C. T. 1993. American tree sparrow. In *The Birds of North America*, No. 37 (A. Poole and F. Gill, eds.). Philadelphia, PA: The Birds of North America, Inc. 12 pp.

Nice, M. M. 1927. Experiences with cardinals at a feeding station in Oklahoma. *Condor* 29:101–103.

Nijhuis, M. 2014. The Audubon Report A storm gathers. *Audubon* 116(5):24–32.

Niven, D., J. R. Sauer, G. S. Butcher and W. S. Link. 2004. Christmas Bird Counts provide insights into population change in land birds that breed in the boreal forest. *American Birds* 58:10–20.

Nolan, V., Jr., E. D. Ketterson, D. A. Cristol, C. M. Rogers, E. D. Clotfelter, R. C. Titus, S. J. Schoech and E. Snajdr. 2003. Dark-eyed junco. In *The Birds of North America*, No. 716 (A. Poole and F. Gill, eds.). Philadelphia, PA: The Birds of North America, Inc. 44 pp.

Norment, C. J., and S. A. Shackleton. 1993. Harris' sparrow. In *The Birds of North America*, No. 64 (A. Poole and F. Gill, eds.). Philadelphia, PA: The Birds of North America, Inc. 20 pp.

North American Bird Conservation Initiative. 2000. Bird Conservation Region descriptions. Washington, DC: U. S. Fish and Wildlife Service.

Odum, E. P. 1941. Annual cycle of the black-capped chickadee. *Auk* 88:314–333, 518–535; 59:499–531.

Ohmart, R. D., and R. C. Lasiewski. 1971. Roadrunners: Energy conservation by hypothermia and absorption of sunlight. *Science* 172:67–69.

Olsen, K. M., and K. Larson. 2004. *Gulls of North America, Europe and Asia.* Princeton, NJ: Princeton University Press.

Orians, G. 1985. *Blackbirds of the Americas.* Seattle, WA: University of Washington Press.

Oring, L. W., E. M. Gray and J. M. Reed. 1997. Spotted sandpiper. In *The Birds of North America*, No. 289 (A. Poole and F. Gill, eds.). Philadelphia, PA: The Birds of North America, Inc. 32 pp.

Osborne, J. 1992. *The Cardinal.* Austin, TX: University of Texas Press.

Owen, M. 1980. *Wild Geese of the World: Their Life History and Ecology.* London, UK: B. T. Bateson, Ltd.

Palmer, R. S. (ed.). 1962. *Handbook of North American Birds: Loons through Flamingos.* Vol. 1. New Haven, CT: Yale University Press.

———. 1978. *Handbook of North American Birds: Waterfowl.* Vols. 2 & 3. New Haven, CT: Yale University Press.

———. 1988. *Handbook of North American Birds: Diurnal Raptors.* Vols. 4 & 5. New Haven, CT: Yale University Press.

Paothong, N. 2012. *Save the Last Dance: A Story of North American Grassland Grouse.* Napppdol Paothong Photography. 203 pp.

Parmelee, D. F. 1992. Snowy owl. In *The Birds of North America*, No. 10 (A. Poole and F. Gill, eds.). Philadelphia, PA: The Birds of North America, Inc. 20 pp.

Peer, B. D., and E. K. Bollinger. 1997. Common grackle. In *The Birds of North America*, No. 271 (A. Poole and F. Gill, eds.). Philadelphia, PA: The Birds of North America, Inc. 20 pp.

Pelikan, M. 2002. Victims, vectors and viruses: Mosquito-borne West Nile virus links humans to bird ecology. *Winging It* 14(10):1–4.

Peters, R. L, and T. E. Lovejoy (eds.). 1992. *Global Warming and Biological Diversity*. New Haven, CT: Yale University Press.

Pettingill, O. S., Jr., and N. R. Whitney, Jr. 1965. *Birds of the Black Hills*. Ithaca, NY: Cornell Laboratory of Ornithology, Special Publication Number 1.

Pierotti, R. J., and T. P. Good. 1994. Herring gull. In *The Birds of North America*, No. 124 (A. Poole and F. Gill, eds.). Philadelphia, PA: The Birds of North America, Inc. 28 pp.

Poole, A. F., R. O. Bierregaard and M. S. Martell. 2003. Osprey. In *The Birds of North America*, No. 683 (A. Poole and F. Gill, eds.). Philadelphia, PA: The Birds of North America, Inc. 44 pp.

Power, H. W., and M. P. Lombardo. 1996. Mountain bluebird. In *The Birds of North America*, No. 222 (A. Poole and F. Gill, eds.). Philadelphia, PA: The Birds of North America, Inc. 24 pp.

Pravosudov, V. V., and T. C. Grubb, Jr. 1993. White-breasted nuthatch. In *The Birds of North America*, No. 54 (A. Poole and F. Gill, eds.). Philadelphia, PA: The Birds of North America, Inc. 16 pp.

Preston, R., and R. D. Beane. 1993. Red-tailed hawk. In *The Birds of North America*, No. 290 (A. Poole and F. Gill, eds.). Philadelphia, PA: The Birds of North America, Inc. 24 pp.

Proudfoot, G. A., D. A. Sherry and S. Johnson. 2000. Cactus wren. In *The Birds of North America*, No. 558 (A. Poole and F. Gill, eds.). Philadelphia, PA: The Birds of North America, Inc. 24 pp.

Ratcliffe, D. A. 1980. *The Peregrine Falcon*. Vermillion, SD: Buteo Books.

Rathke, D., R J. Oelsby, C. M. Rowe and D. A. Wilhite. 2014. *Understanding and Assessing Implications for Climate Change in Nebraska*. Lincoln, NE: University of Nebraska. 72 pp.

Reinking, D. L. (ed.). 2004. *Oklahoma Breeding Bird Atlas*. Norman, OK: University of Oklahoma Press.

Reynolds, T. D., T. D. Rich and D. A. Stephens. 1999. Sage thrasher. In *The Birds of North America*, No. 463 (A. Poole and F. Gill, eds.). Philadelphia, PA: The Birds of North America, Inc. 24 pp.

Rich, T. D., C. J. Beardmore, H. Berlanga, P. J. Blancher, M. S. W. Bradstreet, G. S. Butcher, D. W. Demarest, E. H. Dunn, W. C. Hunter, E. E. Inigo-Elias, J. A. Kennedy, A. M. Martell, A. O. Panjabi, D. N. Pashley, K. V. Rosenberg, C. M. Rustay, J. S. Wendt, and T. C. Will. 2004. *Partners in Flight North American Landbird Conservation Plan*. Ithaca, NY: Cornell Lab of Ornithology.

Rising, J. D. 1974. The status and faunal affinities of the summer birds of western Kansas. *University of Kansas Science Bulletin* 50:347–388.

———, and D. D. Beadle. 1996. *A Guide to the Identification and Natural History of the Sparrows of the United States and Canada*. New York, NY: Academic Press.

Robbins, M. B., and B. C. Dale. 1999. Sprague's pipit. In *The Birds of North America*, No. 439 (A. Poole and F. Gill, eds.). Philadelphia, PA: The Birds of North America, Inc. 16 pp.

Roberts, T. S. 1932. *The Birds of Minnesota*. 2 vols. Minneapolis: Univ. of Minnesota Press.

Robertson, G. I., and J. P. L. Savard. 2002. Long-tailed duck. In *The Birds of North America*, No. 651 (A. Poole and F. Gill, eds.). Philadelphia, PA: The Birds of North America, Inc. 28 pp.

Rodewald, P. G, J. H. Withgott and K. G. Smith. 1999. Pine warbler In *The Birds of North America*, No. 438 (A. Poole and F. Gill, eds.). Philadelphia, PA: The Birds of North America, Inc. 28 pp.

Romagosa, C. M. 2002. Eurasian collared-dove. In *The Birds of North America*, No. 630 (A. Poole and F. Gill, eds.). Philadelphia, PA: The Birds of North America, Inc. 20 pp.

Root, T. 1988. *Atlas of North American Winter Birds*. Chicago, IL: University of Chicago Press.

330        REFERENCES

Rose, P. M., and D. A. Scott. 1997. *Waterfowl Population Estimates*. Publ No. 44, 2nd. Ed. Wageningen, Netherlands: Wetlands International.

Rosenfield, R. N., and J. Bielefeldt. 1993. Cooper's hawk. In *The Birds of North America*, No. 75 (A. Poole and F. Gill, eds.). Philadelphia, PA: The Birds of North America, Inc. 24 pp.

Rusch, D., S. Destafano, M. Reynolds and D. Lauten. 2000. Ruffed Grouse. In *The Birds of North America*, No. 525 (A. Poole and F. Gill, eds.). Philadelphia, PA: The Birds of North America, Inc. 28 pp.

Russell, R. W. 2002. Pacific loon/Arctic loon. In *The Birds of North America*, No. 657 (A. Poole and F. Gill, eds.). The Birds of North America, Inc. Philadelphia, PA. 40 pp.

Russock, H. I. 1979. Observations on the behavior of wintering bald eagles. *Raptor Res.* 13:112–115.

Ryder, J. P. 1993. Ring-billed gull. In *The Birds of North America*, No. 33 (A. Poole and F. Gill, eds.). Philadelphia, PA: The Birds of North America, Inc. 28 pp.

_____. and R. T. Alisauskas. 1995. Ross's goose. In *The Birds of North America*, No. 162 (A. Poole and F. Gill, eds.). Philadelphia, PA: The Birds of North America, Inc. 28 pp.

Sabine, W. S. 1949. Dominance in winter flocks of juncos and tree sparrows. *Physiol. Zool.* 22:261–280.

_____. 1955. The winter society of the Oregon junco: The flock. *Condor* 57:88–111.

_____. 1956. Integrating mechanisms of the junco winter flock. *Condor* 58:338–341.

_____. 1959. The winter society of the Oregon junco: Intolerance, dominance and the pecking order. *Condor* 61:110–134.

Sallabanks, R., and F. C. James. 1999. American robin. In *The Birds of North America*, No. 462 (A. Poole and F. Gill, eds.). Philadelphia, PA: The Birds of North America, Inc. 28 pp.

Sauer, J. R. et al., 2012. The North American Breeding Bird Survey, Results and Analysis 1966 – 2012. Version 02.19.2014. Laurel, MD: USGS Patuxent Wildlife Research Center. See associated web site: http://www.mbr-pwrc.usgs.gov/bbs/bbs2012.html

_____, and W. A. Link. 2002. Using Christmas Bird Count data in analysis of population change. *Am. Birds* 56:10–14.

Sauer, J. R., S. Schwartz, and B. Hoover. 1996. *The Christmas Bird Count Home Page. Version 95.1. Patuxent Wildlife Research Center, Laurel, MD.* See associated website: http://www.mbr-pwrc.usgs.gov/cbc/cbcnew.html

Scheminitz, S. D. 1994. Scaled quail. In *The Birds of North America*, No. 106 (A. Poole and F. Gill, eds.). Philadelphia, PA: The Birds of North America, Inc. 16 pp.

Schroeder, M. A, J. R. Young and C. E. Braun. 1999. Sage grouse. In *The Birds of North America*, No. 425 (A. Poole and F. Gill, eds.). Philadelphia, PA: The Birds of North America, Inc. 28 pp.

_____, and L. A. Robb. 1993. Greater prairie-chicken. In *The Birds of North America*, No. 36 (A. Poole and F. Gill, eds.). Philadelphia, PA: The Birds of North America, Inc. 24 pp.

Schukman, J. M., and B. O. Wolf. 1998. Say's phoebe. In *The Birds of North America*, No. 374 (A. Poole and F. Gill, eds.). Philadelphia, PA: The Birds of North America, Inc. 20 pp.

Schwertner, T. W, H. A. Mathewson, J. A. Robertson, M. Small and G. L. Waggerman. 2003. White-winged dove. In *The Birds of North America*, No. 710 (A. Poole and F. Gill, eds.). Philadelphia, PA: The Birds of North America, Inc. 28 pp.

Seyffert, K, D. 2000. *Birds of the Texas Panhandle: Their Status, Distribution and History.* Natural History series, vol. 29. College Station, TX: Texas A. & M. University Press.

Shackelford, C. E., R. E. Brown and R. N. Conner. 2000. Red-bellied woodpecker. In *The Birds of North America*, No. 500 (A. Poole and F. Gill, eds.). Philadelphia, PA: The Birds of North America, Inc. 24 pp.

Shane, T. G. 1996. The lark bunting: In peril or making progress *C.F.O. Journal* 30:162–168.

———. 2000. Lark bunting. In *The Birds of North America*, No. 542 (A. Poole and F. Gill, eds.). Philadelphia, PA: The Birds of North America, Inc. 28 pp.

Sharpe, R. S., W. R. Silcock and J. G. Jorgensen. 2001. *Birds of Nebraska: Their Distribution and Temporal Occurrence.* Lincoln, NE: University of Nebraska Press.

Short, L. L. 1982. *Woodpeckers of the World.* Greenville, DE: Delaware Museum of Natural History.

Sloane, S. A. 2001. Bushtit. In *The Birds of North America*, No. 598 (A. Poole and F. Gill, eds.). Philadelphia, PA: The Birds of North America, Inc. 20 pp.

Smallwood, J. A., and D. M. Bird. 2002. American kestrel. In *The Birds of North America*, No. 602 (A. Poole and F. Gill, eds.). Philadelphia, PA: The Birds of North America, Inc. 32 pp.

Smith, K., J. Withgott and P. Rodewald. 2000. Red-headed woodpecker. In *The Birds of North America*, No. 518 (A. Poole and F. Gill, eds.). Philadelphia, PA: The Birds of North America, Inc. 28 pp.

Smith, S. M. 1991. *The Black-capped Chickadee: Behavioral Ecology and Natural History.* Ithaca, NY: Cornell University Press.

———. 1993. Black-capped chickadee. In *The Birds of North America*, No. 39 (A. Poole and F. Gill, eds.). Philadelphia, PA: The Birds of North America, Inc. 20 pp.

Snell, R. R. 2003. Iceland gull and Thayer's gull. In *The Birds of North America*, No. 699 (A. Poole and F. Gill, eds.). Philadelphia, PA: The Birds of North America, Inc. 36 pp.

Sodhi, N. S., L. Oliphant, P. James and I. Warkentin. 1993. Merlin. In *The Birds of North America*, No. 44 (A. Poole and F. Gill, eds.). Philadelphia, PA: The Birds of North America, Inc. 20 pp.

Sogge, M K., W. M. Gilbert, and C. Van Riper, III. 1994. Orange-crowned warbler. In *The Birds of North America*, No. 101 (A. Poole and F. Gill, eds.). Philadelphia, PA: The Birds of North America, Inc. 20 pp.

Sprenkle, J. M., and C. R. Blem. 1984. Metabolism and food selection of eastern house finches. *Wilson Bull.* 96:184–195.

Squires, J. R., and R. T. Reynolds. 1997. Northern goshawk. In *The Birds of North America*, No. 298 (A. Poole and F. Gill, eds.). Philadelphia, PA: The Birds of North America, Inc. 32 pp.

Stavy, N. E., K. F. Dybala & M. A. Snyder. 2008. Climate models and ornithology. *Auk* 125:1–10.

Stedman, S. J. 2000. Horned grebe. In *The Birds of North America*, No. 505 (A. Poole and F. Gill, eds.). Philadelphia, PA: The Birds of North America, Inc. 28 pp.

Steenhof, K. 1998. Prairie falcon. In *The Birds of North America*, No. 346 (A. Poole and F. Gill, eds.). Philadelphia, PA: The Birds of North America, Inc. 28 pp.

Stewart, R. E. 1975. *Breeding Birds of North Dakota.* Fargo, ND: Tri-College Center for Environmental Studies.

Strickland, D., and H. Ouellet. 1993. Gray jay. In *The Birds of North America*, No. 40 (A. Poole and F. Gill, eds.). Philadelphia, PA: The Birds of North America, Inc. 24 pp.

Summers-Smith, D. 1963. *The House Sparrow.* London, UK: Collins.

Sutton, G. M. 1967. *Oklahoma Birds: Their Ecology and Distribution, with Comments on the Avifauna of the Southern Great Plains.* Norman, OK: University of Oklahoma Press.

Swenson, J. E., K. C. Jensen, and J. E. Toepfer. 1988. Winter movements by rosy finches in Montana. *J. Field Ornithol.* 59:157–160.

Sydeman, W. J., and M. Guntert. 1983. Winter communal roosting in the pygmy nuthatch. In *Snag Habitat Management: Proceedings of the Symposium.* Fort Collins, CO: U.S. Forest Service General Technology Report. RM-99.

Tacha, T. C., and C. E. Braun (eds.). 1994. *Migratory Shore and Upland Game Management in North America.* , Washington, D.C.: International Assoc. of Fish and Wildlife Agencies.

_____, S. A. Nesbitt, and P. A. Vohs. 1992. Sandhill crane. In *The Birds of North America*, No. 31 (A. Poole and F. Gill, eds.). Philadelphia, PA: The Birds of North America, Inc. 24 pp.

Tallman, D. A. 2001. Fifty years of South Dakota Christmas Counts. *South Dakota Bird Notes* 53:24–29.

_____, D. L. Swanson, and J. S. Palmer. 2002. *Birds of South Dakota.* Aberdeen, SD: South Dakota State University.

Tarvin, K. A., and G. E. Woolfenden. 1999. Blue jay. In *The Birds of North America*, No. 469 (A. Poole and F. Gill, eds.). Philadelphia, PA: The Birds of North America, Inc. 32 pp.

Thompson, M. C., and C. Ely. 1989, 1992. *Birds in Kansas.* 2 vol. Lawrence, KS: University Press of Kansas.

_____, _____, B. Gress, C. Otte, S.T. Patti, D. Seibel, and E. A. Young. 2011. *Birds of Kansas.* Lawrence, KS: University Press of Kansas.

Titman, R. D. 1999. Red-breasted merganser. In *The Birds of North America*, No. 443 (A. Poole and F. Gill, eds.). Philadelphia, PA: The Birds of North America, Inc. 24 pp.

Tobalske, B. W. 1997. Lewis' woodpecker. In *The Birds of North America*, No. 284 (A. Poole and F. Gill, eds.). Philadelphia, PA: The Birds of North America, Inc. 28 pp.

Tomback, D. F. 1998. Clark's nutcracker. In *The Birds of North America*, No. 331 (A. Poole and F. Gill, eds.). Philadelphia, PA: The Birds of North America, Inc. 24 pp

Trost, T. R. 1999. Black-billed magpie In *The Birds of North America*, No. 389 (A. Poole and F. Gill, eds.). Philadelphia, PA: The Birds of North America, Inc. 28 pp.

Twedt, D. J., and R. D. Crawford. 1995. Yellow-headed blackbird. In *The Birds of North America*, No. 192 (A. Poole and F. Gill, eds.). Philadelphia, PA: The Birds of North America, Inc. 28 pp.

Tweit, R. C. 1996. Curve-billed thrasher. In *The Birds of North America*, No. 235 (A. Poole and F. Gill, eds.). Philadelphia, PA: The Birds of North America, Inc. 20 pp.

_____. 1997. Long-billed thrasher. In *The Birds of North America*, No. 317 (A. Poole and F. Gill, eds.). Philadelphia, PA: The Birds of North America, Inc. 12 pp.

_____, and C. W. Thompson. 1999. Pyrrhuloxia. In *The Birds of North America*, No. 391 (A. Poole and F. Gill, eds.). Philadelphia, PA: The Birds of North America, Inc. 20 pp.

United States Fish and Wildlife Service. 2007. Waterfowl Population Status, 2007. Washington, D.C.: Administrative Report, U.S. Dept. of Interior. See associated web site: http://www.fws.gov/migrationbirds/reports/reports.html

United States Fish and Wildlife Service. 2007. Waterfowl Population Status, 2013. Washington, D.C.: Administrative Report, U.S. Dept. of Interior. 64 pp.

United States Fish and Wildlife Service. 2014. Waterfowl Population Status, 2014. Washington, D.C.: Administrative Report, U.S. Dept. of Interior. 57 pp.

United States Geological Survey. 1970. *The National Atlas of the United States of America.* Washington, D.C.: U.S. Printing Office.

U.S. Global Change Research Program (USGCRP), 2014. *Natuonal Climate Assessment.* http://nca2014.globalchange.gov/

Verbeek, N. A. M. 1993. Glaucous-winged gull. In *The Birds of North America*, No. 59 (A. Poole and F. Gill, eds.). Philadelphia, PA: The Birds of North America, Inc. 20 pp.

_____, and C. Caffrey. 2002. American crow. In *The Birds of North America*, No. 647 (A. Poole and F. Gill, eds.). Philadelphia, PA: The Birds of North America, Inc. 36 pp.

_____, and P. Hendricks. 1994. American pipit. In *The Birds of North America*, No. 95 (A. Poole and F. Gill, eds.). Philadelphia, PA: The Birds of North America, Inc. 24 pp.

Vickery, P. D. 1996. Grasshopper sparrow. In *The Birds of North America*, No. 239 (A. Poole and F. Gill, eds.). Philadelphia, PA: The Birds of North America, Inc. 24 pp.

Walters, E. L., E. H. Miller and P. E. Lowther. 2002a. Yellow-bellied sapsucker. In *The Birds of North America*, No. 662 (A. Poole and F. Gill, eds.). Philadelphia, PA: The Birds of North America, Inc. 24 pp.

_____, _____, and _____. 2002b Red-breasted sapsucker and red-naped sapsucker. In *The Birds of North America*, No. 663 (A. Poole and F. Gill, eds.). The Birds of North America, Inc. Philadelphia, PA 32 pp.

Watt, D. J. 1986. Relationship of plumage variability, size, and sex to social dominance in Harris' sparrows. *Anim. Behav.* 34:16–27.

_____, C. J. Ralph, and C. T. Atkinson. 1984. The role of plumage polymorphism in dominance relationships of the white-throated sparrow. *Auk* 101:110–120.

_____, and E. J. Willoughby. 1999. Lesser goldfinch. In *The Birds of North America*, No. 392 (A. Poole and F. Gill, eds.). Philadelphia, PA: The Birds of North America, Inc. 24 pp.

Weckstein, J. D., D. E. Kroodsma, and R. C. Faucett. 2003. Fox sparrow. In *The Birds of North America*, No. 715 (A. Poole and F. Gill, eds.). Philadelphia, PA: The Birds of North America, Inc. 28 pp.

Weeks, H. P., Jr. 1994. Eastern phoebe. In *The Birds of North America*, No. 94 (A. Poole and F. Gill, eds.). Philadelphia, PA: The Birds of North America, Inc. 20 pp.

Weidensaul, S. 2007. CBC: The Climate Bird Count? *American Birds* 61:10-13.

Wheelwright, N. T., and J. D. Rising. 1993. Savannah sparrow. In *The Birds of North America*, No. 45 (A. Poole and F. Gill, eds.). Philadelphia, PA: The Birds of North America, Inc. 28 pp.

White, C. M., N. J. Clum, T. J. Cade, and W. G. Hunt. 2002. Peregrine falcon. In *The Birds of North America*, No. 660 (A. Poole and F. Gill, eds.). Philadelphia, PA: The Birds of North America, Inc. 48 pp.

Winkler, D. W. 1996. California gull. In *The Birds of North America*, No. 259 (A. Poole and F. Gill, eds.). Philadelphia, PA: The Birds of North America, Inc. 28 pp.

Wishart, D. J. (ed.). 2004. *Encyclopedia of the Great Plains*. Lincoln, NE: University of Nebraska Press.

Winkler, H., D. A. Christie, and D. Nurney. 1995. *Woodpeckers of the World*. Boston, MA: Houghton Mifflin Co.

With, K. A. 1994. McCown's longspur. In *The Birds of North America*, No. 96 (A. Poole and F. Gill, eds.). Philadelphia, PA: The Birds of North America, Inc. 24 pp.

Withgott, J. H., and K. G. Smith. 1998. Brown-headed nuthatch. In *The Birds of North America*, No. 349 (A. Poole and F. Gill, eds.). Philadelphia, PA: The Birds of North America, Inc. 24 pp.

Witmer, M. C. 2003. Bohemian waxwing. In *The Birds of North America*, No. 714 (A. Poole and F. Gill, eds.). Philadelphia, PA: The Birds of North America, Inc. 20 pp.

_____, D. J. Mountjoy, and L. Elliot. 1997. Cedar waxwing. In *The Birds of North America*, No. 309 (A. Poole and F. Gill, eds.). Philadelphia, PA: The Birds of North America, Inc. 28 pp.

Wood, D. S., and G. D. Schnell. 1984. *Distribution of Oklahoma Birds*. Norman, OK: University of Oklahoma Press.

Woodin, M. C. and T .C. Michot. 2003. Redhead. In *The Birds of North America*, No. 695 (A. Poole and F. Gill, eds.). Philadelphia, PA: The Birds of North America, Inc. 40 pp.

Woodrey, M. S. 1990. Economics of caching versus immediate consumption by white-breasted nuthatches: the effect of handling time. *Condor* 92:621–624.

_____. 1991. Caching behavior in free-ranging white-breasted nuthatches: The effects of social dominance. *Ornis Scand.* 22:160–166.

Wootton, T. 1996. Purple finch. 1996. In *The Birds of North America*, No. 208 (A. Poole and F. Gill, eds.). Philadelphia, PA: The Birds of North America, Inc. 20 pp.

Wormworth, J., and K. Mullen. Undated. *Bird Species and Climate Change. The Global Status Report.* Version 1.0. A Report to World Wide Fund for Nature. 75 pp. http://assets. panda.org/downloads/birdsclimatereportfinal.pdf

Yasukawa, K. and W. A. Searcy. 1995. Red-winged blackbird. In *The Birds of North America*, No. 184 (A. Poole and F. Gill, eds.). Philadelphia, PA: The Birds of North America, Inc. 28 pp.

Yosef, R. 1996. Loggerhead shrike. In *The Birds of North America*, No. 231 (A. Poole and F. Gill, eds.). Philadelphia, PA: The Birds of North America, Inc. 28 pp.

Zimpfer, N. L., W. E. Rhodes, E. D. Silverman, G. S. Zimmerman, and K. D, Richkus, 2014. Trends in Duck Breeding Populations, 1955–2014. U S Fish & Wildlife Service, Administration Report, July 2, 2014.

# Appendix: Species Abundance Tables

## 1. Greater White-fronted Goose

| Counts No. | Average Number of Birds Reported per Party-hour* | | | | |
|---|---|---|---|---|---|
| | 1968-1977<br>69-78 | 1978-1987<br>79-88 | 1988-1997<br>89-98 | 1998-2007<br>99-108 | 40 yr<br>avg. |
| N. Dakota | 0 | 0 | p | p | p |
| S. Dakota | 0 | p | 0.01 | 0.01 | 0.01 |
| Nebraska | p | 0.06 | 0.01 | 0.03 | 0.02 |
| Kansas | 1.63 | 1.55 | 9.61 | 142.03 | 38.70 |
| Oklahoma | 0.17 | 3.15 | 16.52 | 1.46 | 5.32 |
| 5-state ave.* | 0.57 | 1.24 | 6.95 | 40.11 | 12.22 |
| NW Texas | 0 | 0 | p | 0 | p |

* p = < 0.01 bird/party-hour

## 2. Snow Goose

| Counts No. | Average Number of Birds Reported per Party-hour* | | | | |
|---|---|---|---|---|---|
| | 1968-1977<br>69-78 | 1978-1987<br>79-88 | 1988-1997<br>89-98 | 1998-2007<br>99-108 | 40 yr<br>avg. |
| N. Dakota | p | 0 | p | p | p |
| S. Dakota | p | 0.01 | 0.01 | 0.01 | 0.01 |
| Nebraska | 1.80 | 0.05 | 0.39 | 0.01 | 0.53 |
| Kansas | 6.52 | 15.89 | 4.52 | 144.18 | 49.48 |
| Oklahoma | 3.31 | 36.77 | 12.22 | 9.00 | 15.12 |
| 5-state ave.* | 3.07 | 10.55 | 4.29 | 39.89 | 14.45 |
| NW Texas | 0.16 | 0.07 | 0.02 | p | 0.05 |

* p = < 0.01 bird/party-hour

## 3. Ross's Goose

| Counts No. | Average Number of Birds Reported per Party-hour* | | | | |
|---|---|---|---|---|---|
| | 1968-1977<br>69-78 | 1978-1987<br>79-88 | 1988-1997<br>89-98 | 1998-2007<br>99-108 | 40 yr<br>avg. |
| N. Dakota | 0 | 0 | 0 | 0 | 0 |
| S. Dakota | 0 | p | p | p | p |
| Nebraska | 0 | 0 | p | p | p |
| Kansas | p | p | 0.02 | 3.31 | 0.83 |
| Oklahoma | 0 | 0.02 | 0.54 | 1.81 | 0.59 |
| 5-state ave.* | p | p | 0.11 | 1.33 | 0.36 |
| NW Texas | 0.01 | 0 | p | 0.07 | 0.02 |

* p = < 0.01 bird/party-hour

## 4. Cackling Goose

| | Average Number of Birds Reported per Party-hour* | | | | |
|---|---|---|---|---|---|
| | 1968–1977 | 1978–1987 | 1988–1997 | 1998–2007 | 40 yr |
| Counts No. | 69-78 | 79-88 | 89-98 | 99-108 | avg. |
| N. Dakota | 0 | 0 | 0 | 0.38 | 0.38 |
| S. Dakota | 0 | 0 | 0 | 1.52 | 1.52 |
| Nebraska | 0 | 0 | 0 | 1.28 | 1.28 |
| Kansas | 0 | 0 | 0 | 1.42 | 1.42 |
| Oklahoma | 0 | 0 | 0 | 4.74 | 4.74 |
| 5-state ave.** | 0 | 0 | 0 | 2.01 | 2.01 |
| NW Texas | 0 | 0 | 0 | 0 | 0 |

* Not distinguished from Canada goose until 2005 count; all averages are
  for three years
** 3-year and multi-state averages are unweighted calculations

## 5. Canada Goose

| | Average Number of Birds Reported per Party-hour* | | | | |
|---|---|---|---|---|---|
| | 1968–1977 | 1978–1987 | 1988–1997 | 1998–2007 | 40 yr |
| Counts No. | 69-78 | 79-88 | 89-98 | 99-108 | avg. |
| N. Dakota | 0.02 | 0.65 | 16.73 | 32.82 | 12.56 |
| S. Dakota | 22.78 | 17.37 | 55.73 | 71.165 | 41.76 |
| Nebraska | 0.82 | 11.25 | 63.08 | 81.84 | 39.24 |
| Kansas | 53.27 | 21.61 | 103.35 | 353.12 | 132.89 |
| Oklahoma | 130.25 | 150.98 | 57.29 | 78.28 | 104.19 |
| 5-state ave.* | 52.61 | 46.41 | 63.57 | 145.14 | 76.93 |
| NW Texas | 233.37 | 34.41 | 49.94 | 40.42 | 89.535 |

## 6. Tundra Swan

| | Average Number of Birds Reported per Party-hour* | | | | |
|---|---|---|---|---|---|
| | 1968–1977 | 1978–1987 | 1988–1997 | 1998–2007 | 40 yr |
| Counts No. | 69-78 | 79-88 | 89-98 | 99-108 | avg. |
| N. Dakota | p | 0.01 | 0.02 | 0.01 | 0.01 |
| S. Dakota | 0 | 0.06 | p | 0.13 | 0.04 |
| Nebraska | 0 | 0 | 0 | 0.02 | p |
| Kansas | 0.02 | 0.01 | 0.02 | 0.04 | 0.02 |
| Oklahoma | p | 0.01 | 0.02 | 0.01 | 0.01 |
| 5-state ave.* | 0.01 | 0.02 | 0.02 | 0.04 | 0.02 |
| NW Texas | 0 | 0.04 | 0 | 0 | 0.01 |

* p = < 0.01 bird/party-hour

### 7. Trumpeter Swan

| Counts No. | Average Number of Birds Reported per Party-hour* | | | | |
| --- | --- | --- | --- | --- | --- |
| | 1968–1977 69-78 | 1978–1987 79-88 | 1988–1997 89-98 | 1998–2007 99-108 | 40 yr avg. |
| N. Dakota | 0 | 0 | 0 | p | p |
| S. Dakota | 0.15 | 0.84 | 0.19 | 0 | 0.30 |
| Nebraska | 0 | 0 | p | 0.02 | p |
| Kansas | 0 | 0 | p | p | p |
| Oklahoma | 0 | 0 | p | p | p |
| 5-state ave.* | 0.03 | 0.01 | 0.03 | p | 0.02 |
| NW Texas | 0 | 0 | p | 0 | p |

* p = < 0.01 bird/party-hour

### 8. Wood Duck

| Counts No. | Average Number of Birds Reported per Party-hour* | | | | |
| --- | --- | --- | --- | --- | --- |
| | 1968–1977 69-78 | 1978–1987 79-88 | 1988–1997 89-98 | 1998–2007 99-108 | 40 yr avg. |
| N. Dakota | p | p; | p | p | p |
| S. Dakota | 0.01 | p | p | p | p |
| Nebraska | p | p | p | 0.01 | p |
| Kansas | p | p | 0.05 | 0.04 | 0.03 |
| Oklahoma | 0.03 | 0.47 | 1.30 | 0.02 | 0.50 |
| 5-state ave.* | p | p | 0.34 | 0.06 | 0.13 |
| NW Texas | 0 | p | 0.02 | p | p |

* p = < 0.01 bird/party-hour

### 9. Gadwall

| Counts No. | Average Number of Birds Reported per Party-hour* | | | | |
| --- | --- | --- | --- | --- | --- |
| | 1968–1977 69-78 | 1978–1987 79-88 | 1988–1997 89-98 | 1998–2007 99-108 | 40 yr avg. |
| N. Dakota | p | p | p | p | p |
| S. Dakota | 0.16 | 0.09 | 0.21 | 0.14 | 0.15 |
| Nebraska | p | p | 0.18 | 0.47 | 0.16 |
| Kansas | 0.66 | 1.42 | 0.53 | 1.15 | 0.94 |
| Oklahoma | 2.10 | 5.12 | 2.61 | 5.11 | 3.74 |
| 5-state ave.* | 0.76 | 1.59 | 0.82 | 1.55 | 1.21 |
| NW Texas | 6.32 | 0.93 | 1.08 | 0.98 | 2.32 |

* p = < 0.01 bird/party-hour

10. American Wigeon

| | Average Number of Birds Reported per Party-hour* | | | | |
|---|---|---|---|---|---|
| Counts No. | 1968–1977<br>69-78 | 1978–1987<br>79-88 | 1988–1997<br>89-98 | 1998–2007<br>99-108 | 40 yr<br>avg. |
| N. Dakota | p | p | p | p | p |
| S. Dakota | 0.03 | 0.06 | 0.10 | 0.12 | 0.07 |
| Nebraska | 0.05 | 0.31 | 0.41 | 0.83 | 0.40 |
| Kansas | 5.00 | 2.78 | 0.20 | 1.06 | 2.26 |
| Oklahoma | 24.16 | 12.88 | 1.67 | 1.38 | 10.02 |
| 5-state ave.* | 7.48 | 3.13 | 0.51 | 0.72 | 3.18 |
| NW Texas | 94.05 | 2.50 | 3.76 | 7.80 | 27.02 |

* p = < 0.01 bird/party-hour

11. Mallard

| | Average Number of Birds Reported per Party-hour* | | | | |
|---|---|---|---|---|---|
| Counts No. | 1968–1977<br>69-78 | 1978–1987<br>79-88 | 1988–1997<br>89-98 | 1998–2007<br>99-108 | 40 yr<br>avg. |
| N. Dakota | 0.32 | 13.93 | 17.91 | 8.10 | 10.07 |
| S. Dakota | 376.83 | 60.76 | 42.22 | 55.21 | 134.00 |
| Nebraska | 278.17 | 42.02 | 140.48 | 94.68 | 136.84 |
| Kansas | 338.25 | 223.96 | 86.95 | 217.61 | 216.69 |
| Oklahoma | 359.01 | 192.83 | 66.43 | 21.41 | 159.92 |
| 5-state ave. | 270.52 | 106.70 | 71.00 | 79.40 | 131.90 |
| NW Texas | 140.39 | 72.64 | 26.43 | 27.67 | 66.76 |

12. Northern Shoveler

| | Average Number of Birds Reported per Party-hour* | | | | |
|---|---|---|---|---|---|
| Counts No. | 1968–1977<br>69-78 | 1978–1987<br>79-88 | 1988–1997<br>89-98 | 1998–2007<br>99-108 | 40 yr<br>avg. |
| N. Dakota | 0 | 0 | 0 | p | p |
| S. Dakota | p | p | p | 0.06 | 0.02 |
| Nebraska | 0.01 | p | 0.02 | 0.14 | 0.04 |
| Kansas | 0.38 | 0.07 | 0.12 | 0.87 | 0.36 |
| Oklahoma | 0.07 | 0.26 | 0.37 | 0.92 | 0.40 |
| 5-state ave.* | 0.09 | 0.07 | 0.10 | 0.40 | 0.17 |
| NW Texas | 4.74 | 1.85 | 2.09 | 2.18 | 2.71 |

* p = < 0.01 bird/party-hour

## 13. Northern Pintail

| | Average Number of Birds Reported per Party-hour* | | | | |
|---|---|---|---|---|---|
| Counts No. | 1968–1977 69-78 | 1978–1987 79-88 | 1988–1997 89-98 | 1998–2007 99-108 | 40 yr avg. |
| N. Dakota | p | p | p | p | p |
| S. Dakota | 0.01 | 0.02 | 0.01 | 0.01 | 0.01 |
| Nebraska | 0.10 | 0.04 | 0.08 | 0.18 | 0.10 |
| Kansas | 4.31 | 0.89 | 0.27 | 0.90 | 1.60 |
| Oklahoma | 0.87 | 1.12 | 3.05 | 1.03 | 1.51 |
| 5-state ave.* | 1.06 | 0.41 | 0.68 | 0.43 | 0.64 |
| NW Texas | 354.89 | 23.97 | 15.81 | 31.90 | 106.65 |

* p = < 0.01 bird/party-hour

## 14. Green-winged Teal

| | Average Number of Birds Reported per Party-hour* | | | | |
|---|---|---|---|---|---|
| Counts No. | 1968–1977 69-78 | 1978–1987 79-88 | 1988–1997 89-98 | 1998–2007 99-108 | 40 yr avg. |
| N. Dakota | 0 | 0 | p | p | p |
| S. Dakota | p | p | p | p | p |
| Nebraska | 0.03 | 0.02 | 0.35 | 0.22 | 0.15 |
| Kansas | 5.86 | 0.87 | 0.90 | 1.17 | 2.38 |
| Oklahoma | 0.79 | 3.35 | 7.70 | 2.09 | 3.48 |
| 5-state ave.* | 1.33 | 0.85 | 1.79 | 0.02 | 0.70 |
| NW Texas | 98.57 | 10.98 | 9.85 | 0.82 | 30.06 |

* p = < 0.01 bird/party-hour

## 15. Canvasback

| | Average Number of Birds Reported per Party-hour* | | | | |
|---|---|---|---|---|---|
| Counts No. | 1968–1977 69-78 | 1978–1987 79-88 | 1988–1997 89-98 | 1998–2007 99-108 | 40 yr avg. |
| N. Dakota | 0 | p | p | p | p |
| S. Dakota | p | p | 0.01 | p | p |
| Nebraska | p | p | 0.01 | 0.09 | 0.03 |
| Kansas | 0.17 | 0.06 | 0.18 | 0.27 | 0.17 |
| Oklahoma | 0.55 | 0.14 | 0.21 | 0.19 | 0.28 |
| 5-state ave.* | 0.15 | 0.04 | 0.08 | 0.12 | 0.09 |
| NW Texas | 0.29 | 0.18 | 0.28 | 0.51 | 0.32 |

* p = < 0.01 bird/party-hour

16. Redhead

| Counts No. | Average Number of Birds Reported per Party-hour* | | | | |
| --- | --- | --- | --- | --- | --- |
| | 1968-1977 69-78 | 1978-1987 79-88 | 1988-1997 89-98 | 1998-2007 99-108 | 40 yr avg. |
| N. Dakota | p | p | p | p | p |
| S. Dakota | p | 0.01 | 0.01 | 0.02 | 0.01 |
| Nebraska | p | 0.01 | 0.01 | 0.19 | 0.05 |
| Kansas | 0.06 | 0.08 | 0.10 | 0.19 | 0.10 |
| Oklahoma | 0.16 | 0.09 | 0.14 | 0.29 | 0.17 |
| 5-state ave.* | 0.04 | 0.03 | 0.05 | 0.14 | 0.07 |
| NW Texas | 0.27 | 0.22 | 0.13 | 0.34 | 0.24 |

* p = < 0.01 bird/party-hour

17. Ring-necked Duck

| Counts No. | Average Number of Birds Reported per Party-hour* | | | | |
| --- | --- | --- | --- | --- | --- |
| | 1968-1977 69-78 | 1978-1987 79-88 | 1988-1997 89-98 | 1998-2007 99-108 | 40 yr avg. |
| N. Dakota | 0 | 0 | 0 | p | p |
| S. Dakota | p | p | 0.01 | 0.02 | 0.01 |
| Nebraska | p | p | 0.05 | 0.19 | 0.06 |
| Kansas | 0.11 | 0.38 | 1.13 | 0.51 | 0.53 |
| Oklahoma | 0.19 | 0.33 | 1.03 | 0.57 | 0.53 |
| 5-state ave.* | 0.06 | 0.14 | 044 | 0.26 | 0.22 |
| NW Texas | 0.81 | 0.44 | 1.75 | 1.29 | 1.08 |

* p = < 0.01 bird/party-hour

18. Greater Scaup

| Counts No. | Average Number of Birds Reported per Party-hour* | | | | |
| --- | --- | --- | --- | --- | --- |
| | 1968-1977 69-78 | 1978-1987 79-88 | 1988-1997 89-98 | 1998-2007 99-108 | 40 yr avg. |
| N. Dakota | 0 | 0 | p | p | p |
| S. Dakota | p | p | p | p | p |
| Nebraska | 0 | 0 | p | p | p |
| Kansas | 0 | 0 | p | 0.02 | p |
| Oklahoma | p | p | p | 0.13 | 0.04 |
| 5-state ave.* | p | p | p | 0.03 | 0.01 |
| NW Texas | 0 | p | p | 0 | p |

* p = < 0.01 bird/party-hour

### 19. Lesser Scaup

| Counts No. | Average Number of Birds Reported per Party-hour* | | | | |
|---|---|---|---|---|---|
| | 1968-1977 69-78 | 1978-1987 79-88 | 1988-1997 89-98 | 1998-2007 99-108 | 40 yr avg. |
| N. Dakota | p | p | p | 0.01 | p |
| S. Dakota | 0.17 | 0.02 | 0.02 | 0.07 | 0.07 |
| Nebraska | p | 0.03 | 0.07 | 0.63 | 0.19 |
| Kansas | 0.19 | 0.96 | 0.47 | 0.74 | 0.59 |
| Oklahoma | 0.69 | 0.53 | 1.09 | 0.61 | 0.73 |
| 5-state ave.* | 0.21 | 0.31 | 0.33 | 0.41 | 0.31 |
| NW Texas | 2.16 | 0.19 | 0.67 | 0.63 | 0.91 |

* p = < 0.01 bird/party-hour

### 20. Bufflehead

| Counts No. | Average Number of Birds Reported per Party-hour* | | | | |
|---|---|---|---|---|---|
| | 1968-1977 69-78 | 1978-1987 79-88 | 1988-1997 89-98 | 1998-2007 99-108 | 40 yr avg. |
| N. Dakota | 0 | p | p | 0.01 | p |
| S. Dakota | 0.08 | 0.02 | 0.03 | 0.05 | 0.04 |
| Nebraska | p | p | 0.23 | 0.43 | 0.16 |
| Kansas | 0.03 | 0.03 | 0.15 | 0.25 | 0.12 |
| Oklahoma | 0.18 | 0.15 | 0.41 | 0.70 | 0.36 |
| 5-state ave.* | 0.06 | 0.04 | 0.16 | 0.29 | 0.14 |
| NW Texas | 2.37 | 0.25 | 0.54 | 0.36 | 0.88 |

* p = < 0.01 bird/party-hour

### 21. Common Goldeneye

| Counts No. | Average Number of Birds Reported per Party-hour* | | | | |
|---|---|---|---|---|---|
| | 1968-1977 69-78 | 1978-1987 79-88 | 1988-1997 89-98 | 1998-2007 99-108 | 40 yr avg. |
| N. Dakota | p | 0.45 | 0.49 | 0.50 | 0.36 |
| S. Dakota | 1.35 | 0.88 | 0.33 | 0.35 | 0.74 |
| Nebraska | 0.07 | 0.10 | 2.27 | 3,24 | 1.42 |
| Kansas | 0.16 | 0.75 | 6.24 | 8.64 | 3.95 |
| Oklahoma | 1.17 | 0.98 | 0.72 | 0.55 | 0.86 |
| 5-state ave.* | 0.55 | 0.63 | 2.01 | 2.65 | 1.46 |
| NW Texas | 0.09 | 0.07 | 0.06 | 0.01 | 0,05 |

* p = < 0.01 bird/party-hour

## 22. Hooded Merganser

| Counts No. | Average Number of Birds Reported per Party-hour* | | | | |
|---|---|---|---|---|---|
| | 1968–1977<br>69-78 | 1978–1987<br>79-88 | 1988–1997<br>89-98 | 1998–2007<br>99-108 | 40 yr<br>avg. |
| N. Dakota | p | 0 | p | p | p |
| S. Dakota | p | p | p | 0.01 | p |
| Nebraska | 0 | p | 0.04 | 0.04 | 0.01 |
| Kansas | 0.02 | 0.02 | 0.02 | 0.67 | 0.22 |
| Oklahoma | 0.17 | 0.36 | 0.80 | 0.79 | 0.53 |
| 5-state ave.* | 0.04 | 0.09 | 0.25 | 0.37 | 0.19 |
| NW Texas | 0.04 | 0.06 | 0.05 | 0.09 | 0.06 |

* p = < 0.01 bird/party-hour

## 23. Red-breasted Merganser

| Counts No. | Average Number of Birds Reported per Party-hour* | | | | |
|---|---|---|---|---|---|
| | 1968–1977<br>69-78 | 1978–1987<br>79-88 | 1988–1997<br>89-98 | 1998–2007<br>99-108 | 40 yr<br>avg. |
| N. Dakota | p | p | p | p | p |
| S. Dakota | p | p | p | p | p |
| Nebraska | p | p | 0.01 | 0.02 | 0.01 |
| Kansas | 0 | p | p | 0.02 | 0.01 |
| Oklahoma | p | 0.01 | 0.15 | 0.69 | 0.21 |
| 5-state ave.* | p | p | 0.04 | 0.16 | 0.05 |
| NW Texas | 0.02 | 0 | p | 0 | p |

* p = < 0.01 bird/party-hour

## 24. Common Merganser

| Counts No. | Average Number of Birds Reported per Party-hour* | | | | |
|---|---|---|---|---|---|
| | 1968–1977<br>69-78 | 1978–1987<br>79-88 | 1988–1997<br>89-98 | 1998–2007<br>99-108 | 40 yr<br>avg. |
| N. Dakota | p | 0.25 | 0.37 | 0.21 | 0.21 |
| S. Dakota | 2.91 | 4.75 | 0.95 | 2.62 | 2.59 |
| Nebraska | 0.40 | 0.14 | 63.14 | 21.50 | 21.29 |
| Kansas | 6.35 | 4.98 | 36.33 | 55.54 | 25.80 |
| Oklahoma | 16.39 | 2.14 | 1.97 | 1.07 | 5.39 |
| 5-state ave.* | 6.58 | 2.97 | 17.54 | 18.92 | 11.53 |
| NW Texas | 0.48 | 1.53 | 0.14 | 0.04 | 0.55 |

* p = < 0.01 bird/party-hour

**25. Ruddy Duck**

| Counts No. | Average Number of Birds Reported per Party-hour* | | | | |
| | 1968-1977 69-78 | 1978-1987 79-88 | 1988-1997 89-98 | 1998-2007 99-108 | 40 yr avg. |
|---|---|---|---|---|---|
| N. Dakota | 0 | p | 0 | p | p |
| S. Dakota | p | p | p | p | p |
| Nebraska | p | p | p | 0.08 | 0.02 |
| Kansas | 0.16 | 0.01 | 0.02 | 0.17 | 0.05 |
| Oklahoma | 0.04 | 0.03 | 0.19 | 0.14 | 0.10 |
| 5-state ave.* | 0.02 | 0.01 | 0.05 | 0.09 | 0.04 |
| NW Texas | 2.82 | 0.02 | 0.16 | 0.05 | 0.87 |

* p = < 0.01 bird/party-hour

**26. Gray Partridge**

| Counts No. | Average Number of Birds Reported per Party-hour* | | | | |
| | 1968-1977 69-78 | 1978-1987 79-88 | 1988-1997 89-98 | 1998-2007 99-108 | 40 yr avg. |
|---|---|---|---|---|---|
| N. Dakota | 1.06 | 1.38 | 0.97 | 0.43 | 0.96 |
| S. Dakota | 0.07 | 0.57 | 0.11 | 0.14 | 0.22 |
| Nebraska | 0 | p | 0.02 | p | p |
| Kansas | 0 | 0 | 0 | 0 | 0 |
| Oklahoma | 0 | 0 | 0 | 0 | 0 |
| 5-state ave.* | 0.22 | 0.32 | 0.22 | 0.12 | 0.22 |
| NW Texas | 0 | 0 | 0 | 0 | 0 |

* p = < 0.01 bird/party-hour

**27. Ring-necked Pheasant**

| Counts No. | Average Number of Birds Reported per Party-hour* | | | | |
| | 1968-1977 69-78 | 1978-1987 79-88 | 1988-1997 89-98 | 1998-2007 99-108 | 40 yr avg. |
|---|---|---|---|---|---|
| N. Dakota | 0.98 | 1.93 | 4.31 | 4.54 | 2.94 |
| S. Dakota | 4.49 | 7.64 | 3.85 | 6.00 | 5.42 |
| Nebraska | 0.39 | 0.88 | 1.01 | 0.46 | 0.69 |
| Kansas | 0.46 | 0.44 | 0.39 | 0.69 | 0.50 |
| Oklahoma | 0.01 | 0.02 | 0.01 | 0.04 | 0.02 |
| 5-state ave.* | 1.08 | 1.90 | 1.74 | 2.30 | 1.75 |
| NW Texas | 0.38 | 0.56 | 0.12 | 0.04 | 0.06 |

**28. Sharp-tailed Grouse**

| | Average Number of Birds Reported per Party-hour* | | | | |
|---|---|---|---|---|---|
| Counts No. | 1968–1977 69-78 | 1978–1987 79-88 | 1988–1997 89-98 | 1998–2007 99-108 | 40 yr avg. |
| N. Dakota | 1.57 | 1.70 | 1.28 | 2.03 | 1.65 |
| S. Dakota | 0.16 | 0.35 | 0.61 | 0.93 | 0.51 |
| Nebraska | 0.05 | 0.04 | 0.09 | 0.07 | 0.07 |
| Kansas | 0 | 0 | 0 | 0 | 0 |
| Oklahoma | 0 | 0 | 0 | 0 | 0 |
| 5-state ave.* | 0.34 | 0.36 | 0.37 | 0.61 | 0.42 |
| NW Texas | 0 | 0 | 0 | 0 | 0 |

**29. Greater Prairie-Chicken**

| | Average Number of Birds Reported per Party-hour* | | | | |
|---|---|---|---|---|---|
| Counts No. | 1968–1977 69-78 | 1978–1987 79-88 | 1988–1997 89-98 | 1998–2007 99-108 | 40 yr avg. |
| N. Dakota | 0.01 | 0.02 | p | p | 0.01 |
| S. Dakota | 0.02 | 0.15 | 0.02 | 008 | 0.07 |
| Nebraska | p | 0.51 | 0.83 | 0.42 | 0.44 |
| Kansas | 0.01 | 0.18 | 0.04 | 0.11 | 0.11 |
| Oklahoma | 0.01 | 0.02 | 0.01 | 0.01 | 0.05 |
| 5-state ave.* | 0.04 | 0.13 | 0.09 | 0.09 | 0.09 |
| NW Texas | 0 | 0 | 0 | 0 | 0 |

* p = < 0.01 bird/party-hour

**30. Wild Turkey**

| | Average Number of Birds Reported per Party-hour* | | | | |
|---|---|---|---|---|---|
| Counts No. | 1968–1977 69-78 | 1978–1987 79-88 | 1988–1997 89-98 | 1998–2007 99-108 | 40 yr avg. |
| N. Dakota | 0.04 | 0.39 | 0.92 | 2.08 | 0.86 |
| S. Dakota | 0.08 | 0.38 | 1.16 | 3.02 | 1.16 |
| Nebraska | 0.01 | 0.10 | 0.57 | 1.20 | 0.47 |
| Kansas | 0.04 | 0.20 | 0.59 | 1.70 | 0.64 |
| Oklahoma | 0.47 | 0.60 | 0.39 | 1.06 | 0.63 |
| 5-state ave.* | 0.15 | 0.35 | 0.71 | 1.82 | 0.76 |
| NW Texas | 0.87 | 2.38 | 1.86 | 0.73 | 1.46 |

## 31. Scaled Quail

| Counts No. | Average Number of Birds Reported per Party-hour* | | | | |
|---|---|---|---|---|---|
| | 1968-1977 69-78 | 1978-1987 79-88 | 1988-1997 89-98 | 1998-2007 99-108 | 40 yr avg. |
| N. Dakota | 0 | 0 | 0 | 0 | 0 |
| S. Dakota | 0 | 0 | 0 | 0 | 0 |
| Nebraska | 0 | 0 | 0 | 0 | 0 |
| Kansas | 0 | 0.08 | 0.04 | 0.05 | 0.04 |
| Oklahoma | 0.17 | 0.29 | 0.12 | 0.09 | 0.17 |
| 5-state ave.* | 0.04 | 0.09 | 0.04 | 0.03 | 0.05 |
| NW Texas | 2.85 | 4.81 | 1.59 | 0.71 | 2.49 |

## 32. Northern Bobwhite

| Counts No. | Average Number of Birds Reported per Party-hour* | | | | |
|---|---|---|---|---|---|
| | 1968-1977 69-78 | 1978-1987 79-88 | 1988-1997 89-98 | 1998-2007 99-108 | 40 yr avg. |
| N. Dakota | 0 | 0 | p | 0 | p |
| S. Dakota | 0.01 | 0.08 | 0.03 | 0.02 | 0.03 |
| Nebraska | 0.55 | 0.62 | 0.50 | 0.25 | 0.48 |
| Kansas | 1.78 | 1.46 | 1.50 | 0.74 | 1.37 |
| Oklahoma | 1.79 | 1.48 | 0.91 | 0.54 | 1.18 |
| 5-state ave.* | 1.03 | 0.88 | 0.69 | 0.36 | 0.74 |
| NW Texas | 0.98 | 0.88 | 1.83 | 1.35 | 1.26 |

* p = < 0.01 bird/party-hour

## 33. Pied-billed Grebe

| Counts No. | Average Number of Birds Reported per Party-hour* | | | | |
|---|---|---|---|---|---|
| | 1968-1977 69-78 | 1978-1987 79-88 | 1988-1997 89-98 | 1998-2007 99-108 | 40 yr avg. |
| N. Dakota | 0 | 0 | p | p | p |
| S. Dakota | p | p | 0.01 | p | p |
| Nebraska | 0 | 0 | p | 0.01 | p |
| Kansas | 0.02 | 0.02 | 0.04 | 0.14 | 0.05 |
| Oklahoma | 0.23 | 0.18 | 0.33 | 0.31 | 0.23 |
| 5-state ave.* | 0.06 | 0.04 | 0.06 | 0.11 | 0.07 |
| NW Texas | 0.06 | 0.13 | 0.24 | 0.27 | 0.18 |

* p = < 0.01 bird/party-hour

**34. Horned Grebe**

| | Average Number of Birds Reported per Party-hour* | | | | |
|---|---|---|---|---|---|
| Counts No. | 1968-1977 69-78 | 1978-1987 79-88 | 1988-1997 89-98 | 1998-2007 99-108 | 40 yr avg. |
| N. Dakota | 0 | 0 | p | p | p |
| S. Dakota | 0 | 0 | 0 | p | p |
| Nebraska | 0 | 0 | p | p | p |
| Kansas | 0 | p | p | p | p |
| Oklahoma | 0.13 | 0.13 | 0.10 | 0.37 | 0.15 |
| 5-state ave.* | 003 | 0.03 | 0.02 | 0.08 | 0.04 |
| NW Texas | 0 | 0 | p | p | p |

* p = < 0.01 bird/party-hour

**35. American White Pelican**

| | Average Number of Birds Reported per Party-hour* | | | | |
|---|---|---|---|---|---|
| Counts No. | 1968-1977 69-78 | 1978-1987 79-88 | 1988-1997 89-98 | 1998-2007 99-108 | 40 yr avg. |
| N. Dakota | 0 | 0 | 0 | p | p |
| S. Dakota | 0 | p | p | p | p |
| Nebraska | 0 | 0 | p | p | p |
| Kansas | p | p | 0.05 | 0.04 | 0.02 |
| Oklahoma | 0.01 | 0.01 | 0.01 | 1.69 | 0.65 |
| 5-state ave.* | p | p | p | 0.39 | 0.16 |
| NW Texas | 0 | 0 | 0 | p | p |

* p = < 0.01 bird/party-hour

**36. Double-crested Cormorant**

| | Average Number of Birds Reported per Party-hour* | | | | |
|---|---|---|---|---|---|
| Counts No. | 1968-1977 69-78 | 1978-1987 79-88 | 1988-1997 89-98 | 1998-2007 99-108 | 40 yr avg. |
| N. Dakota | 0 | 0 | 0 | p | p |
| S. Dakota | 0 | 0 | 0 | p | p |
| Nebraska | p | p | p | 0.01 | p |
| Kansas | p | 0.06 | 0.07 | 0.21 | 0.09 |
| Oklahoma | 0.05 | 0.73 | 7.89 | 10.79 | 4.87 |
| 5-state ave.* | 0.01 | 0.10 | 1.93 | 2.50 | 1.16 |
| NW Texas | 0 | 0.01 | 001 | 0.01 | 0.01 |

* p = < 0.01 bird/party-hour

## 37. Great Blue Heron

| Counts No. | Average Number of Birds Reported per Party-hour* | | | | |
|---|---|---|---|---|---|
| | 1968-1977 69-78 | 1978-1987 79-88 | 1988-1997 89-98 | 1998-2007 99-108 | 40 yr avg. |
| N. Dakota | 0 | p | p | p | p |
| S. Dakota | p | p | p | p | p |
| Nebraska | p | 0.02 | 0.04 | 0.05 | 0.03 |
| Kansas | 0.06 | 0.10 | 0.18 | 0.22 | 0.14 |
| Oklahoma | 0.28 | 0.50 | 0.65 | 0.54 | 0.49 |
| 5-state ave.* | 0.08 | 0.15 | 0.21 | 0.19 | 0.16 |
| NW Texas | 0.07 | 0.13 | 0.25 | 0.29 | 0.19 |

* p = < 0.01 bird/party-hour

## 38. Black Vulture

| Counts No. | Average Number of Birds Reported per Party-hour* | | | | |
|---|---|---|---|---|---|
| | 1968-1977 69-78 | 1978-1987 79-88 | 1988-1997 89-98 | 1998-2007 99-108 | 40 yr avg. |
| N. Dakota | 0 | 0 | 0 | 0 | 0 |
| S. Dakota | 0 | 0 | 0 | 0 | 0 |
| Nebraska | 0 | 0 | 0 | 0 | 0 |
| Kansas | 0 | 0 | 0 | 0 | 0 |
| Oklahoma | 0.07 | 0.15 | 0.28 | 0.58 | 0.27 |
| 5-state ave.* | 0.01 | 0.04 | 0.06 | 0.13 | 0.06 |
| NW Texas | 0 | 0 | 0 | 0 | 0 |

## 39. Turkey Vulture

| Counts No. | Average Number of Birds Reported per Party-hour* | | | | |
|---|---|---|---|---|---|
| | 1968-1977 69-78 | 1978-1987 79-88 | 1988-1997 89-98 | 1998-2007 99-108 | 40 yr avg. |
| N. Dakota | 0 | 0 | 0 | p | p |
| S. Dakota | p | 0 | 0 | 0 | p |
| Nebraska | 0 | 0 | p | p | p |
| Kansas | p | p | p | p | p |
| Oklahoma | 0.18 | 0.52 | 0.80 | 1.05 | 0.63 |
| 5-state ave.* | 0.05 | 0.12 | 0.19 | 0.23 | 0.14 |
| NW Texas | 0 | 0 | 0 | 0 | 0 |

* p = < 0.01 bird/party-hour

**40. Bald Eagle**

| | Average Number of Birds Reported per Party-hour* | | | | |
| --- | --- | --- | --- | --- | --- |
| Counts No. | 1968–1977<br>69-78 | 1978–1987<br>79-88 | 1988–1997<br>89-98 | 1998–2007<br>99-108 | 40 yr<br>avg. |
| N. Dakota | p | 0.02 | 0.04 | 0.09 | 0.04 |
| S. Dakota | 0.34 | 0.37 | 0.23 | 0.36 | 0.33 |
| Nebraska | 0.03 | 0.11 | 0.29 | 0.55 | 0.25 |
| Kansas | 0.12 | 0.11 | 0.30 | 0.50 | 0.25 |
| Oklahoma | 0.11 | 0.18 | 0.23 | 0.17 | 0.17 |
| 5-state ave.* | 0.12 | 0.16 | 0.22 | 0.32 | 0.21 |
| NW Texas | 0.11 | 0.25 | 0.31 | 0.19 | 0.22 |

* p = < 0.01 bird/party-hour

**41. Northern Harrier**

| | Average Number of Birds Reported per Party-hour* | | | | |
| --- | --- | --- | --- | --- | --- |
| Counts No. | 1968–1977<br>69-78 | 1978–1987<br>79-88 | 1988–1997<br>89-98 | 1998–2007<br>99-108 | 40 yr<br>avg. |
| N. Dakota | p | p | p | 0.01 | p |
| S. Dakota | 0.03 | 0.03 | 0.02 | 0.03 | 0.03 |
| Nebraska | 0.13 | 0.09 | 0.12 | 0.12 | 0.12 |
| Kansas | 0.67 | 0.45 | 0.63 | 0.87 | 0.64 |
| Oklahoma | 0.49 | 0.42 | 0.43 | 0.46 | 0.45 |
| 5-state ave.* | 0.34 | 0.25 | 0.29 | 0.37 | 0.31 |
| NW Texas | 0.32 | 0.43 | 0.58 | 0.61 | 0.36 |

* p = < 0.01 bird/party-hour

**42. Red-tailed Hawk**

| | Average Number of Birds Reported per Party-hour* | | | | |
| --- | --- | --- | --- | --- | --- |
| Counts No. | 1968–1977<br>69-78 | 1978–1987<br>79-88 | 1988–1997<br>89-98 | 1998–2007<br>99-108 | 40 yr<br>avg. |
| N. Dakota | p | p | 0.01 | 0.03 | 0.01 |
| S. Dakota | 0.04 | 0.07 | 0.06 | 0.17 | 0.09 |
| Nebraska | 0.20 | 0.30 | 0.50 | 0.64 | 0.41 |
| Kansas | 0.92 | 1.11 | 1.87 | 1.98 | 1.45 |
| Oklahoma | 0.72 | 0.86 | 1.15 | 1.00 | 0.93 |
| 5-state ave.* | 0.49 | 0.59 | 0.87 | 0.88 | 0.71 |
| NW Texas | 0.20 | 0.38 | 0.41 | 0.56 | 0.39 |

* p = < 0.01 bird/party-hour

## 43. Ferruginous Hawk

| Counts No. | Average Number of Birds Reported per Party-hour* | | | | |
|---|---|---|---|---|---|
| | 1968–1977 69-78 | 1978–1987 79-88 | 1988–1997 89-98 | 1998-2007 99-108 | 40 yr avg. |
| N. Dakota | 0 | 0 | 0 | 0- | 0 |
| S. Dakota | p | 0.01 | 0.01 | 0.01 | 0.01 |
| Nebraska | 0 | 0.02 | 0.02 | 0.01 | 0.01 |
| Kansas | 0.03 | 0.03 | 0.06 | 0.05` | 0.04 |
| Oklahoma | p | 0.02 | 0.02 | 0.01 | 0.01 |
| 5-state ave.* | 0.01 | 0.02 | 0.02 | 0.02 | 0.02 |
| NW Texas | 0.06 | 0.11 | 0.24 | 0.16 | 0.14 |

* p = < 0.01 bird/party-hour

## 44. Rough-legged Hawk

| Counts No. | Average Number of Birds Reported per Party-hour* | | | | |
|---|---|---|---|---|---|
| | 1968–1977 69-78 | 1978–1987 79-88 | 1988–1997 89-98 | 1998-2007 99-108 | 40 yr avg. |
| N. Dakota | 0.01 | ˋ | 0.02 | 0.01 | 0.07 |
| 0.03 | | | | | |
| S. Dakota | 0.07 | 0.11 | 0.13 | 0.21 | 0.13 |
| Nebraska | 0.06 | 0.15 | 0.17 | 0.12 | 0.13 |
| Kansas | 0.24 | 0.13 | 0.08 | 0.09 | 0.14 |
| Oklahoma | 0.03 | 0.03 | 0.03 | 0.03 | 0.04 |
| 5-state ave.* | 0.10 | 0.09 | 0.07 | 0.10 | 0.09 |
| NW Texas | 0.05 | 0.05 | 0.06 | 0.07 | 0.06 |

## 45. Golden Eagle

| Counts No. | Average Number of Birds Reported per Party-hour* | | | | |
|---|---|---|---|---|---|
| | 1968–1977 69-78 | 1978–1987 79-88 | 1988–1997 89-98 | 1998-2007 99-108 | 40 yr avg. |
| N. Dakota | 0.02 | 0.03 | 0.04 | 0.05 | 0.03 |
| S. Dakota | 0.04 | 0.04 | 0.05 | 0.07 | 0.05 |
| Nebraska | 0.17 | 0.02 | 0.03 | 0.02 | 0.02 |
| Kansas | 0.03 | 0.01 | 0.01 | 0.01 | 0.02 |
| Oklahoma | 0.04 | 0.01 | 0.01 | 0.01 | 0.01 |
| 5-state ave.* | 0.03 | 0.02 | 0.03 | 0.03 | 0.03 |
| NW Texas | 0.12 | 0.09 | 0.06 | 0.04 | 0.08 |

**46. Virginia Rail**

| Counts No. | Average Number of Birds Reported per Party-hour* | | | | |
|---|---|---|---|---|---|
| | 1968–1977 69-78 | 1978–1987 79-88 | 1988–1997 89-98 | 1998–2007 99-108 | 40 yr avg. |
| N. Dakota | 0 | 0 | 0 | 0 | 0 |
| S. Dakota | 0 | 0 | 0 | p | p |
| Nebraska | p | 0 | p | 0.08 | 0.02 |
| Kansas | p | 0 | p | 0.09 | 0.02 |
| Oklahoma | 0.02 | 0.02 | 0.09 | 0.10 | 0.08 |
| 5-state ave.* | p | 0.01 | 0.02 | 0.06 | 0.02 |
| NW Texas | 0 | p | p | 0.07 | 0.02 |

\* p = < 0.01 bird/party-hour

**47. American Coot**

| Counts No. | Average Number of Birds Reported per Party-hour* | | | | |
|---|---|---|---|---|---|
| | 1968–1977 69-78 | 1978–1987 79-88 | 1988–1997 89-98 | 1998–2007 99-108 | 40 yr avg. |
| N. Dakota | 0 | 0 | 0 | p | p |
| S. Dakota | 0.02 | 0.05 | 0.72 | 1.20 | 0.50 |
| Nebraska | p | p | 0.02 | 0.66 | 0.17 |
| Kansas | 0.25 | 0.94 | 0.50 | 1.92 | 0.90 |
| Oklahoma | 5.03 | 3.57 | 3.47 | 4.86 | 4.23 |
| 5-state ave.* | 1.32 | 0.92 | 0.98 | 1.82 | 1.16 |
| NW Texas | 7.86 | 2.35 | 3.50 | 2.90 | 4.15 |

\* p = < 0.01 bird/party-hour

**48. Sandhill Crane**

| Counts No. | Average Number of Birds Reported per Party-hour* | | | | |
|---|---|---|---|---|---|
| | 1968–1977 69-78 | 1978–1987 79-88 | 1988–1997 89-98 | 1998–2007 99-108 | 40 yr avg. |
| N. Dakota | 0 | 0 | p | p | p |
| S. Dakota | p | 0 | 0 | 0 | p |
| Nebraska | 0 | p | 0 | 0 | p |
| Kansas | p | 0 | 7.22 | 16.32 | 5.88 |
| Oklahoma | 0.87 | 5.63 | 0.68 | 6.78 | 3.49 |
| 5-state ave.* | 0.21 | 1.29 | 2.33 | 4.62 | 2.11 |
| NW Texas | 826.63 | 1134.68 | 154.44 | 139.55 | 564.32 |

\* p = < 0.01 bird/party-hour

**49. Killdeer**

| | Average Number of Birds Reported per Party-hour* | | | | |
| Counts No. | 1968–1977 69-78 | 1978–1987 79-88 | 1988–1997 89-98 | 1998–2007 99-108 | 40 yr avg. |
|---|---|---|---|---|---|
| N. Dakota | 0 | 0 | 0 | 0 | 0 |
| S. Dakota | p | 0.01 | p | p | p |
| Nebraska | 0.03 | 0.01 | 0.02 | 0.02 | 0.02 |
| Kansas | 0.05 | 0.05 | 0.04 | 0.03 | 0.04 |
| Oklahoma | 0.26 | 0.25 | 0.28 | 0.32 | 0.28 |
| 5-state ave.* | 0.08 | 0.08 | 0.08 | 0.08 | 0.08 |
| NW Texas | 0.09 | 0.09 | 0.07 | 0.12 | 0.09 |

* p = < 0.01 bird/party-hour


**50. Least Sandpiper**

| | Average Number of Birds Reported per Party-hour* | | | | |
| Counts No. | 1968–1977 69-78 | 1978–1987 79-88 | 1988–1997 89-98 | 1998–2007 99-108 | 40 yr avg. |
|---|---|---|---|---|---|
| N. Dakota | 0 | 0 | 0 | 0 | 0 |
| S. Dakota | 0 | 0 | 0 | 0 | 0 |
| Nebraska | 0 | 0 | 0 | p | p |
| Kansas | 0 | 0 | 0 | p | p |
| Oklahoma | 0.03 | 0.01 | 0.05 | 0.14 | 0.06 |
| 5-state ave.* | p | p | 0.01 | 003 | 0.01 |
| NW Texas | 0.01 | 0.02 | p | p | p |

* p = < 0.01 bird/party-hour


**51. Wilson's Snipe**

| | Average Number of Birds Reported per Party-hour* | | | | |
| Counts No. | 1968–1977 69-78 | 1978–1987 79-88 | 1988–1997 89-98 | 1998–2007 99-108 | 40 yr avg. |
|---|---|---|---|---|---|
| N. Dakota | 0 | p | 0 | 0 | p |
| S. Dakota | 0.01 | 0.03 | 0.03 | 0.02 | 0.02 |
| Nebraska | 0.03 | 0.02 | 0.05 | 0.04 | 0.03 |
| Kansas | 0.05 | 0.03 | 0.02 | 0.01 | 0.03 |
| Oklahoma | 0.09 | 0.10 | 0.13 | 0.12 | 0.10 |
| 5-state ave.* | 0.04 | 0.04 | 0.05 | 0.05 | 0.04 |
| NW Texas | 0.02 | 0.01 | p | p | 0.01 |

* p = < 0.01 bird/party-hour

**52. Bonaparte's Gull**

| Counts No. | Average Number of Birds Reported per Party-hour* | | | | |
| --- | --- | --- | --- | --- | --- |
| | 1968-1977 69-78 | 1978-1987 79-88 | 1988-1997 89-98 | 1998-2007 99-108 | 40 yr avg. |
| N. Dakota | 0 | 0 | 0 | 0 | 0 |
| S. Dakota | 0 | p | p | 0.05 | 0.01 |
| Nebraska | 0 | 0 | p | 0.06 | 0.01 |
| Kansas | p | p | 0.07 | 0.53 | 0.15 |
| Oklahoma | 0.12 | 0.31 | 2.25 | 1.93 | 1.15 |
| 5-state ave.* | 0.03 | 0.07 | 0.06 | 0.06 | 0.31 |
| NW Texas | 0.16 | 0 | 0 | p | 0.04 |

* p = < 0.01 bird/party-hour

**53. Ring-billed Gull**

| Counts No. | Average Number of Birds Reported per Party-hour* | | | | |
| --- | --- | --- | --- | --- | --- |
| | 1968-1977 69-78 | 1978-1987 79-88 | 1988-1997 89-98 | 1998-2007 99-108 | 40 yr avg. |
| N. Dakota | 0 | 0.35 | 0.40 | 0.66 | 0.36 |
| S. Dakota | 0.07 | 0.32 | 0.07 | 0.12 | 0.14 |
| Nebraska | p | 0.09 | 0.50 | 0.62 | 0.30 |
| Kansas | 0.03 | 0.47 | 0.85 | 2.04 | 0.85 |
| Oklahoma | 8.82 | 10.63 | 22.89 | 11.96 | 13.58 |
| 5-state ave.* | 2.47 | 3.62 | 8.45 | 6.23 | 5.20 |
| NW Texas | 1.20 | 0.47 | 0.75 | 1.35 | 0.94 |

**54. Herring Gull**

| Counts No. | Average Number of Birds Reported per Party-hour* | | | | |
| --- | --- | --- | --- | --- | --- |
| | 1968-1977 69-78 | 1978-1987 79-88 | 1988-1997 89-98 | 1998-2007 99-108 | 40 yr avg. |
| N. Dakota | 0 | 0.06 | 0.10 | 0.10 | 0.065 |
| S. Dakota | 0.025 | 0.07 | 0.07 | 0.07 | 0.06 |
| Nebraska | p | p | 0.02 | 0.25 | 0.07 |
| Kansas | 0.02 | 0.11 | 0.19 | 0.32 | 0.16 |
| Oklahoma | 0.35 | 0.12 | 0.87 | 0.17 | 0.18 |
| 5-state ave.* | 0.08 | 0.07 | 0.09 | 0.18 | 0.10 |
| NW Texas | 0.01 | p | p | 0.02 | 0.01 |

* p = < 0.01 bird/party-hour

## 55. Rock Pigeon

| | Average Number of Birds Reported per Party-hour* | | | | |
| --- | --- | --- | --- | --- | --- |
| | 1968-1977 | 1978-1987 | 1988-1997 | 1998-2007 | 40 yr |
| Counts No. | 69-78 | 79-88 | 89-98 | 99-108 | avg. |
| N. Dakota | 2.32 | 5.21 | 5.56 | 6.56 | 4.91 |
| S. Dakota | 1.77 | 3.82 | 4.40 | 5.98 | 3.99 |
| Nebraska | 1.90 | 4.14 | 4.40 | 3.05 | 3.37 |
| Kansas | 1.87 | 4.22 | 4.92 | 4.24 | 3.81 |
| Oklahoma | 0.80 | 1.09 | 1.27 | 1.42 | 1.14 |
| 5-state ave.* | 1.73 | 3.70 | 4.11 | 4.25 | 3.45 |
| NW Texas | 0.12 | 0.52 | 0.39 | 0.70 | 0.43 |

## 56. Eurasian Collared-Dove

| | Average Number of Birds Reported per Party-hour* | | | | |
| --- | --- | --- | --- | --- | --- |
| | 1968-1977 | 1978-1987 | 1988-1997 | 1998-2007 | 40 yr |
| Counts No. | 69-78 | 79-88 | 89-98 | 99-108 | avg. |
| N. Dakota | 0 | 0 | 0 | 0.03 | 0.01 |
| S. Dakota | 0 | 0 | 0 | 0.10 | 0.02 |
| Nebraska | 0 | 0 | 0 | 0.18 | 0.05 |
| Kansas | 0 | 0 | 0 | 0.46 | 0.11 |
| Oklahoma | 0 | 0 | p | 0.06 | 0.01 |
| 5-state ave.* | 0 | 0 | p | 0.18 | 0.04 |
| NW Texas | 0 | 0 | 0 | 0.44 | 0.11 |

* p = < 0.01 bird/party-hour

## 57. Mourning Dove

| | Average Number of Birds Reported per Party-hour* | | | | |
| --- | --- | --- | --- | --- | --- |
| | 1968-1977 | 1978-1987 | 1988-1997 | 1998-2007 | 40 yr |
| Counts No. | 69-78 | 79-88 | 89-98 | 99-108 | avg. |
| N. Dakota | 0.02 | 0.08 | 0.02 | 0.07 | 0.05 |
| S. Dakota | 0.08 | 0.06 | 0.02 | 0.05 | 0.06 |
| Nebraska | 1.12 | 0.37 | 0.24 | 0.28 | 0.50 |
| Kansas | 1.06 | 0.92 | 0.97 | 1.56 | 1.13 |
| Oklahoma | 1.07 | 0.98 | 0.84 | 1.39 | 1.07 |
| 5-state ave.* | 0.71 | 0.58 | 0.51 | 0.80 | 0.65 |
| NW Texas | 3.48 | 1.88 | 3.32 | 3.10 | 2.94 |

**58. Greater Roadrunner**

| Counts No. | Average Number of Birds Reported per Party-hour* | | | | |
|---|---|---|---|---|---|
| | 1968-1977 69-78 | 1978-1987 79-88 | 1988-1997 89-98 | 1998-2007 99-108 | 40 yr avg. |
| N. Dakota | 0 | 0 | 0 | 0 | 0 |
| S. Dakota | 0 | 0 | 0 | 0 | 0 |
| Nebraska | 0 | 0 | 0 | 0 | 0 |
| Kansas | p | p | p | p | p |
| Oklahoma | 0.02 | 0.01 | 0.01 | 0.02 | 0.02 |
| 5-state ave.* | 0.01 | p | p | p | p |
| NW Texas | 0.06 | 0.05 | 0.11 | 0.12 | 0.09 |

\* p = < 0.01 bird/party-hour

**59. Great Horned Owl**

| Counts No. | Average Number of Birds Reported per Party-hour* | | | | |
|---|---|---|---|---|---|
| | 1968-1977 69-78 | 1978-1987 79-88 | 1988-1997 89-98 | 1998-2007 99-108 | 40 yr avg. |
| N. Dakota | 0.08 | 0.14 | 0.14 | 0.12 | 0.12 |
| S. Dakota | 0.17 | 0.18 | 0.15 | 0.13 | 0.16 |
| Nebraska | 0.11 | 0.14 | 0.18 | 0.10 | 0.13 |
| Kansas | 0.20 | 0.20 | 0.20 | 0.15 | 0.19 |
| Oklahoma | 0.09 | 0.10 | 0.09 | 0.08 | 0.09 |
| 5-state ave. | 0.14 | 0.16 | 0.15 | 0.12 | 0.14 |
| NW Texas | 0.01 | 0.10 | 0.08 | 0.12 | 0.10 |

**60. Belted Kingfisher**

| Counts No. | Average Number of Birds Reported per Party-hour* | | | | |
|---|---|---|---|---|---|
| | 1968-1977 69-78 | 1978-1987 79-88 | 1988-1997 89-98 | 1998-2007 99-108 | 40 yr avg. |
| N. Dakota | p | 0 | p | p | p |
| S. Dakota | p | p | p | p | p |
| Nebraska | 0.08 | 0.08 | 0.07 | 0.04 | 0.07 |
| Kansas | 0.32 | 0.35 | 0.24 | 0.20 | 0.28 |
| Oklahoma | 0.18 | 0.14 | 0.16 | 0.16 | 0.16 |
| 5-state ave.* | 0.08 | 0.07 | 0.08 | 0.07 | 0.07 |
| NW Texas | 0.04 | 0.05 | 0.06 | 0.06 | 0.05 |

\* p = < 0.01 bird/party-hour

## 61. Red-headed Woodpecker

| | Average Number of Birds Reported per Party-hour* | | | | |
|---|---|---|---|---|---|
| Counts No. | 1968–1977 69-78 | 1978–1987 79-88 | 1988–1997 89-98 | 1998–2007 99-108 | 40 yr avg. |
| N. Dakota | p | 0 | p | p | p |
| S. Dakota | p | p | p | p | p |
| Nebraska | 0.09 | 0.08 | 0.07 | 0.04 | 0.07 |
| Kansas | 0.32 | 0.35 | 0.24 | 0.20 | 0.28 |
| Oklahoma | 0.52 | 0.55 | 0.43 | 0.39 | 047 |
| 5-state ave.* | 0.23 | 0.25 | 0.18 | 0.15 | 0.20 |
| NW Texas | 0 | 0 | p | 0.01 | p |

* p = < 0.01 bird/party-hour

## 62. Lewis's Woodpecker

| | Average Number of Birds Reported per Party-hour* | | | | |
|---|---|---|---|---|---|
| Counts No. | 1968–1977 69-78 | 1978–1987 79-88 | 1988–1997 89-98 | 1998–2007 99-108 | 40 yr avg. |
| N. Dakota | 0 | 0 | 0 | 0 | 0 |
| S. Dakota | 0 | 0.10 | 0.11 | 0.14 | 0.08 |
| Nebraska | 0 | 0 | 0 | 0 | 0 |
| Kansas | 0 | 0 | 0 | 0 | 0 |
| Oklahoma | 0 | 0.05 | 0.09 | p | 0.04 |
| 5-state ave.* | 0 | 0.03 | 0.04 | 0.02 | 0.02 |
| NW Texas | 0 | 0 | 0 | 0 | 0 |

* p = < 0.01 bird/party-hour

## 63. Golden-fronted Woodpecker

| | Average Number of Birds Reported per Party-hour* | | | | |
|---|---|---|---|---|---|
| Counts No. | 1968–1977 69-78 | 1978–1987 79-88 | 1988–1997 89-98 | 1998–2007 99-108 | 40 yr avg. |
| N. Dakota | 0 | 0 | 0 | 0 | 0 |
| S. Dakota | 0 | 0 | 0 | 0 | 0 |
| Nebraska | 0 | 0 | 0 | 0 | 0 |
| Kansas | 0 | 0 | 0 | 0 | 0 |
| Oklahoma | p | p | 0 | 0 | p |
| 5-state ave.* | p | p | 0 | 0 | p |
| NW Texas | 0.33 | 0.69 | 0.49 | 039 | 0.47 |

* p = < 0.01 bird/party-hour

**64. Red-bellied Woodpecker**

| Counts No. | Average Number of Birds Reported per Party-hour* | | | | |
| --- | --- | --- | --- | --- | --- |
| | 1968–1977 69-78 | 1978–1987 79-88 | 1988–1997 89-98 | 1998–2007 99-108 | 40 yr avg. |
| N. Dakota | p | p | p | p | p |
| S. Dakota | 0.01 | 0.04 | 0.03 | 0.04 | 0.03 |
| Nebraska | 0.25 | 0.18 | 0.26 | 0.44 | 0.28 |
| Kansas | 0.65 | 0.66 | 0.86 | 0.92 | 0.78 |
| Oklahoma | 0.55 | 0.59 | 0.68 | 0.85 | 0.67 |
| 5-state ave.* | 0.36 | 0.37 | 0.44 | 0.50 | 35.2 |
| NW Texas | 0.01 | 0.02 | 0.02 | 0.02 | 0.02 |

* p = < 0.01 bird/party-hour

**65. Yellow-bellied Sapsucker**

| Counts No. | Average Number of Birds Reported per Party-hour* | | | | |
| --- | --- | --- | --- | --- | --- |
| | 1968–1977 69-78 | 1978–1987 79-88 | 1988–1997 89-98 | 1998–2007 99-108 | 40 yr avg. |
| N. Dakota | p | 0 | 0 | 0 | p |
| S. Dakota | p | p | p | p | p |
| Nebraska | 0.01 | 0.01 | 0.05 | 0.01 | 0.01 |
| Kansas | 0.03 | 0.04 | 0.03 | 0.04 | 0.04 |
| Oklahoma | 0.26 | 0.35 | 0.32 | 0.33 | 0.31 |
| 5-state ave.* | 0.04 | 0.03 | 0.04 | 0.04 | 0.04 |
| NW Texas | 0.25 | 0.41 | 0.32 | 0.34 | 0.33 |

* p = < 0.01 bird/party-hour

**66. Downy Woodpecker**

| Counts No. | Average Number of Birds Reported per Party-hour* | | | | |
| --- | --- | --- | --- | --- | --- |
| | 1968–1977 69-78 | 1978–1987 79-88 | 1988–1997 89-98 | 1998–2007 99-108 | 40 yr avg. |
| N. Dakota | 0.31 | 0.43 | 0.47 | 0.52 | 0.43 |
| S. Dakota | 0.55 | 0.65 | 0.48 | 0.44 | 0.53 |
| Nebraska | 0.68 | 0.77 | 0.76 | 0.82 | 0.75 |
| Kansas | 0.61 | 0.83 | 0.98 | 0.83 | 0.82 |
| Oklahoma | 0.48 | 0.62 | 0.60 | 0.72 | 0.61 |
| 5-state ave.* | 0.53 | 0.66 | 0.66 | 0.67 | 0.63 |
| NW Texas | 0.04 | 0.05 | 0.07 | 0.05 | 0. 05 |

## 67. Hairy Woodpecker

| Counts No. | Average Number of Birds Reported per Party-hour* | | | | |
|---|---|---|---|---|---|
| | 1968-1977 69-78 | 1978-1987 79-88 | 1988-1997 89-98 | 1998-2007 99-108 | 40 yr avg. |
| N. Dakota | 0.31 | 0.40 | 0.51 | 0.35 | 0.39 |
| S. Dakota | 0.22 | 0.26 | 0.22 | 0.18 | 0.22 |
| Nebraska | 0.17 | 0.19 | 0.17 | 0.17 | 0.18 |
| Kansas | 0.17 | 0.17 | 0.17 | 0.17 | 0.17 |
| Oklahoma | 0.12 | 0.11 | 0.09 | 0.09 | 0.10 |
| 5-state ave.* | 0.20 | 0.22 | 0.23 | 0.19 | 0.2 |
| NW Texas | 0 | 0.1 | 0 | 0 | 0 |

## 68. Northern Flicker

| Counts No. | Average Number of Birds Reported per Party-hour* | | | | |
|---|---|---|---|---|---|
| | 1968-1977 69-78 | 1978-1987 79-88 | 1988-1997 89-98 | 1998-2007 99-108 | 40 yr avg. |
| N. Dakota | p | p | 0.01 | 0.01 | 0.01 |
| S. Dakota | 0.02 | 0.09 | 0.05 | 0.10 | 0.07 |
| Nebraska | 0.13 | 0.06 | 0.14 | 0.31 | 0.16 |
| Kansas | p | 0.15 | 0.59 | 0.66 | 0.36 |
| Oklahoma | 0.20 | 0.52 | 0.59 | 0.64 | 0.49 |
| 5-state ave.* | 0.07 | 0.16 | 0.16 | 0.34 | 0.18 |
| NW Texas | 0.19 | 0.96 | 0.46 | 0.50 | 0.53 |

* p = < 0.01 bird/party-hour

## 69. Pileated Woodpecker

| Counts No. | Average Number of Birds Reported per Party-hour* | | | | |
|---|---|---|---|---|---|
| | 1968-1977 69-78 | 1978-1987 79-88 | 1988-1997 89-98 | 1998-2007 99-108 | 40 yr avg. |
| N. Dakota | p | p | 0.01 | 0.02 | 0.01 |
| S. Dakota | 0 | 0 | 0 | 0 | 0 |
| Nebraska | 0 | 0 | 0 | 0 | 0 |
| Kansas | p | 0.01 | 0.02 | 0.03 | 0.02 |
| Oklahoma | 0.11 | 0.12 | 0.11 | 0.13 | 0.12 |
| 5-state ave.* | 0.2 | 0.02 | 0.02 | 0.03 | 0.02 |
| NW Texas | 0 | 0 | 0 | 0 | 0 |

* p = < 0.01 bird/party-hour

**70. American Kestrel**

| Counts No. | Average Number of Birds Reported per Party-hour* | | | | |
|---|---|---|---|---|---|
| | 1968–1977 69-78 | 1978–1987 79-88 | 1988–1997 89-98 | 1998–2007 99-108 | 40 yr avg. |
| N. Dakota | 0.01 | 0.01 | 0.01 | 0.01 | 0.01 |
| S. Dakota | 0.03 | 0.06 | 0.07 | 0,06 | 0.06 |
| Nebraska | 0.13 | 0.36 | 0.36 | 0.33 | 0.30 |
| Kansas | 0.45 | 0.55 | 0.66 | 0.67 | 0.58 |
| Oklahoma | 0.28 | 0.30 | 0.43 | 0.43 | 0.36 |
| 5-state ave.* | 0.22 | 0.29 | 0.34 | 0.33 | 0.30 |
| NW Texas | 0.36 | 0.78 | 0.78 | 1.11 | 0.76 |

**71. Eastern Phoebe**

| Counts No. | Average Number of Birds Reported per Party-hour* | | | | |
|---|---|---|---|---|---|
| | 1968–1977 69-78 | 1978–1987 79-88 | 1988–1997 89-98 | 1998–2007 99-108 | 40 yr avg. |
| N. Dakota | 0 | 0 | 0 | 0 | 0 |
| S. Dakota | 0 | 0 | 0 | 0 | 0 |
| Nebraska | 0 | p | 0 | 0 | p |
| Kansas | p | 0 | 0 | p | p |
| Oklahoma | p | p | 0.10 | 0.06 | 0.04 |
| 5-state ave.* | p | p | p | p | p |
| NW Texas | 0 | 0 | 0 | 0 | 0 |

* p = < 0.01 bird/party-hour

**72. Loggerhead Shrike**

| Counts No. | Average Number of Birds Reported per Party-hour* | | | | |
|---|---|---|---|---|---|
| | 1968–1977 69-78 | 1978–1987 79-88 | 1988–1997 89-98 | 1998–2007 99-108 | 40 yr avg. |
| N. Dakota | p | p | p | p | p |
| S. Dakota | p | p | p | p | p |
| Nebraska | p | p | p | p | p |
| Kansas | 0.19 | 0.15 | 0.14 | 0.06 | 0.13 |
| Oklahoma | 0.43 | 0.29 | 0.26 | 0.15 | 0.28 |
| 5-state ave.* | 0.12 | 0.08 | 0.08 | 0.04 | 0.08 |
| NW Texas | 0.57 | 0.33 | 0.34 | 0.22 | 0.36 |

* p = < 0.01 bird/party-hour

73. Northern Shrike

| Counts No. | Average Number of Birds Reported per Party-hour* | | | | |
|---|---|---|---|---|---|
| | 1968–1977 69-78 | 1978–1987 79-88 | 1988–1997 89-98 | 1998–2007 99-108 | 40 yr avg. |
| N. Dakota | 0.38 | 0.08 | 0.06 | 0.06 | 0.15 |
| S. Dakota | 0.04 | 0.06 | 0.08 | 0.06 | 0.06 |
| Nebraska | 0.03 | 0.05 | 0.06 | 0.07 | 0.06 |
| Kansas | p | p | 0.01 | 0.01 | p |
| Oklahoma | p | p | p | p | p |
| 5-state ave.* | p | p | p | p | p |
| NW Texas | p | p | p | p | p |

* p = < 0.01 bird/party-hour

74. Blue Jay

| Counts No. | Average Number of Birds Reported per Party-hour* | | | | |
|---|---|---|---|---|---|
| | 1968–1977 69-78 | 1978–1987 79-88 | 1988–1997 89-98 | 1998–2007 99-108 | 40 yr avg. |
| N. Dakota | 0.38 | 0.79 | 0.67 | 0.72 | 0.64 |
| S. Dakota | 0.37 | 0.66 | 0.82 | 1.01 | 0.72 |
| Nebraska | 0.94 | 0.98 | 1.21 | 1.01 | 1.03 |
| Kansas | 1.87 | 2.00 | 2.22 | 1.51 | 1.90 |
| Oklahoma | 2.75 | 2.80 | 3.20 | 2.70 | 2.86 |
| 5-state ave.* | 1.26 | 1.45 | 1.62 | 1.39 | 1.43 |
| NW Texas | 0.01 | 0.13 | 0.15 | 0.13 | 0.11 |

75. Western Scrub-Jay

| Counts No. | Average Number of Birds Reported per Party-hour* | | | | |
|---|---|---|---|---|---|
| | 1968–1977 69-78 | 1978–1987 79-88 | 1988–1997 89-98 | 1998–2007 99-108 | 40 yr avg. |
| N. Dakota | 0 | 0 | 0 | 0 | 0 |
| S. Dakota | 0 | 0 | 0 | 0 | 0 |
| Nebraska | 0 | 0 | 0 | 0 | 0 |
| Kansas | 0 | 0 | 0 | p | 0 |
| Oklahoma | 0 | 0 | p | 0.02 | p |
| 5-state ave.* | 0 | 0 | p | p | p |
| NW Texas | 0 | 0 | 0.03 | 0.13 | 0.03 |

* p = < 0.01 bird/party-hour

### 76. Pinyon Jay

| Counts No. | Average Number of Birds Reported per Party-hour* | | | | |
|---|---|---|---|---|---|
| | 1968–1977 69-78 | 1978–1987 79-88 | 1988–1997 89-98 | 1998–2007 99-108 | 40 yr avg. |
| N. Dakota | 0 | p | 0 | 0 | 0 |
| S. Dakota | 0.55 | 0.38 | 0.03 | 0.02 | 0.24 |
| Nebraska | 0.32 | 0.23 | p | p | 0.07 |
| Kansas | p | 0 | 0 | 0 | 0 |
| Oklahoma | p | p | 0.07 | 0.05 | 0 |
| 5-state ave.* | 0.17 | 0.12 | 0.02 | 0.01 | 0.06 |
| NW Texas | 0 | p | 0 | 0 | 0 |

* p = < 0.01 bird/party-hour

### 77. Black-billed Magpie

| Counts No. | Average Number of Birds Reported per Party-hour* | | | | |
|---|---|---|---|---|---|
| | 1968–1977 69-78 | 1978–1987 79-88 | 1988–1997 89-98 | 1998–2007 99-108 | 40 yr avg. |
| N. Dakota | 0.53 | 0.53 | 0.76 | 0.49 | 0.57 |
| S. Dakota | 0.28 | 0.19 | 0.17 | 0.23 | 0.22 |
| Nebraska | 0.42 | 0.57 | 0.77 | 0.52 | 0.58 |
| Kansas | 0.13 | 0.15 | 0.39 | 0.33 | 0.25 |
| Oklahoma | 0.06 | 0.05 | 0.03 | 0.03 | 0.04 |
| 5-state ave.* | 0.24 | 0.24 | 0.38 | 0.30 | 0.29 |
| NW Texas | 0 | 0 | 0 | 0 | 0 |

### 78. American Crow

| Counts No. | Average Number of Birds Reported per Party-hour* | | | | |
|---|---|---|---|---|---|
| | 1968–1977 69-78 | 1978–1987 79-88 | 1988–1997 89-98 | 1998–2007 99-108 | 40 yr avg. |
| N. Dakota | 0.09 | 0.25 | 0.60 | 1.75 | 0.67 |
| S. Dakota | 1.03 | 1.70 | 2.58 | 3.68 | 2.25 |
| Nebraska | 2.24 | 3.80 | 7.68 | 16.48 | 7.55 |
| Kansas | 14.11 | 13.26 | 19.95 | 19.73 | 16.76 |
| Oklahoma | 4.62 | 4.89 | 7.91 | 9.67 | 6.77 |
| 5-state ave.* | 5.81 | 6.03 | 8.94 | 10.45 | 7.84 |
| NW Texas | 1.62 | 9.85 | 0.77 | 0.47 | 3.18 |

**79. Horned Lark**

| | Average Number of Birds Reported per Party-hour* | | | | |
|---|---|---|---|---|---|
| Counts No. | 1968–1977<br>69-78 | 1978–1987<br>79-88 | 1988–1997<br>89-98 | 1998–2007<br>99-108 | 40 yr<br>avg. |
| N. Dakota | 2.81 | 2.38 | 4.37 | 8.30 | 4.45 |
| S. Dakota | 7.77 | 6.93 | 3.87 | 6.04 | 6.15 |
| Nebraska | 19.54 | 9.05 | 10.81 | 3.01 | 10.58 |
| Kansas | 26.81 | 37.71 | 36.08 | 46.03 | 36.73 |
| Oklahoma | 1.08 | 2.36 | 1.17 | 1.12 | 1.43 |
| 5-state ave.* | 12.13 | 15.27 | 13.29 | 15.80 | 14.12 |
| NW Texas | 7.69 | 10.86 | 10.12 | 5.13 | 8.45 |

**80. Carolina Chickadee**

| | Average Number of Birds Reported per Party-hour* | | | | |
|---|---|---|---|---|---|
| Counts No. | 1968–1977<br>69-78 | 1978–1987<br>79-88 | 1988–1997<br>89-98 | 1998–2007<br>99-108 | 40 yr<br>avg. |
| N. Dakota | 0 | 0 | 0 | 0 | 0 |
| S. Dakota | 0 | 0 | 0 | 0 | 0 |
| Nebraska | 0 | 0 | 0 | 0 | 0 |
| Kansas | 0.24 | 0.62 | 0.53 | 0.40 | 0.45 |
| Oklahoma | 2.16 | 2.92 | 3.06 | 2.90 | 2.76 |
| 5-state ave.* | 0.81 | 0.87 | 0.88 | 0.75 | 0.78 |
| NW Texas | 0.13 | 0.31 | 0.33 | 0.05 | 0.20 |

* p = < 0.01 bird/party-hour

**81. Black-capped Chickadee**

| | Average Number of Birds Reported per Party-hour* | | | | |
|---|---|---|---|---|---|
| Counts No. | 1968–1977<br>69-78 | 1978–1987<br>79-88 | 1988–1997<br>89-98 | 1998–2007<br>99-108 | 40 yr<br>avg. |
| N. Dakota | 1.62 | 2.77 | 3.32 | 2.75 | 2.61 |
| S. Dakota | 2.14 | 3.03 | 2.70 | 2.10 | 2.49 |
| Nebraska | 2.39 | 2.76 | 3.04 | 2.02 | 2.55 |
| Kansas | 2.67 | 3.33 | 2.93 | 1.67 | 2.65 |
| Oklahoma | 0 | p | p | p | p |
| 5-state ave.* | 1.68 | 2.34 | 2.26 | 1.64 | 1.98 |
| NW Texas | p | 0 | 0 | 0 | p |

* p = < 0.01 bird/party-hour

**82. Tufted Titmouse**

| Counts No. | Average Number of Birds Reported per Party-hour* | | | | |
|---|---|---|---|---|---|
| | 1968–1977 69-78 | 1978–1987 79-88 | 1988–1997 89-98 | 1998–2007 99-108 | 40 yr avg. |
| N. Dakota | 0 | 0 | 0 | 0 | 0 |
| S. Dakota | 0 | 0 | 0 | 0 | 0 |
| Nebraska | 0.10 | 0.05 | 0.07 | 0.08 | 0.07 |
| Kansas | 0.75 | 0.64 | 0.81 | 0.64 | 0.71 |
| Oklahoma | 1.20 | 1.13 | 1.30 | 1.37 | 1.25 |
| 5-state ave.* | 0.53 | 0.47 | 0.55 | 0.50 | 0.51 |
| NW Texas | 0 | 0- | 0.04 | 0.08 | 0.03 |

* p = < 0.01 bird/party-hour

**83. Black-crested Titmouse**

| Counts No. | Average Number of Birds Reported per Party-hour* | | | | |
|---|---|---|---|---|---|
| | 1968–1977 69-78 | 1978–1987 79-88 | 1988–1997 89-98 | 1998–2007 99-108 | 40 yr avg. |
| N. Dakota | 0 | 0 | 0 | 0 | 0 |
| S. Dakota | 0 | 0 | 0 | 0 | 0 |
| Nebraska | 0 | 0 | 0 | 0 | 0 |
| Kansas` | 0 | 0 | 0 | 0 | 0 |
| Oklahoma | p | 0.01 | 0 | 0 | p |
| 5-state ave.* | p | p | 0 | 0 | p |
| NW Texas | 0.12 | 0.14 | 0.18 | 0.38 | 0.20 |

* p = < 0.01 bird/party-hour

**84. Bushtit**

| Counts No. | Average Number of Birds Reported per Party-hour* | | | | |
|---|---|---|---|---|---|
| | 1968–1977 69-78 | 1978–1987 79-88 | 1988–1997 89-98 | 1998–2007 99-108 | 40 yr avg. |
| N. Dakota | 0 | 0 | 0 | 0 | 0 |
| S. Dakota | 0 | 0 | 0 | 0 | 0 |
| Nebraska | 0 | 0 | 0 | 0 | 0 |
| Kansas | p | 0 | 0.01 | p | p |
| Oklahoma | 0.02 | 0.13 | 0.06 | 0.04 | 0.06 |
| 5-state ave.* | p | 0.03 | 0.01 | 0.01 | 0.01 |
| NW Texas | 0.32 | 0.26 | 0.43 | 0.22 | 0.31 |

* p = < 0.01 bird/party-hour

## 85. Red-breasted Nuthatch

| | Average Number of Birds Reported per Party-hour* | | | | |
|---|---|---|---|---|---|
| | 1968–1977 | 1978–1987 | 1988–1997 | 1998–2007 | 40 yr |
| Counts No. | 69-78 | 79-88 | 89-98 | 99-108 | avg. |
| N. Dakota | 0.01 | 0.06 | 0.13 | 0.18 | 0.10 |
| S. Dakota | 0.08 | 0.15 | 0.21 | 0.27 | 0.18 |
| Nebraska | 0.12 | 0.08 | 0.07 | 0.10 | 0.09 |
| Kansas | 0.05 | 0.06 | 0.07 | 0.11 | 0.07 |
| Oklahoma | 0.03 | 0.02 | 0.04 | 0.03 | 0.03 |
| 5-state ave.* | 0.05 | 0.07 | 0.10 | 0.13 | 0.09 |
| NW Texas | 0.01 | 0.02 | 0.02 | 0.01 | 0.01 |

## 86. White-breasted Nuthatch

| | Average Number of Birds Reported per Party-hour* | | | | |
|---|---|---|---|---|---|
| | 1968–1977 | 1978–1987 | 1988–1997 | 1998–2007 | 40 yr |
| Counts No. | 69-78 | 79-88 | 89-98 | 99-108 | avg. |
| N. Dakota | 0.35 | 0.63 | 0.85 | 0.70 | 0.63 |
| S. Dakota | 0.32 | 0.43 | 0.43 | 0.37 | 0.39 |
| Nebraska | 0.44 | 0.49 | 0.64 | 0.83 | 0.60 |
| Kansas | 0.17 | 0.35 | 0.47 | 0.46 | 0.36 |
| Oklahoma | 0.17 | 0.22 | 0.23 | 0.27 | 0.22 |
| 5-state ave.* | 0.25 | 0.40 | 0.50 | 0.50 | 0.41 |
| NW Texas | p | 0.01 | p | p | p |

* p = < 0.01 bird/party-hour

## 87. Brown Creeper

| | Average Number of Birds Reported per Party-hour* | | | | |
|---|---|---|---|---|---|
| | 1968–1977 | 1978–1987 | 1988–1997 | 1998–2007 | 40 yr |
| Counts No. | 69-78 | 79-88 | 89-98 | 99-108 | avg. |
| N. Dakota | 0.04 | 0.04 | 0.02 | 0.04 | 0.03 |
| S. Dakota | 0.07 | 0.07 | 0.06 | 0.07 | 0.07 |
| Nebraska | 0.14 | 0.12 | 0.10 | 0.12 | 0.12 |
| Kansas | 0.13 | 0.15 | 0.11 | 0.12 | 0.13 |
| Oklahoma | 0.12 | 0.12 | 0.13 | 0.11 | 0.12 |
| 5-state ave.* | 0.10 | 0.10 | 0.09 | 0.09 | 010 |
| NW Texas | 0.06 | 0.05 | 0.03 | 0.01 | 0.04 |

## 88. Carolina Wren

| Counts No. | Average Number of Birds Reported per Party-hour* | | | | |
| --- | --- | --- | --- | --- | --- |
| | 1968-1977 69-78 | 1978-1987 79-88 | 1988-1997 89-98 | 1998-2007 99-108 | 40 yr avg. |
| N. Dakota | 0 | 0 | 0 | 0 | 0 |
| S. Dakota | 0 | 0 | 0 | 0 | 0 |
| Nebraska | p | p | 0.01 | 0.02 | 0.01 |
| Kansas | 0.12 | 0.01 | 0.19 | 0.27 | 0.15 |
| Oklahoma | 0.21 | 0.15 | 0.52 | 0.66 | 0.39 |
| 5-state ave.* | 0.10 | 0.04 | 0.18 | 0.23 | 0.14 |
| NW Texas | 0 | 0 | p | 0.01 | 0.01 |

* p = < 0.01 bird/party-hour

## 89. Bewick's Wren

| Counts No. | Average Number of Birds Reported per Party-hour* | | | | |
| --- | --- | --- | --- | --- | --- |
| | 1968-1977 69-78 | 1978-1987 79-88 | 1988-1997 89-98 | 1998-2007 99-108 | 40 yr avg. |
| N. Dakota | 0 | 0 | 0 | 0 | 0 |
| S. Dakota | 0 | 0 | 0 | 0 | 0 |
| Nebraska | 0` | 0 | 0 | 0` | 0 |
| Kansas | 0.01 | p | 0.01 | 0.01 | 0.01 |
| Oklahoma | 0.12 | 0.08 | 0.08 | 0.08 | 0.09 |
| 5-state ave.* | 0.03 | 0.02 | 0.02 | 0.02 | 0.02 |
| NW Texas | 0.39 | 0.42 | 0.28 | 0.24 | 0.33 |

* p = < 0.01 bird/party-hour

## 90. Marsh Wren

| Counts No. | Average Number of Birds Reported per Party-hour* | | | | |
| --- | --- | --- | --- | --- | --- |
| | 1968-1977 69-78 | 1978-1987 79-88 | 1988-1997 89-98 | 1998-2007 99-108 | 40 yr avg. |
| N. Dakota | 0 | 0 | 0 | 0 | 0 |
| S. Dakota | p | p | 0 | p | p |
| Nebraska | 0 | p | 0.01 | 0.01 | 0.01 |
| Kansas | p | 0.01 | 0.01 | 0.04 | 0.02 |
| Oklahoma | 0.01 | 0.02 | 0.02 | 0.02 | 0.02 |
| 5-state ave.* | p | 0.01 | 0.01 | 0.02 | 0.01 |
| NW Texas | 0.06 | 0.11 | 0.13 | 0.14 | 0.11 |

* p = < 0.01 bird/party-hour

### 91. Golden-crowned Kinglet

| Counts No. | Average Number of Birds Reported per Party-hour* | | | | |
|---|---|---|---|---|---|
| | 1968-1977 69-78 | 1978-1987 79-88 | 1988-1997 89-98 | 1998-2007 99-108 | 40 yr avg. |
| N. Dakota | p | 0.02 | 0.02 | 0.03 | 0.02 |
| S. Dakota | 0.04 | 0.05 | 0.06 | 0.06 | 0.05 |
| Nebraska | 0.13 | 0.12 | 0.20 | 0.15 | 0.15 |
| Kansas | 0.20 | 0.14 | 0.21 | 0.21 | 0.19 |
| Oklahoma | 0.27 | 0.13 | 0.19 | 0.17 | 0.19 |
| 5-state ave.* | 0.15 | 0.10 | 0.14 | 0.13 | 0.13 |
| NW Texas | 0.11 | 0.10 | 0.11 | 0.06 | 0.09 |

### 92. Ruby-crowned Kinglet

| Counts No. | Average Number of Birds Reported per Party-hour* | | | | |
|---|---|---|---|---|---|
| | 1968-1977 69-78 | 1978-1987 79-88 | 1988-1997 89-98 | 1998-2007 99-108 | 40 yr avg. |
| N. Dakota | p | 0 | p | p | p |
| S. Dakota | p | 0 | 0 | p | p |
| Nebraska | 0.02 | p | p | p | p |
| Kansas | 0.02 | 0.02 | 0.03 | 0.07 | 0.03 |
| Oklahoma | 0.25 | 0.18 | 0.22 | 0.30 | 0.24 |
| 5-state ave.* | 0.07 | 0.05 | 0.06 | 0.06 | 0.07 |
| NW Texas | 0.08 | 0.21 | 0.21 | 0.37 | 0.22 |

* p = < 0.01 bird/party-hour

### 93. Eastern Bluebird

| Counts No. | Average Number of Birds Reported per Party-hour* | | | | |
|---|---|---|---|---|---|
| | 1968-1977 69-78 | 1978-1987 79-88 | 1988-1997 89-98 | 1998-2007 99-108 | 40 yr avg. |
| N. Dakota | 0 | 0 | 0 | p | p |
| S. Dakota | p | p | 0.01 | 0.02 | 0.01 |
| Nebraska | p | 0.05 | 0.05 | 0.22 | 0.9 |
| Kansas | 0.59 | 0.39 | 0.98 | 1.61 | 0.89 |
| Oklahoma | 1.54 | 1.19 | 2.14 | 2.51 | 1.84 |
| 5-state ave.* | 0.56 | 0.41 | 0.80 | 1.04 | 0.70 |
| NW Texas | 0.36 | 0.87 | 0.55 | 0.41 | 0.54 |

* p = < 0.01 bird/party-hour

### 94. Mountain Bluebird

| Counts No. | Average Number of Birds Reported per Party-hour* | | | | |
| | 1968–1977 69-78 | 1978–1987 79-88 | 1988–1997 89-98 | 1998–2007 99-108 | 40 yr avg. |
|---|---|---|---|---|---|
| N. Dakota | 0 | 0 | 0 | 0 | 0 |
| S. Dakota | p | 0 | 0 | 0 | p |
| Nebraska | 0 | p | p | 0.01 | p |
| Kansas | p | p | 0.13 | 0.63 | 0.19 |
| Oklahoma | 0.50 | 0.67 | 1.46 | 1.32 | 0.99 |
| 5-state ave.* | 0.13 | 0.16 | 0.39 | 0.46 | 0.28 |
| NW Texas | 2.23 | 2.42 | 2.45 | 2.47 | 2.39 |

* p = < 0.01 bird/party-hour

### 95. Townsend's Solitaire

| Counts No. | Average Number of Birds Reported per Party-hour* | | | | |
| | 1968–1977 69-78 | 1978–1987 79-88 | 1988–1997 89-98 | 1998–2007 99-108 | 40 yr avg. |
|---|---|---|---|---|---|
| N. Dakota | 0.01 | 0.02 | 0.02 | 0.03 | 0.02 |
| S. Dakota | 0.10 | 0.13 | 0.11 | 0.11 | 0.11 |
| Nebraska | 0.04 | 0.05 | 0.18 | 0.18 | 0.11 |
| Kansas | 0.02 | 0.01 | 0.03 | 0.05 | 0.03 |
| Oklahoma | 0.05 | 0.09 | 0.07 | 0.02 | 0.06 |
| 5-state ave.* | 0.04 | 0.05 | 0.07 | 0.06 | 0.06 |
| NW Texas | 0.19 | 0.29 | 0.16 | 0.5 | 0.17 |

### 96. American Robin

| Counts No. | Average Number of Birds Reported per Party-hour* | | | | |
| | 1968–1977 69-78 | 1978–1987 79-88 | 1988–1997 89-98 | 1998–2007 99-108 | 40 yr avg. |
|---|---|---|---|---|---|
| N. Dakota | 0.27 | 0.95 | 0.62 | 0.89 | 0.68 |
| S. Dakota | 23.03 | 2.13 | 1.76 | 24.01 | 12.73 |
| Nebraska | 2.50 | 7.14 | 17.04 | 25.97 | 13.16 |
| Kansas | 4.48 | 9.01 | 5.25 | 39.32 | 14.52 |
| Oklahoma | 11.22 | 9.65 | 8.41 | 22.50 | 12.94 |
| 5-state ave.* | 8.39 | 6.44 | 5.65 | 23.13 | 10.91 |
| NW Texas | 11.07 | 11.84 | 7.48 | 8.43 | 9.70 |

## 97. Northern Mockingbird

| Counts No. | Average Number of Birds Reported per Party-hour* | | | | |
|---|---|---|---|---|---|
| | 1968-1977 69-78 | 1978-1987 79-88 | 1988-1997 89-98 | 1998-2007 99-108 | 40 yr avg. |
| N. Dakota | p | 0 | p | p | p |
| S. Dakota | p | p | p | p | p |
| Nebraska | p | p | P | 0 | P |
| Kansas | 0.23 | 0.10 | 0.14 | 0.15 | 0.16 |
| Oklahoma | 0.42 | 0.44 | 0.56 | 0.61 | 0.51 |
| 5-state ave.* | 0.17 | 0.14 | 0.17 | 0.18 | 0.17 |
| NW Texas | 0.43 | 0.32 | 0.22 | 0.22 | 0.30 |

* p = < 0.01 bird/party-hour

## 98. Brown Thrasher

| Counts No. | Average Number of Birds Reported per Party-hour* | | | | |
|---|---|---|---|---|---|
| | 1968-1977 69-78 | 1978-1987 79-88 | 1988-1997 89-98 | 1998-2007 99-108 | 40 yr avg. |
| N. Dakota | p | p | p | p | p |
| S. Dakota | p | p | p | p | p |
| Nebraska | p | p | p | p | p |
| Kansas | 0.02 | 0.01 | 0.01 | 0.01 | 0.01 |
| Oklahoma | 0.27 | 0.08 | 0.09 | 0.08 | 0.13 |
| 5-state ave.* | 0.07 | 0.02 | 0.02 | 0.02 | 0.04 |
| NW Texas | 0.03 | 0.04 | 0.05 | 0.03 | 0.04 |

* p = < 0.01 bird/party-hour

## 99. Curve-billed Thrasher

| Counts No. | Average Number of Birds Reported per Party-hour* | | | | |
|---|---|---|---|---|---|
| | 1968-1977 69-78 | 1978-1987 79-88 | 1988-1997 89-98 | 1998-2007 99-108 | 40 yr avg. |
| N. Dakota`` | 0 | 0 | 0 | 0 | 0 |
| S. Dakota | 0 | 0 | 0 | 0 | 0 |
| Nebraska | 0 | 0 | 0 | 0 | 0 |
| Kansas | p | p | p | p | p |
| Oklahoma | 0.01 | 0.01 | 0.01 | p | p |
| 5-state ave.* | p | p | p | p | p |
| NW Texas | 0.11 | 0.18 | 0.11 | 0.14 | 0.14 |

* p = < 0.01 bird/party-hour

**100. European Starling**

| Counts No. | Average Number of Birds Reported per Party-hour* | | | | |
|---|---|---|---|---|---|
| | 1968–1977 69-78 | 1978–1987 79-88 | 1988–1997 89-98 | 1998–2007 99-108 | 40 yr avg. |
| N. Dakota | 2.0 | 3.5 | 2.6 | 6.1 | 3.6 |
| S. Dakota | 28.7 | 11.3 | 10.4 | 19.7 | 17.5 |
| Nebraska | 95.8 | 42.6 | 38.5 | 46.1 | 55.7 |
| Kansas | 986.8 | 105.1 | 123.5 | 542.3 | 439.4 |
| Oklahoma | 153.6 | 194.7 | 75.8 | 127.8 | 138.0 |
| 5-state ave.* | 369.6 | 88.2 | 60.3 | 189.6 | 176.9 |
| NW Texas | 4.8 | 10.7 | 5.4 | 6.4 | 6.8 |

**101. American Pipit**

| Counts No. | Average Number of Birds Reported per Party-hour* | | | | |
|---|---|---|---|---|---|
| | 1968–1977 69-78 | 1978–1987 79-88 | 1988–1997 89-98 | 1998–2007 99-108 | 40 yr avg. |
| N. Dakota | 0 | 0 | 0 | 0 | 0 |
| S. Dakota | 0 | p | 0 | 0 | p |
| Nebraska | 0 | p | 0 | 0 | p |
| Kansas | 0.01 | p | 0.01 | 0.04 | 0.02 |
| Oklahoma | 0.83 | 0.08 | 0.12 | 0.29 | 0.33 |
| 5-state ave.* | 0.02 | 0.02 | 0.03 | 0.07 | 0.04 |
| NW Texas | 0.04 | 0.01 | 0.09 | 0.01 | 0.04 |

* p = < 0.01 bird/party-hour

**102. Bohemian Waxwing**

| Counts No. | Average Number of Birds Reported per Party-hour* | | | | |
|---|---|---|---|---|---|
| | 1968–1977 69-78 | 1978–1987 79-88 | 1988–1997 89-98 | 1998–2007 99-108 | 40 yr avg. |
| N. Dakota | 2.8 | 3.9 | 2.1 | 0.50 | 2.3 |
| S. Dakota | 1.4 | 1.2 | 1.1 | 1.7 | 1.2 |
| Nebraska | 0.23 | 0.16 | p | 0.20 | 0.15 |
| Kansas | 0.01 | p | p | p | p |
| Oklahoma | 0 | 0 | 0 | 0 | 0 |
| 5-state ave.* | 0.08 | 1.00 | 0.60 | 0.41 | 0.70 |
| NW Texas | 0 | 0 | 0 | 0 | 0 |

* p = < 0.01 bird/party-hour

## 103. Cedar Waxwing

| | Average Number of Birds Reported per Party-hour* | | | | |
|---|---|---|---|---|---|
| Counts No. | 1968-1977 69-78 | 1978-1987 79-88 | 1988-1997 89-98 | 1998-2007 99-108 | 40 yr avg. |
| N. Dakota | 0.34 | 0.67 | 1.07 | 1.88 | 0.99 |
| S. Dakota | 0.22 | 0.87 | 1.66 | 2.64 | 1.34 |
| Nebraska | 0.65 | 0.77 | 1.16 | 2.92 | 1.38 |
| Kansas | 0.87 | 1.39 | 1.31 | 2.20 | 1.45 |
| Oklahoma | 2.93 | 3.25 | 4.38 | 6.25 | 4.20 |
| 5-state ave.* | 1.00 | 1.39 | 1.91 | 3.18 | 1.87 |
| NW Texas | 0.69 | 1.41 | 1.08 | 1.30 | 1.12 |

## 104. Chestnut-collared Longspur

| | Average Number of Birds Reported per Party-hour* | | | | |
|---|---|---|---|---|---|
| Counts No. | 1968-1977 69-78 | 1978-1987 79-88 | 1988-1997 89-98 | 1998-2007 99-108 | 40 yr avg. |
| N. Dakota | 0 | p | 0 | 0 | p |
| S. Dakota | 0.01 | 0 | 0.01 | 0 | p |
| Nebraska | 0 | 0 | p | 0 | p |
| Kansas | 0.60 | 0.18 | 0.17 | 0.11 | 0.26 |
| Oklahoma | 0.24 | 0.51 | 0.31 | 0.31 | 0.35 |
| 5-state ave.* | 0.47 | 0.15 | 0.23 | 0.11 | 0.19 |
| NW Texas | 0.03 | 0.17 | 0.13 | 0.10 | 0.11 |

* p = < 0.01 bird/party-hour

## 105. Smith's Longspur

| | Average Number of Birds Reported per Party-hour* | | | | |
|---|---|---|---|---|---|
| Counts No. | 1968-1977 69-78 | 1978-1987 79-88 | 1988-1997 89-98 | 1998-2007 99-108 | 40 yr avg. |
| N. Dakota | 0 | 0 | 0 | 0 | 0 |
| S. Dakota | 0 | p | p | p | p |
| Nebraska | 0 | p | p | p | p |
| Kansas | 0.07 | p | p | 0.03 | 0.03 |
| Oklahoma | 0.11 | 0.31 | 0.58 | 0.49 | 0.38 |
| 5-state ave.* | 0.04 | 0.07 | 0.15 | 0.12 | 0.09 |
| NW Texas | 0 | p | 0 | 0 | p |

* p = < 0.01 bird/party-hour

106. McCown's Longspur

| Counts No. | Average Number of Birds Reported per Party-hour* | | | | |
| | 1968-1977 69-78 | 1978-1987 79-88 | 1988-1997 89-98 | 1998-2007 99-108 | 40 yr avg. |
|---|---|---|---|---|---|
| N. Dakota | 0 | 0 | 0 | 0 | 0 |
| S. Dakota | 0 | 0 | 0 | 0 | 0 |
| Nebraska | 0 | 0 | 0 | 0 | 0 |
| Kansas | p | 0.01 | 0.06 | 0.04 | 0.03 |
| Oklahoma | p | p | p | 0.01 | p |
| 5-state ave.* | p | p | 0.01 | 0.01 | 0.01 |
| NW Texas | 0.76 | 1.58 | 7.79 | 2.62 | 3.18 |

* p = < 0.01 bird/party-hour

107. Lapland Longspur

| Counts No. | Average Number of Birds Reported per Party-hour* | | | | |
| | 1968-1977 69-78 | 1978-1987 79-88 | 1988-1997 89-98 | 1998-2007 99-108 | 40 yr avg. |
|---|---|---|---|---|---|
| N. Dakota | 1.20 | 2.90 | 1.62 | 5.94 | 2.91 |
| S. Dakota | 2.90 | 2.12 | 0.45 | 0.87 | 1.58 |
| Nebraska | 4.46 | 409 | 0.36 | 3.02 | 2.98 |
| Kansas | 46.60 | 15.99 | 44.70 | 66.92 | 43.80 |
| Oklahoma | 0.34 | 0.09 | 0.20 | 0.59 | 0.31 |
| 5-state ave.* | 14.79 | 6.59 | 13.5 | 20.54 | 13.66 |
| NW Texas | 0.32 | 0.07 | 0.46 | 1.20 | 0.52 |

* p = < 0.01 bird/party-hour

108. Snow Bunting

| Counts No. | Average Number of Birds Reported per Party-hour* | | | | |
| | 1968-1977 69-78 | 1978-1987 79-88 | 1988-1997 89-98 | 1998-2007 99-108 | 40 yr avg. |
|---|---|---|---|---|---|
| N. Dakota | 15.37 | 8.43 | 9.97 | 8.46 | 10.56 |
| S. Dakota | 1.27 | 2.71 | 0.50 | 1.31 | 1.45 |
| Nebraska | p | 0.01 | 0.10 | 0.04 | 0.04 |
| Kansas | p | p | 0.01 | p | p |
| Oklahoma | 0 | 0 | p | 0 | 0 |
| 5-state ave.* | 3.08 | 2.04 | 2.14 | 2.06 | 2.33 |
| NW Texas | 0 | 0 | 0 | 0 | 0 |

* p = < 0.01 bird/party-hour

## 109. Yellow-rumped Warbler

| Counts No. | Average Number of Birds Reported per Party-hour* | | | | |
|---|---|---|---|---|---|
| | 1968–1977 69-78 | 1978–1987 79-88 | 1988–1997 89-98 | 1998–2007 99-108 | 40 yr avg. |
| N. Dakota | 0 | 0 | 0 | p | p |
| S. Dakota | p | 0.01 | 0.02 | p | 0.01 |
| Nebraska | p | p | p | p | p |
| Kansas | 0.04 | 0.05 | 0.08 | 0.36 | 0.14 |
| Oklahoma | 0.36 | 0.52 | 0.84 | 1.40 | 0.78 |
| 5-state ave.* | 0.10 | 0.14 | 0.22 | 0.42 | 0.22 |
| NW Texas | 0.01 | 0.07 | 0.14 | 0.11 | 0.08 |

* p = < 0.01 bird/party-hour

## 110. Pine Warbler

| Counts No. | Average Number of Birds Reported per Party-hour* | | | | |
|---|---|---|---|---|---|
| | 1968–1977 69-78 | 1978–1987 79-88 | 1988–1997 89-98 | 1998–2007 99-108 | 40 yr avg. |
| N. Dakota | 0 | 0 | 0 | 0 | 0 |
| S. Dakota | 0 | 0 | 0 | 0 | 0 |
| Nebraska | 0 | 0 | 0 | 0 | 0 |
| Kansas | 0 | 0 | 0 | 0 | 0 |
| Oklahoma | 0.04 | 0.02 | 0.11 | 0.06 | 0.06 |
| 5-state ave.* | 0.01 | p | 0.03 | 0.01 | 0.01 |
| NW Texas | 0 | 0 | 0 | 0 | 0 |

* p = < 0.01 bird/party-hour

## 111. Spotted Towhee

| Counts No. | Average Number of Birds Reported per Party-hour* | | | | |
|---|---|---|---|---|---|
| | 1968–1977 69-78 | 1978–1987 79-88 | 1988–1997 89-98 | 1998–2007 99-108 | 40 yr avg. |
| N. Dakota | p | 0 | 0 | 0 | 0 |
| S. Dakota | p | p | p | p | p |
| Nebraska | p | p | 0.01 | 0.01 | 0.01 |
| Kansas | 0.14 | 0.01 | 0.09 | 0.11 | 0.09 |
| Oklahoma | 0.19 | 0.05 | 0.18 | 0.26 | 0.17 |
| 5-state ave.* | 0.07 | 0.01 | 0.07 | 0.09 | 0.06 |
| NW Texas | 0.26 | 0.13 | 0.38 | 0.47 | 0.31 |

* p = < 0.01 bird/party-hour
** Not distinguished from eastern towhee during between counts 69 and 86;
   numbers shown for 1968-1977 may include some eastern towhees

## 112. Canyon Towhee

| Counts No. | Average Number of Birds Reported per Party-hour* | | | | |
|---|---|---|---|---|---|
| | 1968-1977 69-78 | 1978-1987 79-88 | 1988-1997 89-98 | 1998-2007 99-108 | 40 yr avg. |
| N. Dakota | 0 | 0 | 0 | 0 | 0 |
| S. Dakota | 0 | 0 | 0 | 0 | 0 |
| Nebraska | 0 | 0 | 0 | 0 | 0 |
| Kansas | p | 0 | p | p | p |
| Oklahoma | 0.27 | 0.22 | 0.11 | 0.09 | 0.18 |
| 5-state ave.* | 0.06 | 0.05 | 0.03 | 0.02 | 0.01 |
| NW Texas | 0.02 | 0.08 | 0.05 | 0.10 | 0.04 |

* p = < 0.01 bird/party-hour

## 113. Rufous-crowned Sparrow

| Counts No. | Average Number of Birds Reported per Party-hour* | | | | |
|---|---|---|---|---|---|
| | 1968-1977 69-78 | 1978-1987 79-88 | 1988-1997 89-98 | 1998-2007 99-108 | 40 yr avg. |
| N. Dakota | 0 | 0 | 0 | 0 | 0 |
| S. Dakota | 0 | 0 | 0 | 0 | 0 |
| Nebraska | 0 | 0 | 0 | 0 | 0 |
| Kansas | 0 | p | p | p | p |
| Oklahoma | 0.04 | 0.02 | 0.02 | 0.01 | 0.03 |
| 5-state ave.* | 0.01 | p | p | p | 0.01 |
| NW Texas | 0.23 | 0.12 | 0.13 | 0.13 | 0.15 |

* p = < 0.01 bird/party-hour

## 114. Chipping Sparrow

| Counts No. | Average Number of Birds Reported per Party-hour* | | | | |
|---|---|---|---|---|---|
| | 1968-1977 69-78 | 1978-1987 79-88 | 1988-1997 89-98 | 1998-2007 99-108 | 40 yr avg. |
| N. Dakota | p | 0 | 0.04 | 0 | 0.01 |
| S. Dakota | 1.43 | 0.01 | 0.02 | 0.09 | 0.36 |
| Nebraska | 0.20 | 0.01 | p | 0.01 | 0.01 |
| Kansas | 0.09 | 0.01 | 0.01 | 0.01 | 0.01 |
| Oklahoma | 2.18 | 0.49 | 1.26 | 3.41 | 1.29 |
| 5-state ave.* | 0.85 | 0.12 | 0.31 | 0.79 | 0.31 |
| NW Texas | 0.29 | 0.31 | 0.43 | 0.19 | 0.23 |

* p = < 0.01 bird/party-hour

## 115. Field Sparrow

| Counts No. | Average Number of Birds Reported per Party-hour* | | | | |
|---|---|---|---|---|---|
| | 1968-1977 69-78 | 1978-1987 79-88 | 1988-1997 89-98 | 1998-2007 99-108 | 40 yr avg. |
| N. Dakota | 0 | 0.01 | 0 | 0 | 0 |
| S. Dakota | 0.02 | 0.04 | p | 0.01 | 0.02 |
| Nebraska | 0.01 | 0.05 | p | 0.01 | .02 |
| Kansas | 0.16 | 0.11 | 0.09 | 0.11 | 0.09 |
| Oklahoma | 1.12 | 0.82 | 0.94 | 0.93 | 0.72 |
| 5-state ave.* | 0.33 | 0.24 | 0.25 | 0.25 | 0.27 |
| NW Texas | 0.17 | 0.13 | 0.14 | 0.10 | 0.14 |

* p = < 0.01 bird/party-hour

## 116. American Tree Sparrow

| Counts No. | Average Number of Birds Reported per Party-hour* | | | | |
|---|---|---|---|---|---|
| | 1968-1977 69-78 | 1978-1987 79-88 | 1988-1997 89-98 | 1998-2007 99-108 | 40 yr avg. |
| N. Dakota | 0.24 | 0.69 | 0.98 | 0.47 | 0.47 |
| S. Dakota | 3.84 | 3.33 | 2.63 | 2.13 | 2.98 |
| Nebraska | 7.96 | 9.09 | 11.39 | 6.72 | 8.79 |
| Kansas | 24.43 | 20.42 | 18.54 | 20.02 | 20.85 |
| Oklahoma | 5.02 | 3.77 | 2.35 | 1.33 | 3.12 |
| 5-state ave.* | 10.10 | 9.03 | 7.61 | 7.05 | 8.45 |
| NW Texas | 1.84 | 0.94 | 0.21 | 0.01 | 0.75 |

## 117. Vesper Sparrow

| Counts No. | Average Number of Birds Reported per Party-hour* | | | | |
|---|---|---|---|---|---|
| | 1968-1977 69-78 | 1978-1987 79-88 | 1988-1997 89-98 | 1998-2007 99-108 | 40 yr avg. |
| N. Dakota | P | 0 | 0 | 0 | 0 |
| S. Dakota | 0 | p | 0 | p | p |
| Nebraska | 0 | 0.01 | 0 | 0 | p |
| Kansas | p | p | p | p | p |
| Oklahoma | 0.03 | 0.05 | 0.03 | 0.06 | 0.04 |
| 5-state ave.* | 0.01 | 0.01 | 0.01 | 0.01 | 0.01 |
| NW Texas | 0.02 | 0.11 | 0.03 | 0.12 | 0.07 |

* p = < 0.01 bird/party-hour

### 118. Lark Bunting

| Counts No. | Average Number of Birds Reported per Party-hour* | | | | |
|---|---|---|---|---|---|
|  | 1968–1977 69-78 | 1978–1987 79-88 | 1988–1997 89-98 | 1998–2007 99-108 | 40 yr avg. |
| N. Dakota | 0 | 0 | 0 | 0 | 0 |
| S. Dakota | 0 | 0 | 0 | 0 | 0 |
| Nebraska | 0 |  | 0 | 0 | 0 |
| 0 |  |  |  |  |  |
| Kansas | p | 0.07 | 0.08 | 0.03 | 0.06 |
| Oklahoma | p | 0.45 | 0.01 | p | 0.11 |
| 5-state ave.* | 0.01 | 0.12 | 0.02 | 0.01 | 0.04 |
| NW Texas | 0.09 | 0.75 | 4.34 | 0.57 | 1.62 |

* p = < 0.01 bird/party-hour

### 119. Savannah Sparrow

| Counts No. | Average Number of Birds Reported per Party-hour* | | | | |
|---|---|---|---|---|---|
|  | 1968–1977 69-78 | 1978–1987 79-88 | 1988–1997 89-98 | 1998–2007 99-108 | 40 yr avg. |
| N. Dakota | 0 | 0 | 0 | 0 | 0 |
| S. Dakota | 0 | p | 0 | 0 | p |
| Nebraska | 0 | 0 | p | p | p |
| Kansas | 0.01 | 0.01 | 0.01 | 0.15 | 0.04 |
| Oklahoma | 0.12 | 0.17 | 0.30 | 0.97 | 0.40 |
| 5-state ave.* | 0.03 | 0.04 | 0.07 | 0.26 | 0.10 |
| NW Texas | 0.09 | 0.58 | 0.10 | 0.30 | 0.27 |

* p = < 0.01 bird/party-hour

### 120. Fox Sparrow

| Counts No. | Average Number of Birds Reported per Party-hour* | | | | |
|---|---|---|---|---|---|
|  | 1968–1977 69-78 | 1978–1987 79-88 | 1988–1997 89-98 | 1998–2007 99-108 | 40 yr avg. |
| N. Dakota | p | p | p | p | p |
| S. Dakota | p | p | 0 | p | p |
| Nebraska | p | p | 0 | p | p |
| Kansas | 0.01 | 0.01 | 0.01 | 0.02 | 0.01 |
| Oklahoma | 0.14 | 0.21 | 0.18 | 0.33 | 0.21 |
| 5-state ave.* | 0.04 | 0.05 | 0.05 | 0.09 | 0.06 |
| NW Texas | 0.01 | 0.03 | 0.02 | 0.02 | 0.02 |

* p = < 0.01 bird/party-hour

### 121. Song Sparrow

| Counts No. | Average Number of Birds Reported per Party-hour* | | | | |
|---|---|---|---|---|---|
| | 1968–1977 69-78 | 1978–1987 79-88 | 1988–1997 89-98 | 1998–2007 99-108 | 40 yr avg. |
| N. Dakota | 0.09 | p | p | 0.01 | 0.03 |
| S. Dakota | 0.02 | 0.05 | 0.07 | 0.02 | 0.04 |
| Nebraska | 0.30 | 0.13 | 0.14 | 0.15 | 0.18 |
| Kansas | 1.01 | 0.86 | 1.10 | 1.48 | 1.11 |
| Oklahoma | 1.34 | 1.39 | 1.43 | 1.90 | 1.43 |
| 5-state ave.* | 0.68 | 0.62 | 0.68 | 0.87 | 0.71 |
| NW Texas | 0.82 | 1.30 | 1.14 | 0.87 | 1.03 |

* p = < 0.01 bird/party-hour

### 122. Lincoln's Sparrow

| Counts No. | Average Number of Birds Reported per Party-hour* | | | | |
|---|---|---|---|---|---|
| | 1968–1977 69-78 | 1978–1987 79-88 | 1988–1997 89-98 | 1998–2007 99-108 | 40 yr avg. |
| N. Dakota | 0 | 0 | 0 | 0 | 0 |
| S. Dakota | 0 | 0 | p | p | p |
| Nebraska | p | p | p | p | p |
| Kansas | 0.02 | 0.03 | 0.03 | 0.02 | 0.03 |
| Oklahoma | 0.05 | 0.04 | 0.03 | 0.05 | 0.04 |
| 5-state ave.* | 0.02 | 0.02 | 0.01 | 0.02 | 0.02 |
| NW Texas | 0.01 | 0.01 | 0.02 | 0.05 | 0.02 |

* p = < 0.01 bird/party-hour

### 123. Swamp Sparrow

| Counts No. | Average Number of Birds Reported per Party-hour* | | | | |
|---|---|---|---|---|---|
| | 1968–1977 69-78 | 1978–1987 79-88 | 1988–1997 89-98 | 1998–2007 99-108 | 40 yr avg. |
| N. Dakota | 0 | p | p | 0 | p |
| S. Dakota | p | 0 | 0 | p | p |
| Nebraska | p | p | p | p | p |
| Kansas | 0.01 | 0.05 | 0.08 | 0.09 | 0.06 |
| Oklahoma | 0.08 | 0.11 | 0.14 | 0.17 | 12.5 |
| 5-state ave.* | 0.02 | 0.04 | 0.05 | 0.07 | 0.05 |
| NW Texas | 0.04 | 0.02 | 0.02 | 0.1 | 0.02 |

* p = < 0.01 bird/party-hour

### 124. White-throated Sparrow

| Counts No. | Average Number of Birds Reported per Party-hour* | | | | |
| | 1968-1977 69-78 | 1978-1987 79-88 | 1988-1997 89-98 | 1998-2007 99-108 | 40 yr avg. |
|---|---|---|---|---|---|
| N. Dakota | p | p | p | p | p |
| S. Dakota | p | p | p | p | p |
| Nebraska | 0.01 | 0.02 | 0.02 | 0.03 | 0.02 |
| Kansas | 0.07 | 0.08 | 0.10 | 0.25 | 0.13 |
| Oklahoma | 1.02 | 0.86 | 1.99 | 3.05 | 1.73 |
| 5-state ave.* | 0.27 | 0.23 | 0.50 | 0.76 | 0.44 |
| NW Texas | 0.04 | 0.28 | 0.09 | 0.06 | 0.11 |

* p = < 0.01 bird/party-hour

### 125. Harris's Sparrow

| Counts No. | Average Number of Birds Reported per Party-hour* | | | | |
| | 1968-1977 69-78 | 1978-1987 79-88 | 1988-1997 89-98 | 1998-2007 99-108 | 40 yr avg. |
|---|---|---|---|---|---|
| N. Dakota | p | p | | p | 0.01 p |
| S. Dakota | 0.11 | 0.06 | 0.01 | 0.02 | 0.05 |
| Nebraska | 3.01 | 1.72 | 0.52 | 0.36 | 1.40 |
| Kansas | 14.15 | 9.63 | 0.85 | 8.99 | 8.40 |
| Oklahoma | 3.66 | 3.04 | 2.73 | 2.88 | 3.08 |
| 5-state ave.* | 4.18 | 3.69 | 0.82 | 2.36 | 2.59 |
| NW Texas | 0.12 | 0.10 | 0.04 | 0.01 | 0.07 |

### 126. White-crowned Sparrow

| Counts No. | Average Number of Birds Reported per Party-hour* | | | | |
| | 1968-1977 69-78 | 1978-1987 79-88 | 1988-1997 89-98 | 1998-2007 99-108 | 40 yr avg. |
|---|---|---|---|---|---|
| N. Dakota | p | p | p | p | p |
| S. Dakota | P | p | p | p | p |
| Nebraska | 0.06 | 0.14 | 0.16 | 0.17 | 0.13 |
| Kansas | 0.29 | 0.68 | 1.49 | 1.52 | 0.99 |
| Oklahoma | 0.32 | 1.07 | 0.65 | 1.01 | 0.76 |
| 5-state ave.* | 0.18 | 0.48 | 0.59 | 0.68 | 0.48 |
| NW Texas | 4.35 | 6.43 | 6.71 | 7.20 | 6.17 |

* p = < 0.01 bird/party-hour

## 127. Dark-eyed Junco

| | Average Number of Birds Reported per Party-hour* | | | | |
|---|---|---|---|---|---|
| Counts No. | 1968–1977 69-78 | 1978–1987 79-88 | 1988–1997 89-98 | 1998–2007 99-108 | 40 yr avg. |
| N. Dakota | 0 | 0.01 | 0.02 | 0.18 | 0.05 |
| S. Dakota | 0 | 0.08 | 0.09 | 0.96 | 0.28 |
| Nebraska | 0 | 1.84 | 3.64 | 4.44 | 2.48 |
| Kansas | 0.01 | 1.64 | 5.21 | 6.88 | 3.44 |
| Oklahoma | 0.04 | 1.49 | 2.99 | 6.57 | 2.78 |
| 5-state ave.* | 0.02 | 1.07 | 2.50 | 4.08 | 1.91 |
| NW Texas | 2.12 | 5.71 | 8.85 | 6.68 | 5.84 |

* p = < 0.01 bird/party-hour

## 128. Northern Cardinal

| | Average Number of Birds Reported per Party-hour* | | | | |
|---|---|---|---|---|---|
| Counts No. | 1968–1977 69-78 | 1978–1987 79-88 | 1988–1997 89-98 | 1998–2007 99-108 | 40 yr avg. |
| N. Dakota | 0.01 | p | p | p | p |
| S. Dakota | 0.08 | 0.09 | 0.08 | 0.10 | 0.09 |
| Nebraska | 1.04 | 0.82 | 0.75 | 0.91 | 0.88 |
| Kansas | 3.16 | 316 | 3.85 | 2.85 | 3.25 |
| Oklahoma | 3.23 | 3.12 | 4.45 | 4.48 | 3.81 |
| 5-state ave.* | 1.86 | 1.82 | 2.23 | 1.92 | 1.96 |
| NW Texas | 0.69 | 0.96 | 1.01 | 1.01 | 0.92 |

* p = < 0.01 bird/party-hour

## 129. Red-winged Blackbird

| | Average Number of Birds Reported per Party-hour* | | | | |
|---|---|---|---|---|---|
| Counts No. | 1968–1977 69-78 | 1978–1987 79-88 | 1988–1997 89-98 | 1998–2007 99-108 | 40 yr avg. |
| N. Dakota | 0.19 | 051 | 0.37 | 0.24 | 0.32 |
| S. Dakota | 5.68 | 2.14 | 1.24 | 1.87 | 2.73 |
| Nebraska | 7.83 | 270.34 | 57.01 | 15.12 | 87.58 |
| Kansas | 7862.85 | 608.80 | 404.44 | 3791.71 | 3166.95 |
| Oklahoma | 895.25 | 148.67 | 183.93 | 1459.63 | 673.36 |
| 5-state ave.* | 2804.03 | 253.12 | 172.94 | 1428.94 | 1164.76 |
| NW Texas | 34.99 | 24.69 | 1283.74 | 136.12 | 369.89 |

130. Eastern Meadowlark

| | Average Number of Birds Reported per Party-hour* | | | | |
|---|---|---|---|---|---|
| Counts No. | 1968-1977 69-78 | 1978-1987 79-88 | 1988-1997 89-98 | 1998-2007 99-108 | 40 yr avg. |
| N. Dakota | 0 | 0 | 0 | 0 | 0 |
| S. Dakota | 0.01 | p | p | p | 0.14 |
| Nebraska | 0.56 | p | p | p | p |
| Kansas | 4.04 | 2.10 | 2.07 | 0.98 | 2.30 |
| Oklahoma | 5.79 | 3.05 | 3.35 | 2.53 | 3.68 |
| 5-state ave.* | 2.72 | 1.38 | 1.39 | 0.85 | 1.58 |
| NW Texas | 0.14 | p | 0.02 | 0.08 | 0.06 |

* p = < 0.01 bird/party-hour

131. Western Meadowlark

| | Average Number of Birds Reported per Party-hour* | | | | |
|---|---|---|---|---|---|
| Counts No. | 1968-1977 69-78 | 1978-1987 79-88 | 1988-1997 89-98 | 1998-2007 99-108 | 40 yr avg. |
| N. Dakota | 0.01 | p | p | p | p |
| S. Dakota | 0.08 | 0.07 | 0.01 | 0.06 | 0.05 |
| Nebraska | 2.34 | 0.62 | 0.26 | 0.08 | 0.82 |
| Kansas | 4.57 | 2.77 | 3.57 | 4.09 | 3.75 |
| Oklahoma | 1.07 | 0.82 | 0.36 | 0.23 | 0.62 |
| 5-state ave.* | 1.90 | 1.16 | 1.14 | 1.22 | 1.35 |
| NW Texas | 9.37 | 5.61 | 6.48 | 7.18 | 7.16 |

* p = < 0.01 bird/party-hour

132. Yellow-headed Blackbird

| | Average Number of Birds Reported per Party-hour* | | | | |
|---|---|---|---|---|---|
| Counts No. | 1968-1977 69-78 | 1978-1987 79-88 | 1988-1997 89-98 | 1998-2007 99-108 | 40 yr avg. |
| N. Dakota | p | p | p | p | p |
| S. Dakota | p | 0.03 | 0.03 | 0.01 | 0.02 |
| Nebraska | 0 | p | p | p | p |
| Kansas | p | p | p | p | p |
| Oklahoma | 0.20 | p | p | p | 0.05 |
| 5-state ave.* | 0.04 | p | p | p | 0.01 |
| NW Texas | p | p | p | p | p |

### 133. Brewer's Blackbird

| Counts No. | Average Number of Birds Reported per Party-hour* | | | | |
|---|---|---|---|---|---|
| | 1968–1977 69-78 | 1978–1987 79-88 | 1988–1997 89-98 | 1998–2007 99-108 | 40 yr avg. |
| N. Dakota | 0.02 | 0.01 | 0.01 | 0.01 | 0.01 |
| S. Dakota | 0.10 | 0.35 | 0.05 | 0.03 | 0.13 |
| Nebraska | 0.23 | 0.11 | 0.01 | 0.05 | 0.10 |
| Kansas | 1.48 | 1.54 | 1.41 | 5.99 | 2.60 |
| Oklahoma | 12.97 | 4.07 | 8.06 | 6.85 | 7.98 |
| 5-state ave.* | 3.67 | 1.51 | 2.31 | 3.25 | 2.69 |
| NW Texas | 4.97 | 0.73 | 2.07 | 4.37 | 3.10 |

### 134. Rusty Blackbird

| Counts No. | Average Number of Birds Reported per Party-hour* | | | | |
|---|---|---|---|---|---|
| | 1968–1977 69-78 | 1978–1987 79-88 | 1988–1997 89-98 | 1998–2007 99-108 | 40 yr avg. |
| N. Dakota | 0.15 | 0.11 | 0.03 | 0.04 | 0.08 |
| S. Dakota | 0.20 | 0.09 | 0.13 | 0.04 | 0.12 |
| Nebraska | 0.41 | 0.43 | 0.09 | 0.06 | 0.03 |
| Kansas | 1.61 | 2.46 | 0.76 | 6.95 | 2.94 |
| Oklahoma | 1.67 | 0.78 | 0.83 | 0.60 | 0.97 |
| 5-state ave.* | 1.02 | 1.06 | 0.46 | 2.18 | 1.83 |
| NW Texas | 0.05 | 0 | 0 | 0.02 | 0.01 |

### 135. Common Grackle

| Counts No. | Average Number of Birds Reported per Party-hour* | | | | |
|---|---|---|---|---|---|
| | 1968–1977 69-78 | 1978–1987 79-88 | 1988–1997 89-98 | 1998–2007 99-108 | 40 yr avg. |
| N. Dakota | 0.03 | 0.04 | 0.05 | 0.08 | 0.05 |
| S. Dakota | 0.11 | 0.14 | 0.23 | 0.11 | 0.15 |
| Nebraska | 0.91 | 1.97 | 1.98 | 0.06 | 0.78 |
| Kansas | 5.81 | 1.63 | 2.68 | 57.04 | 16.79 |
| Oklahoma | 0.98 | 4.16 | 227.94 | 23.97 | 64.14 |
| 5-state ave.* | 2.06 | 1.68 | 53.16 | 22.08 | 19.94 |
| NW Texas | 0 | 0.02 | 0.10 | 0.09 | 0.06 |

## 136. Great-tailed Grackle

| Counts No. | Average Number of Birds Reported per Party-hour* | | | | |
|---|---|---|---|---|---|
| | 1968–1977 69-78 | 1978–1987 79-88 | 1988–1997 89-98 | 1998–2007 99-108 | 40 yr avg. |
| N. Dakota | 0 | 0 | 0 | 0 | 0 |
| S. Dakota | 0 | p | 0 | p | p |
| Nebraska | 0 | p | 0.02 | 0.04 | 0.01 |
| Kansas | 0.02 | 0.42 | 2.85 | 9.05 | 2.33 |
| Oklahoma | 0.90 | 1.98 | 7.85 | 1.44 | 3.05 |
| 5-state ave.* | 0.23 | 0.60 | 2.78 | 2.07 | 1.42 |
| NW Texas | 0 | 0 | 0.02 | 0.33 | 0.09 |

* p = < 0.01 bird/party-hour

## 137. Brown-headed Cowbird

| Counts No. | Average Number of Birds Reported per Party-hour* | | | | |
|---|---|---|---|---|---|
| | 1968–1977 69-78 | 1978–1987 79-88 | 1988–1997 89-98 | 1998–2007 99-108 | 40 yr avg. |
| N. Dakota | p | 0.01 | p | p | p |
| S. Dakota | 0.02 | 0.02 | 0.01 | 0.03 | 0.02 |
| Nebraska | 1.06 | 0.14 | 0.11 | 0.13 | 0.36 |
| Kansas | 12.35 | 25.66 | 2.37 | 79.97 | 30.09 |
| Oklahoma | 6.80 | 2.72 | 4.33 | 3.49 | 4.34 |
| 5-state ave.* | 5.72 | 9.14 | 1.74 | 24.02 | 10.16 |
| NW Texas | 0.11 | 0.24 | 0.29 | 0.33 | 0.24 |

* p = < 0.01 bird/party-hour

## 138. Gray-crowned Rosy-finch

| Counts No. | Average Number of Birds Reported per Party-hour* | | | | |
|---|---|---|---|---|---|
| | 1968–1977 69-78 | 1978–1987 79-88 | 1988–1997 89-98 | 1998–2007 99-108 | 40 yr avg. |
| N. Dakota | p | 0.42 | p | p | 0.01 |
| S. Dakota | 0.07 | p | 0.02 | p | 0.02 |
| Nebraska | 0 | p | p | 0 | p |
| Kansas | 0 | 0 | 0 | 0 | 0 |
| Oklahoma | 0 | 0 | 0 | 0 | 0 |
| 5-state ave.* | p | p | p | p | p |
| NW Texas | 0 | 0 | 0 | 0 | 0 |

* p = < 0.01 bird/party-hour

## 139. Pine Grosbeak

| Counts No. | Average Number of Birds Reported per Party-hour* | | | | |
|---|---|---|---|---|---|
| | 1968-1977 69-78 | 1978-1987 79-88 | 1988-1997 89-98 | 1998-2007 99-108 | 40 yr avg. |
| N. Dakota | 0.16 | 0.41 | 0.04 | 0.01 | 0.16 |
| S. Dakota | p | 0.04 | 0.01 | p | 0.01 |
| Nebraska | 0.01 | p | 0 | p | p |
| Kansas | p | p | p | p | p |
| Oklahoma | 0 | 0 | 0 | 0 | 0 |
| 5-state ave.* | 0.03 | 0.08 | 0.01 | p | 0.03 |
| NW Texas | 0 | 0 | 0 | 0 | 0 |

* p = < 0.01 bird/party-hour

## 140. Purple Finch

| Counts No. | Average Number of Birds Reported per Party-hour* | | | | |
|---|---|---|---|---|---|
| | 1968-1977 69-78 | 1978-1987 79-88 | 1988-1997 89-98 | 1998-2007 99-108 | 40 yr avg. |
| N. Dakota | 0.13 | 0.52 | 0.22 | 0.30 | 0.29 |
| S. Dakota | 0.27 | 0.34 | 0.27 | 0.18 | 0.26 |
| Nebraska | 0.10 | 0.16 | 0.06 | 0.11 | 0.11 |
| Kansas | 0.18 | 0.10 | 0.06 | 0.07 | 0.10 |
| Oklahoma | 0.65 | 0.24 | 0.19 | 0.07 | 0.29 |
| 5-state ave.* | 0.29 | 0.27 | 0.15 | 0.14 | 0.21 |
| NW Texas | 0.01 | P | 0 | P | P |

* p = < 0.01 bird/party-hour

## 141. House Finch

| Counts No. | Average Number of Birds Reported per Party-hour* | | | | |
|---|---|---|---|---|---|
| | 1968-1977 69-78 | 1978-1987 79-88 | 1988-1997 89-98 | 1998-2007 99-108 | 40 yr avg. |
| N. Dakota | 0 | 0 | 0.72 | 2.02 | 0.68 |
| S. Dakota | p | p | 0.72` | 1.42 | 0.53 |
| Nebraska | 0.32 | 0.85 | 1.24 | 1.82 | 1.06 |
| Kansas | 0.12 | 0.28 | 2.51 | 2.24 | 1.29 |
| Oklahoma | 0.10 | 0.07 | 0.56 | 0.64 | 0.34 |
| 5-state ave.* | 0.09 | 0.02 | 1.02 | 1.09 | 0.92 |
| NW Texas | 0.41 | 1.18 | 1.02 | 1.09 | 0.92 |

* p = < 0.01 bird/party-hour

## 142. Red Crossbill

| Counts No. | 1968–1977 69-78 | 1978–1987 79-88 | 1988–1997 89-98 | 1998–2007 99-108 | 40 yr avg. |
|---|---|---|---|---|---|
| N. Dakota | 0.13 | 0.13 | 0.06 | 0.08 | 0.10 |
| S. Dakota | 0.09 | 0.18 | 0.12 | 0.07 | 0.12 |
| Nebraska | 0.14 | 0.03 | 0.01 | 0.07 | 0.06 |
| Kansas | 0.02 | 0.02 | 0.01 | p | 0.01 |
| Oklahoma | 0.01 | p | p | p | p |
| 5-state ave.* | 0.06 | 0.06 | 0.04 | 0.03 | 0.05 |
| NW Texas | 0 | p | 0 | 0 | p |

Average Number of Birds Reported per Party-hour*

* p = < 0.01 bird/party-hour

## 143. Evening Grosbeak

| Counts No. | 1968–1977 69-78 | 1978–1987 79-88 | 1988–1997 89-98 | 1998–2007 99-108 | 40 yr avg. |
|---|---|---|---|---|---|
| N. Dakota | 0.51 | 0.43 | 0.56 | 0.01 | 0.38 |
| S. Dakota | 0.37 | 0.95 | 1.29 | 0.01 | 0.68 |
| Nebraska | 0.08 | 0.09 | 0.15 | 0 | 0.08 |
| Kansas | p | 0.02 | 0.01 | p | 0.01 |
| Oklahoma | p | 0.03 | 0.01 | 0 | 0.01 |
| 5-state ave.* | 0.17 | 0.26 | 0.36 | 0.02 | 0.20 |
| NW Texas | 0.01 | 0.04 | p | 0 | 0.01 |

Average Number of Birds Reported per Party-hour*

* p = < 0.01 bird/party-hour

## 144. Common Redpoll

| Counts No. | 1968–1977 69-78 | 1978–1987 79-88 | 1988–1997 89-98 | 1998–2007 99-108 | 40 yr avg. |
|---|---|---|---|---|---|
| N. Dakota | 8.38 | 13.81 | 5.54 | 4.83 | 8.14 |
| S. Dakota | 2.52 | 2.05 | 0.66 | 0.69 | 1.48 |
| Nebraska | 0.06 | 0.23 | 0.05 | p | 0.08 |
| Kansas | 0.11 | p | p | p | 0.03 |
| Oklahoma | 0 | 0 | 0 | 0 | 0 |
| 5-state ave.* | 2.04 | 3.21 | 1.27 | 1.15 | 1.92 |
| NW Texas | 0 | 0 | 0 | 0 | 0 |

Average Number of Birds Reported per Party-hour*

* p = < 0.01 bird/party-hour

## 145. Pine Siskin

| Counts No. | Average Number of Birds Reported per Party-hour* | | | | |
| --- | --- | --- | --- | --- | --- |
| | 1968-1977<br>69-78 | 1978-1987<br>79-88 | 1988-1997<br>89-98 | 1998-2007<br>99-108 | 40 yr<br>avg. |
| N. Dakota | 0.18 | 4.42 | 3.22 | 2.02 | 2.46 |
| S. Dakota | 0.48 | 1.55 | 2.06 | 0.96 | 1.26 |
| Nebraska | 1.41 | 1.00 | 1.39 | 0.82 | 1.16 |
| Kansas | 0.68 | 1.05 | 1.00 | 0.23 | 0.74 |
| Oklahoma | 0.85 | 0.72 | 0.42 | 0.17 | 0.54 |
| 5-state ave.* | 0.67 | 1.75 | 1.55 | 0.79 | 1.19 |
| NW Texas | 3.97 | 1.26 | 0.78 | 0.59 | 1.65 |

* p = < 0.01 bird/party-hour

## 146. American Goldfinch

| Counts No. | Average Number of Birds Reported per Party-hour* | | | | |
| --- | --- | --- | --- | --- | --- |
| | 1968-1977<br>69-78 | 1978-1987<br>79-88 | 1988-1997<br>89-98 | 1998-2007<br>99-108 | 40 yr<br>avg. |
| N. Dakota | 0.11 | 0.88 | 0.68 | 1.43 | 0.77 |
| S. Dakota | 0.67 | 1.09 | 1.66 | 2.04 | 1.37 |
| Nebraska | 2.49 | 2.67 | 4.12 | 4.88 | 3.54 |
| Kansas | 2.42 | 3.49 | 4.21 | 2.82 | 3.24 |
| Oklahoma | 4.41 | 3.01 | 3.45 | 2.55 | 3.36 |
| 5-state ave.* | 2.18 | 2.41 | 2.87 | 2.56 | 3.36 |
| NW Texas | 1.91 | 2.51 | 2.19 | 2.65 | 2.32 |

## 147. House Sparrow

| Counts No. | Average Number of Birds Reported per Party-hour* | | | | |
| --- | --- | --- | --- | --- | --- |
| | 1968-1977<br>69-78 | 1978-1987<br>79-88 | 1988-1997<br>89-98 | 1998-2007<br>99-108 | 40 yr<br>avg. |
| N. Dakota | 36.13 | 40.06 | 31.96 | 17.44 | 31.40 |
| S. Dakota | 27.73 | 21.54 | 19.98 | 15.62 | 21.22 |
| Nebraska | 39.41 | 27.74 | 16.43 | 8.47 | 23.01 |
| Kansas | 26.36 | 21.78 | 17.59 | 2.11 | 19.46 |
| Oklahoma | 6.25 | 5.94 | 3.28 | 2.79 | 4.56 |
| 5-state ave.* | 24.52 | 21.87 | 17.40 | 11.39 | 18.79 |
| NW Texas | 2.50 | 3.79 | 13.11 | 2.15 | 5.39 |

www.ingramcontent.com/pod-product-compliance
Lightning Source LLC
Chambersburg PA
CBHW020654270326

41928CB00005B/108